An Introduction to
Gödel's Theorems

Second edition

Peter Smith
University of Cambridge

CAMBRIDGE
UNIVERSITY PRESS

CAMBRIDGE
UNIVERSITY PRESS

University Printing House, Cambridge CB2 8BS, United Kingdom

Published in the United States of America by Cambridge University Press, New York

Cambridge University Press is part of the University of Cambridge.

It furthers the University's mission by disseminating knowledge in the pursuit of education, learning and research at the highest international levels of excellence.

www.cambridge.org
Information on this title: www.cambridge.org/9781107022843

First published 2007
Second edition 2013
Reprinted 2014

Printed in the United Kingdom by Clays, St Ives plc.

A catalogue record for this publication is available from the British Library

ISBN 978-1-107-02284-3 hardback
ISBN 978-1-107-60675-3 paperback

Additional resources for this publication at www.godelbook.net

An Introduction to Gödel's Theorems

In 1931, the young Kurt Gödel published his First Incompleteness Theorem, which tells us that, for any sufficiently rich theory of arithmetic, there are some arithmetical truths the theory cannot prove. This remarkable result is among the most intriguing (and most misunderstood) in logic. Gödel also outlined an equally significant Second Incompleteness Theorem. How are these Theorems established, and why do they matter? Peter Smith answers these questions by presenting an unusual variety of proofs for the First Theorem, showing how to prove the Second Theorem, and exploring a family of related results (including some not easily available elsewhere). The formal explanations are interwoven with discussions of the wider significance of the two Theorems. This book – extensively rewritten for its second edition – will be accessible to philosophy students with a limited formal background. It is equally suitable for mathematics students taking a first course in mathematical logic.

PETER SMITH was formerly Senior Lecturer in Philosophy at the University of Cambridge. His books include *Explaining Chaos* (1998) and *An Introduction to Formal Logic* (2003), and he is also a former editor of the journal *Analysis*.

For Patsy, as ever

Contents

Contents

Contents

Contents

Contents

Vagueness and the idea of computability · Formal proofs and informal demonstrations · Squeezing arguments – the very idea · Kreisel's squeezing argument · The first premiss for a squeezing argument · The other premisses, thanks to Kolmogorov and Uspenskii · The squeezing argument defended · To summarize

Preface

In 1931, the young Kurt Gödel published his First and Second Incompleteness Theorems; very often, these are referred to simply as 'Gödel's Theorems' (even though he proved many other important results). These Incompleteness Theorems settled – or at least, seemed to settle – some of the crucial questions of the day concerning the foundations of mathematics. They remain of the greatest significance for the philosophy of mathematics, though just what that significance is continues to be debated. It has also frequently been claimed that Gödel's Theorems have a much wider impact on very general issues about language, truth and the mind.

This book gives proofs of the Theorems and related formal results, and touches – necessarily briefly – on some of their implications. Who is the book for? Roughly speaking, for those who want a lot more fine detail than you get in books for a general audience (the best of those is Franzén, 2005), but who find the rather forbidding presentations in classic texts in mathematical logic (like Mendelson, 1997) too short on explanatory scene-setting. I assume only a modest amount of background in logic. So I hope philosophy students will find the book useful, as will mathematicians who want a more accessible exposition.

But don't be misled by the relatively relaxed style; don't try to browse through too quickly. We do cover a lot of ground in quite a bit of detail, and new ideas often come thick and fast. Take things slowly!

I originally intended to write a shorter book, leaving many of the formal details to be filled in from elsewhere. But while that plan might have suited some readers, I soon realized that it would seriously irritate others to be sent hither and thither to consult a variety of textbooks with different terminologies and different notations. So, in the end, I have given more or less full proofs of most of the key results we cover ('⊠' serves as our end-of-proof marker, as we want the more usual '□' for another purpose).

However, my original plan shows through in two ways. First, some proofs are still only partially sketched in. Second, I try to signal very clearly when the detailed proofs I do give can be skipped without much loss of understanding. With judicious skimming, you should be able to follow the main formal themes of the book even if you have limited taste for complex mathematical arguments. For those who want to fill in more details and test their understanding there are exercises on the book's website at www.godelbook.net, where there are also other supplementary materials.

As we go through, there is also an amount of broadly philosophical commentary. I follow Gödel in believing that our formal investigations and our general reflections on foundational matters should illuminate and guide each other. I hope that these brief philosophical discussions – relatively elementary though certainly not always uncontentious – will also be reasonably widely accessible. Note however that I am more interested in patterns of ideas and arguments than in being historically very precise when talking e.g. about logicism or about Hilbert's Programme.

Writing a book like this presents many problems of organization. For example, we will need to call upon a number of ideas from the general theory of computation – we will make use of both the notion of a 'primitive recursive function' and the more general notion of a 'μ-recursive function'. Do we explain these related ideas all at once, up front? Or do we give the explanations many chapters apart, when the respective notions first get put to use?

I've mostly adopted the second policy, introducing new ideas as and when needed. This has its costs, but I think that there is a major compensating benefit, namely that the way the book is organized makes it clearer just what depends on what. It also reflects something of the historical order in which ideas emerged.

How does this second edition differ from the first? This edition is over twenty pages longer, but that isn't because there is much new material. Rather, I have mostly used the extra pages to make the original book more reader-friendly; there has been a lot of rewriting and rearrangement, particularly in the opening chapters. Perhaps the single biggest change is in using a more traditional line of proof for the adequacy of Robinson Arithmetic (Q) for capturing all the primitive recursive functions. I will probably have disappointed some readers by still resisting the suggestion that I provide a full-blown, warts-and-all, proof of the Second Theorem, though I do say rather more than before. But after all, this *is* supposed to be a relatively introductory book.

Below, I acknowledge the help that I have so generously been given by so many. But here I must express thanks of a quite different order to Patsy Wilson-Smith, without whose continuing love and support neither edition of this book would ever have been written. This book is for her.

Thanks

'Acknowledgements' is far too cold a word. I have acquired many intellectual debts in the course of writing this book: with great kindness, a lot of people have given me comments, suggestions and corrections. As a result, the book – whatever its remaining shortcomings – is so much better than it might have been.

My colleague Michael Potter has been an inspiring presence ever since I returned to Cambridge. Many thanks are due to him and to all those who gave me advice on draft extracts while I was writing the first edition, including the late Torkel Franzén, Tim Button, Luca Incurvati, Jeffrey Ketland, Aatu Koskensilta, Christopher Leary, Mary Leng, Toby Ord, Alex Paseau, Jacob Plotkin, José F. Ruiz, Kevin Scharp, Hartley Slater, and Tim Storer. I should especially mention Richard Zach, whose comments were particularly extensive and particularly helpful.

I no longer have a list of all those who found errors in the first printing. But Arnon Avron, Peter Milne, Saeed Salehi, and Adil Sanaulla prompted the most significant changes in content that made their way into corrected reprints of the first edition. Orlando May then spotted some still remaining technical errors, as well as a distressing number of residual typos. Jacob Plotkin, Tony Roy and Alfredo Tomasetta found further substantive mistakes. Again, I am very grateful.

It is an oddity that the books which are read the most – texts aimed at students – are reviewed the least in the journals: but Arnon Avron and Craig Smoryński did write friendly reviews of the first edition, and I have now tried to meet at least some of their expressed criticisms.

When I started working on this second edition, I posted some early parts on my blog Logic Matters, and received very useful suggestions from a number of people. Encouraged by that, at a late stage in the writing I experimentally asked for volunteers to proof-read thirty-page chunks of the book: the bribe I offered was tiny, just a mention here! But over forty more people took up the invitation, so every page was looked at again three or four times. Many of these readers applied themselves to the task with quite extraordinary care and attention, telling me not just about the inevitable typos, but about ill-phrased sentences, obscurities, phrases that puzzled a non-native speaker of English, and more besides. A handful of readers – I report this with mixed feelings – also found small technical errors *still* lurking in the text. The experiment, then, was a resounding success. So warm thanks are due to, among others, Sama Agahi, Amir Anvari, Bert Baumgaertner, Alex Blum, Seamus Bradley, Matthew Brammall, Benjamin Briggs,

Catrin Campbell-Moore, Jordan Collins, Irena Cronin, Matthew Dentith, Neil Dewar, Jan Eißfeldt, Özge Ekin, Nicolas Fillion, Stefan Daniel Firth, Joshua P. Fry, Marc Alcobé García, André Gargoura, Jaime Gaspar, Jake Goritski, Scott Hendricks, Harrison Hibbert, Daniel Hoek, Anil Joshi, Abolfazl Karimi, Amir Khamseh, Baldur Arnviður Kristinsson, Jim Laird, S. P. Lam, Jim Lippard, Carl Mummett, Fredrick Nelson, Robert Rynasiewicz, Noah David Schweber, Alin Soare, Shane Steinert-Threlkeld, Andrew Stephenson, Alexander auf der Straße, S. P. Suresh, Andrew Tedder, Arhat Virdi, Benjamin Wells, John Wigglesworth, and Andrew Withy. I am very grateful to them all!

Three more readers require special mention. David Auerbach's many suggestions have uniformly made for significant improvements. Matthew Donald has also given me extremely helpful comments on the whole book. Frank Scheld has proof-read successive versions with enthusiasm and an eagle's eye. Again, very warm thanks.

Finally, I must thank Hilary Gaskin at Cambridge University Press, who initially accepted the book for publication and then offered me the chance to write a second edition.

1 What Gödel's Theorems say

1.1 Basic arithmetic

It is child's play to grasp the fundamental notions involved in the arithmetic of addition and multiplication. Starting from zero, there is a sequence of 'counting' numbers, each having exactly one immediate successor. This sequence of numbers – officially, *the natural numbers* – continues without end, never circling back on itself; and there are no 'stray' natural numbers, lurking outside this sequence. Adding n to m is the operation of starting from m in the number sequence and moving n places along. Multiplying m by n is the operation of (starting from zero and) repeatedly adding m, n times. It's as simple as that.

Once these fundamental notions are in place, we can readily define many more arithmetical concepts in terms of them. Thus, for any natural numbers m and n, $m < n$ iff there is a number $k \neq 0$ such that $m + k = n$. m is a factor of n iff $0 < m$ and there is some number k such that $0 < k$ and $m \times k = n$. m is even iff it has 2 as a factor. m is prime iff $1 < m$ and m's only factors are 1 and itself. And so on.[1]

Using our basic and defined concepts, we can then frame various general claims about the arithmetic of addition and multiplication. There are obvious truths like 'addition is commutative', i.e. for any numbers m and n, $m + n = n + m$. There are also some very unobvious claims, yet to be proved, like Goldbach's conjecture that every even number greater than two is the sum of two primes.

That second example illustrates the truism that it is one thing to understand what we'll call *the language of basic arithmetic* (i.e. the language of the addition and multiplication of natural numbers, together with the standard first-order logical apparatus), and it is quite another thing to be able to evaluate claims that can be framed in that simple language.

Still, it is extremely plausible to suppose that, whether the answers are readily available to us or not, questions posed in the language of basic arithmetic do *have* entirely determinate answers. The structure of the natural number sequence, with each number having a unique successor and there being no repetitions, is (surely) simple and clear. The operations of addition and multiplication are (surely) entirely well-defined; their outcomes are fixed by the school-room rules. So what more could be needed to fix the truth or falsity of propositions that – perhaps via a chain of definitions – amount to claims of basic arithmetic?

To put it fancifully: God lays down the number sequence and specifies how the operations of addition and multiplication work. He has then done all he needs

---

[1] 'Iff' is, of course, the standard logicians' shorthand for 'if and only if'.

to do to make it the case e.g. that Goldbach's conjecture is true (or false, as the case may be).

Of course, that way of putting it is rather *too* fanciful for comfort. We may indeed find it compelling to think that the sequence of natural numbers has a definite structure, and that the operations of addition and multiplication are entirely nailed down by the familiar basic rules. But what is the real content of the thought that the truth-values of all basic arithmetic propositions are thereby 'fixed'?

Here's one appealing way of giving non-metaphorical content to that thought. The idea is that we can specify a bundle of fundamental assumptions or *axioms* which somehow pin down the structure of the natural number sequence, and which also characterize addition and multiplication (after all, it is pretty natural to suppose that we *can* give a reasonably simple list of true axioms to encapsulate the fundamental principles so readily grasped by the successful learner of school arithmetic). So now suppose that φ is a proposition which can be formulated in the language of basic arithmetic. Then, the appealing suggestion continues, the assumed truth of our axioms 'fixes' the truth-value of any such φ in the following sense: either φ is logically deducible from the axioms, and so φ is true; or its negation $\neg\varphi$ is deducible from the axioms, and so φ is false. We may not, of course, actually stumble on a proof one way or the other: but the proposal is that such a proof is always possible, since the axioms contain enough information to enable the truth-value of any basic arithmetical proposition to be deductively extracted by deploying familiar step-by-step logical rules of inference.

Logicians say that a theory T is *negation-complete* if, for every sentence φ in the language of the theory, either φ or $\neg\varphi$ can be derived in T's proof system. So, put into that jargon, the suggestion we are considering is this: we should be able to specify a reasonably simple bundle of true axioms which, together with some logic, give us a *negation-complete* theory of basic arithmetic – i.e. we could in principle use the theory to prove or disprove any claim which is expressible in the language of basic arithmetic. If that's right, truth in basic arithmetic could just be equated with provability in this complete theory.

It is tempting to say rather more. For what will the axioms of basic arithmetic look like? Here's one candidate: 'For every natural number, there's a unique next one'. This is evidently true; but evident *how*? As a first thought, you might say 'we can just see, using mathematical intuition, that this axiom is true'. But the idea of mathematical intuition is obscure, to say the least. Maybe, on second thoughts, we don't need to appeal to it. Perhaps the axiom is evidently true because it is some kind of definitional triviality. Perhaps it is just part of what we *mean* by talk of the natural numbers that we are dealing with an ordered sequence where each member of the sequence has a unique successor. And, plausibly, other candidate axioms are similarly true by definition.

If those tempting second thoughts are right, then true arithmetical claims are *analytic* in the philosophers' sense of the word; that is to say, the truths of basic arithmetic will all flow deductively from logic plus axioms which are trivially true-by-definition.[2] This so-called 'logicist' view would then give us a very neat explanation of the special certainty and the necessary truth of correct claims of basic arithmetic.

1.2 Incompleteness

But now, in headline terms, *Gödel's First Incompleteness Theorem shows that the entirely natural idea that we can give a complete theory of basic arithmetic with a tidy set of axioms is wrong.*

Suppose we try to specify a suitable axiomatic theory T to capture the structure of the natural number sequence and pin down addition and multiplication (and maybe a lot more besides). We want T to have a nice set of true axioms and a reliably truth-preserving deductive logic. In that case, everything T proves must be true, i.e. T is a *sound* theory. But now Gödel gives us a recipe for coming up with a corresponding sentence G_T, couched in the language of basic arithmetic, such that – assuming T really is sound – (i) we can show that G_T can't be derived in T, and yet (ii) we can recognize that G_T must be true.

This is surely quite astonishing. Somehow, it seems, the truths of basic arithmetic must elude our attempts to pin them down by giving a nice set of fundamental assumptions from which we can deduce everything else. So how does Gödel show this in his great 1931 paper which presents the Incompleteness Theorems?

Well, note how we can use numbers and numerical propositions to encode facts about all sorts of things. For a trivial example, students in the philosophy department might be numbered off in such a way that the first digit encodes information about whether a student is an undergraduate or postgraduate, the next two digits encode year of admission, and so on. Much more excitingly, Gödel notes that we can use numbers and numerical propositions to encode facts about *theories*, e.g. facts about what can be derived in a theory T.[3] And what he then did is find a general method that enabled him to take any theory T strong enough to capture a modest amount of basic arithmetic and construct a corresponding arithmetical sentence G_T which encodes the claim 'The sentence G_T itself is unprovable in theory T'. So G_T is true if and only if T can't prove it.

---

[2]Thus Gottlob Frege, writing in his wonderful *Grundlagen der Arithmetik*, urges us to seek the proof of a mathematical proposition by 'following it up right back to the primitive truths. If, in carrying out this process, we come only on general logical laws and on definitions, then the truth is an analytic one.' (Frege, 1884, p. 4)

[3]By the way, it is absolutely standard for logicians to talk of a theory T as *proving* a sentence φ when there is a logically correct derivation of φ from T's assumptions. But T's assumptions may be contentious or plain false or downright absurd. So, T's proving φ in this logician's sense does not mean that φ is proved in the sense that it is established as true. It is far too late in the game to kick against the logician's usage, and in most contexts it is harmless. But our special concern in this book is with the connections and contrasts between being true and being provable in this or that theory T. So we need to be on our guard. And to help emphasize that proving-in-T is not always proving-as-true, I'll often talk of 'deriving' rather than 'proving' sentences when it is the logician's notion which is in play.

Suppose then that T is sound. If T were to prove G_T, G_T would be false, and T would then prove a falsehood, which it can't do. Hence, if T is sound, G_T is unprovable in T. Which makes G_T *true*. Hence $\neg G_T$ is false. And so that too can't be proved by T, because T only proves truths. In sum, still assuming T is sound, neither G_T nor its negation will be provable in T. Therefore T can't be negation-complete.

But in fact we don't need to assume that T is *sound*; we can make do with very significantly less. Gödel's official version of the First Theorem shows that T's mere *consistency* is enough to guarantee that a suitably constructed G_T is true-but-unprovable-in-T. And we only need a little more to show that $\neg G_T$ is not provable either (we won't pause now to fuss about the needed extra assumption).

We said: the sentence G_T encodes the claim that that very sentence is un-provable. But doesn't this make G_T rather uncomfortably reminiscent of the Liar sentence 'This very sentence is false' (which is false if it is true, and true if it is false)? You might well wonder whether Gödel's argument doesn't lead to a cousin of the Liar paradox rather than to a theorem. But not so. As we will see, there really is nothing at all suspect or paradoxical about Gödel's First Theorem as a technical result about formal axiomatized systems (a result which in any case can be proved without appeal to 'self-referential' sentences).

'Hold on! If we locate G_T, a Gödel sentence for our favourite nicely axiomatized sound theory of arithmetic T, and can argue that G_T is true-though-unprovable-in-T, why can't we just patch things up by adding it to T as a new axiom?' Well, to be sure, if we start off with the sound theory T (from which we can't deduce G_T), and add G_T as a new axiom, we will get an expanded sound theory $U = T + G_T$ from which we *can* quite trivially derive G_T. But we can now just re-apply Gödel's method to our improved theory U to find a new true-but-unprovable-in-U arithmetic sentence G_U that encodes 'I am unprovable in U'. So U again is incomplete. Thus T is not only incomplete but, in a quite crucial sense, is *incompletable*.

Let's emphasize this key point. There's nothing at all mysterious about a the-ory's failing to be negation-complete. Imagine the departmental administrator's 'theory' D which records some basic facts about the course selections of a group of students. The language of D, let's suppose, is very limited and can only be used to tell us about who takes what course in what room when. From the 'ax-ioms' of D we'll be able, let's suppose, to deduce further facts – such as that Jack and Jill take a course together, and that ten people are taking the advanced logic course. But if there's currently no relevant axiom in D about their classmate Jo, we might not be able to deduce either J = 'Jo takes logic' or $\neg J$ = 'Jo doesn't take logic'. In that case, D isn't yet a negation-complete story about the course selections of students.

However, that's just boring: for the 'theory' about course selection is no doubt completable (i.e. it can readily be expanded to settle every question that can be posed in its very limited language). By contrast, what gives Gödel's First Theorem its real bite is that it shows that any nicely axiomatized and sound theory of basic arithmetic must *remain* incomplete, however many new true axioms we give it.[4] (And again, we can weaken the soundness condition, and can – more or less – just require consistency for incompletability.)

1.3 More incompleteness

Incompletability does not just affect theories of basic arithmetic. Consider set theory, for example. Start with the empty set $\varnothing$. Form the set $\{\varnothing\}$ containing $\varnothing$ as its sole member. Then form the set containing the empty set we started off with plus the set we've just constructed. Keep on going, at each stage forming the set of all the sets so far constructed. We get the sequence

$$\varnothing, \{\varnothing\}, \{\varnothing, \{\varnothing\}\}, \{\varnothing, \{\varnothing\}, \{\varnothing, \{\varnothing\}\}\}, \ldots$$

This sequence has the structure of the natural numbers. We can pick out a first member (corresponding to zero); each member has one and only one successor; it never repeats. We can go on to define analogues of addition and multiplication. Now, any standard set theory allows us to define this sequence. So if we could have a negation-complete and sound axiomatized set theory, then we could, in particular, have a negation-complete theory of the fragment of set theory which provides us with an analogue of arithmetic. Adding a simple routine for translating the results for this fragment into the familiar language of basic arithmetic would then give us a complete sound theory of arithmetic. But Gödel's First Incompleteness Theorem tells us there can't be such a theory. So there cannot be a sound negation-complete set theory.

The point evidently generalizes: any sound axiomatized mathematical theory T that can define (an analogue of) the natural-number sequence and replicate enough of the basic arithmetic of addition and multiplication must be incomplete and incompletable.

1.4 Some implications?

Gödelian incompleteness immediately challenges what otherwise looks to be a really rather attractive suggestion about the status of basic arithmetic – namely the logicist idea that it all flows deductively using simple logic from a simple bunch of definitional truths that articulate the very ideas of the natural numbers, addition and multiplication.

But then, how *do* we manage somehow to latch on to the nature of the unending number sequence and the operations of addition and multiplication in a way that outstrips whatever rules and principles can be captured in definitions? At this point it can begin to seem that we must have a rule-transcending cognitive grasp of the numbers which underlies our ability to recognize certain 'Gödel

---

[4]What makes for being a 'nicely' axiomatized theory is the topic of Section 4.3.

sentences' as correct arithmetical propositions. And if you are tempted to think so, then you may well be further tempted to conclude that minds such as ours, capable of such rule-transcendence, can't be machines (supposing, reasonably enough, that the cognitive operations of anything properly called a machine can be fully captured by rules governing the machine's behaviour).

So already there's apparently a quick route from reflections about Gödel's First Theorem to some conclusions about the nature of arithmetical truth and the nature of the minds that grasp it. Whether those conclusions really follow will emerge later. For the moment, we have an initial idea of what the First Theorem says and why it might matter – enough, I hope, already to entice you to delve further into the story that unfolds in this book.

1.5   The unprovability of consistency

If we can derive even a modest amount of basic arithmetic in theory T, then we'll be able to derive $0 \neq 1$.[5] So if T *also* proves $0 = 1$, it is inconsistent. Conversely, if T is inconsistent, then – since we can derive anything in an inconsistent theory[6] – it can prove $0 = 1$. But we said that we can use numerical propositions to encode facts about what can be derived in T. So there will in particular be a *numerical* proposition Con_T that encodes the claim that we can't derive $0 = 1$ in T, i.e. encodes in a natural way the claim that T is consistent.

We know, however, that there is a numerical proposition which encodes the claim that G_T is unprovable: we have already said that it is G_T itself.

So this means that (half of) the conclusion of Gödel's official First Theorem, namely the claim that if T is consistent then G_T is unprovable, can *itself* be encoded by a numerical proposition, namely $\mathsf{Con}_T \rightarrow \mathsf{G}_T$. And now for another wonderful Gödelian insight. It turns out that the informal reasoning that we use, outside T, to show 'if T is consistent, then G_T is unprovable' is elementary enough to be mirrored by reasoning inside T (i.e. by reasoning with numerical propositions which encode facts about T-proofs). Or at least that's true so long as T satisfies conditions just a bit stronger than the official First Theorem assumes. So, again on modest assumptions, we can derive $\mathsf{Con}_T \rightarrow \mathsf{G}_T$ inside T.

But the official First Theorem has already shown that if T is consistent we can't derive G_T in T. So it immediately follows that if T is consistent it can't prove Con_T. *And that is Gödel's Second Incompleteness Theorem.* Roughly interpreted: nice theories that include enough basic arithmetic can't prove their own consistency.[7]

---

[5]We'll allow ourselves to abbreviate expressions of the form $\neg\sigma = \tau$ as $\sigma \neq \tau$.

[6]There are, to be sure, deviant non-classical logics in which this principle doesn't hold. In this book, however, we aren't going to say much more about them, if only because of considerations of space.

[7]That *is* rough. The Second Theorem shows that, if T is consistent, T can't prove Con_T, which is certainly *one* natural way of expressing T's consistency inside T. But couldn't there be some *other* sentence, Con'_T, which also in some good sense expresses T's consistency, where

1.6 More implications?

Suppose that there's a genuine issue about whether T is consistent. Then even before we'd ever heard of Gödel's Second Theorem, we wouldn't have been convinced of its consistency by a derivation of Con_T inside T. For we'd just note that if T were in fact inconsistent, we'd be able to derive any T-sentence we like in the theory – including a false statement of its own consistency!

The Second Theorem now shows that we would indeed be right not to trust a theory's announcement of its own consistency. For (assuming T includes enough arithmetic), if T entails Con_T, then the theory must in fact be *inconsistent*.

However, the real impact of the Second Theorem isn't in the limitations it places on a theory's proving its own consistency. The key point is this. If a nice arithmetical theory T can't even prove *itself* to be consistent, it certainly can't prove that a *richer* theory T^+ is consistent (since if the richer theory is consistent, then any cut-down part of it is consistent). Hence we can't use 'safe' reasoning of the kind we can encode in ordinary arithmetic to prove that other more 'risky' mathematical theories are in good shape. For example, we can't use unproblematic arithmetical reasoning to convince ourselves of the consistency of set theory (with its postulation of a universe of wildly infinite sets).

And *that* is a very interesting result, for it seems to sabotage what is called Hilbert's Programme, which is precisely the project of trying to defend the wilder reaches of infinitistic mathematics by giving consistency proofs which use only 'safe' methods. A great deal more about this in due course.

1.7 What's next?

What we've said so far, of course, has been extremely sketchy and introductory. We must now start to do better. After preliminaries in Chapter 2 (including our first example of a 'diagonalization' argument), we go on in Chapter 3 to introduce the notions of effective computability, decidability and enumerability, notions we are going to need in what follows. Then in Chapter 4, we explain more carefully what we mean by talking about an 'axiomatized theory' and prove some elementary results about axiomatized theories in general. In Chapter 5, we introduce some concepts relating specifically to axiomatized theories of arithmetic. Then in Chapters 6 and 7 we prove a pair of neat and relatively easy results – first that any sound and 'sufficiently expressive' axiomatized theory of arithmetic is negation incomplete, and then similarly for any consistent and 'sufficiently strong' axiomatized theory. For reasons that we will explain in Chapter 8, these informal results fall some way short of Gödel's own First Incompleteness Theorem. But they do provide a very nice introduction to some key ideas that we'll be developing more formally in the ensuing chapters.

T *does* prove Con'_T (and we avoid trouble because T doesn't prove $\mathsf{Con}'_T \to \mathsf{G}_T$)? We'll return to this question in Sections 31.6 and 36.1.

2 Functions and enumerations

We start by fixing some entirely standard notation and terminology for talking about functions (worth knowing anyway, quite apart from the occasional use we make of it in coming chapters). We next introduce the useful little notion of a 'characteristic function'. Then we explain the idea of enumerability and give our first example of a 'diagonalization argument' – an absolutely crucial type of argument which will feature repeatedly in this book.

2.1 Kinds of function

(a) Functions, and in particular functions from natural numbers to natural numbers, will feature pivotally in everything that follows.

Note though that our concern will be with *total* functions. A total one-place function maps each and every element of its *domain* to some unique corresponding value in its *codomain*. Similarly for many-place functions: for example, the total two-place addition function maps any two numbers to their unique sum.

For certain wider mathematical purposes, especially in the broader theory of computation, the more general idea of a *partial* function can take centre stage. This is a mapping f which does not necessarily have an output for each argument in its domain (for a simple example, consider the function mapping a natural number to its natural number square root, if it has one). However, we won't need to say much about partial functions in this book, and hence – by default – plain 'function' will henceforth always mean 'total function'.

(b) The conventional notation to indicate that the one-place total function f maps elements of the domain Δ to values in the codomain Γ is, of course, $f \colon \Delta \to \Gamma$. Let f be such a function. Then we say

1. The *range* of f is $\{f(x) \mid x \in \Delta\}$, i.e. the set of elements in Γ that are values of f for arguments in Δ. Note, the range of a function need not be the whole codomain.

2. f is *surjective* iff the range of f indeed *is* the whole codomain Γ – i.e. just if for every $y \in \Gamma$ there is some $x \in \Delta$ such that $f(x) = y$. (If you prefer that in plainer English, you can say that such a function is *onto*, since it maps Δ onto the whole of Γ.)

3. f is *injective* iff f maps different elements of Δ to different elements of Γ – i.e. just if, whenever $x \neq y$, $f(x) \neq f(y)$. (In plainer English, you can say that such a function is *one-to-one*.)

4. f is *bijective* iff it is both surjective and injective. (Or if you prefer, f is then a *one-one correspondence* between Δ and Γ.)[1]

These definitions generalize in natural ways to many-place functions that map two or more objects to values: but we needn't pause over this.

(c)   Our special concern, we said, is going to be with numerical functions. It is conventional to use '$\mathbb{N}$' for the set of natural numbers (which includes zero, remember). So $f\colon \mathbb{N} \to \mathbb{N}$ is a one-place total function, defined for all natural numbers, with number values. While $c\colon \mathbb{N} \to \{0, 1\}$, for example, is a one-place numerical function whose values are restricted to 0 and 1.

'$\mathbb{N}^2$' standardly denotes the set of ordered pairs of numbers, so $f\colon \mathbb{N} \to \mathbb{N}^2$ is a one-place function that maps numbers to ordered pairs of numbers. Note, $g\colon \mathbb{N}^2 \to \mathbb{N}$ is another *one*-place function which this time maps an ordered pair (which is one thing!) to a number. So, to be really pernickety, if we want to indicate a function like addition which maps *two* numbers to a number, we really need a notation such as $h\colon \mathbb{N}, \mathbb{N} \to \mathbb{N}$.

2.2   Characteristic functions

As well as talking about numerical *functions*, we will also be talking a lot about numerical *properties* and *relations*. But discussion of these can be tied back to discussion of functions using the following idea:

> The *characteristic function* of the numerical property P is the one-place function $c_P\colon \mathbb{N} \to \{0, 1\}$ such that if n is P, then $c_P(n) = 0$, and if n isn't P, then $c_P(n) = 1$. (So if P is the property of being even, then c_P maps even numbers to 0 and odd numbers to 1.)
>
> The characteristic function of the two-place numerical relation R is the two-place function $c_R\colon \mathbb{N}, \mathbb{N} \to \{0, 1\}$ such that if m is R to n, then $c_R(m, n) = 0$, and if m isn't R to n, then $c_R(m, n) = 1$.

The notion evidently generalizes to many-place relations in the obvious way.

The choice of values for the characteristic function is, of course, pretty arbitrary; any pair of distinct objects would do as the set of values. Our choice is supposed to be reminiscent of the familiar use of 0 and 1, one way round or the other, to stand in for *true* and *false*. And our selection of 0 rather than 1 for *true* – not the usual choice, but it was Gödel's – is merely for later neatness.

Now, the numerical property P partitions the numbers into two sets, the set of numbers that have the property and the set of numbers that don't. Its corresponding characteristic function c_P also partitions the numbers into two sets, the set of numbers the function maps to the value 0, and the set of numbers the function maps to the value 1. And these are of course exactly the *same*

---

[1] If these notions really *are* new to you, it will help to look at the on-line exercises.

partition both times. So in a good sense, *P and its characteristic function c_P encapsulate just the same information about a partition.* That's why we can typically move between talk of a property and talk of its characteristic function without loss of relevant information. Similarly, of course, for relations.

We will be making use of characteristic functions a lot, starting in the next chapter. But the rest of this chapter discusses something else, namely ...

2.3   Enumerable sets

Suppose that Σ is some set of items: its members might be natural numbers, computer programs, infinite binary strings, complex numbers or whatever. Then, as a suggestive first shot, we can say

> The set Σ is *enumerable* iff its members can – at least in principle – be listed off in some numerical order (a zero-th, first, second, ...) with every member appearing on the list; repetitions are allowed, and the list may be infinite.

It is tidiest to think of the empty set as the limiting case of an enumerable set; after all, it is enumerated by the empty list.

One issue with this rough definition is that, if we are literally to 'list off' elements of Σ, then we need to be dealing with elements which are either things that themselves can be written down (like finite strings of symbols), or which at least have standard representations that can be written down (in the way that natural numbers have numerals which denote them). That condition will be satisfied in most of the cases that interest us in this book; but we need the idea of enumerability to apply more widely.

A more immediate problem is that it is of course careless to talk about 'listing off' *infinite* sets as if we can complete the job. What we really mean is that any member of Σ will eventually appear on this list, if we go on long enough.

Let's give a more rigorous definition, then, that doesn't presuppose that we have a way of writing down the members of Σ, and doesn't imagine us actually making a list. So officially we will now say

> The set Σ is enumerable iff either Σ is empty or else there is a surjective function $f\colon \mathbb{N} \to \Sigma$ (so Σ is the range of f: we can say that such a function enumerates Σ).

This is equivalent to our original informal definition (at least in the cases we are most interested in, when it makes sense to talk of listing the members of Σ).

Proof   Both definitions trivially cover the case where Σ is empty. So concentrate on the non-empty cases.

Pretend we can list off all the members of Σ in some order, repetitions allowed. Count off the members of the list from zero, and define the function f as follows:

$f(n) =$ the n-th member of the list, if the list goes on that far, or $f(n) = f(0)$ otherwise. Then $f \colon \mathbb{N} \to \Sigma$ is evidently a surjection.

Suppose conversely that $f \colon \mathbb{N} \to \Sigma$ is a surjection. Then, if we evaluate f for the arguments $0, 1, 2, \ldots$ in turn, we generate a list of the corresponding values $f(0), f(1), f(2), \ldots$. This list eventually contains any given element of Σ, with repetitions allowed. $\boxtimes$

Note, however, that our official definition of enumerability *doesn't* put any restriction on the 'niceness' of the enumerating surjection $f \colon \mathbb{N} \to \Sigma$. It could be a function that a computer could readily handle; but equally, it could – as far as our current definition is concerned – be a 'wild' function making a quite arbitrary association between members of $\mathbb{N}$ and members of Σ. All that matters for enumerability is that there is *some* function f, however whacky, which pairs up numbers with elements of Σ in such a way that we don't run out of numbers before we've matched every element.

2.4   Enumerating pairs of numbers

(a) It is immediate that every set of natural numbers $\Sigma \subseteq \mathbb{N}$, is enumerable, whether it is finite or infinite. That is trivial if Σ is empty. Otherwise, suppose $s \in \Sigma$. Now define the function f such that $f(n) = n$ if $n \in \Sigma$, and $f(n) = s$ otherwise. Then f evidently enumerates Σ.

(b)   Here's a rather less obvious example of an enumerable set:

> **Theorem 2.1** *The set of ordered pairs of natural numbers $\langle i, j \rangle$ is enumerable.*

Proof   The idea is simple. Systematically arrange the ordered pairs as shown and zig-zag through them, going down successive northeast to southwest diagonals:

$$
\begin{array}{llll}
\langle 0,0 \rangle \to \langle 0,1 \rangle & \langle 0,2 \rangle & \langle 0,3 \rangle \ldots \\
\swarrow \quad \swarrow \quad \swarrow \quad \swarrow \\
\langle 1,0 \rangle \quad \langle 1,1 \rangle & \langle 1,2 \rangle & \langle 1,3 \rangle \ldots \\
\swarrow \quad \swarrow \quad \swarrow \\
\langle 2,0 \rangle \quad \langle 2,1 \rangle & \langle 2,2 \rangle & \langle 2,3 \rangle \ldots \\
\swarrow \quad \swarrow \\
\langle 3,0 \rangle \quad \langle 3,1 \rangle & \langle 3,2 \rangle & \langle 3,3 \rangle \ldots \\
\vdots \quad \swarrow \quad \vdots & \vdots & \vdots
\end{array}
$$

This procedure is entirely mechanical, runs through all the pairs, and evidently defines a bijection $f \colon \mathbb{N} \to \mathbb{N}^2$, a one-one correspondence mapping n to the n-th pair on the zig-zag path (so f is a fortiori a surjection). $\boxtimes$

11

It is easy to see that the ordered pair $\langle i, j \rangle$ will appear at the location $pair(i, j) = \{(i + j)^2 + 3i + j\}/2$ on the list (counting from zero). And if you have a taste for puzzles, you can likewise write down two readily calculable functions $fst(n)$ and $snd(n)$ which return, respectively, the first member i and the second member j of the n-th pair in the zig-zag enumeration. (Exercise!)

2.5 An indenumerable set: Cantor's theorem

(a) Some sets, however, are 'too big' to enumerate. We can readily find an example and thereby prove *Cantor's Theorem*:[2]

Theorem 2.2 *There are infinite sets that are not enumerable.*

Proof Consider the powerset of $\mathbb{N}$, in other words the collection $\mathcal{P}$ whose members are all the sets of numbers (so $X \in \mathcal{P}$ iff $X \subseteq \mathbb{N}$).

Suppose for reductio that there is a function $f \colon \mathbb{N} \to \mathcal{P}$ which enumerates $\mathcal{P}$, and consider what we'll call the diagonal set $D \subseteq \mathbb{N}$ such that $n \in D$ iff $n \notin f(n)$.

Since $D \in \mathcal{P}$ and f by hypothesis enumerates all the members of $\mathcal{P}$, there must be some number d such that $f(d) = D$.

So we have, for all numbers n, $n \in f(d)$ iff $n \notin f(n)$. Hence in particular $d \in f(d)$ iff $d \notin f(d)$. Contradiction!

There therefore cannot be such an enumerating function as f. Hence the power set $\mathcal{P}$ cannot be enumerated. In a word, it is *indenumerable*. ⊠

Which is very neat, though perhaps a little mysterious. For a start, why did we call D a 'diagonal' set? Let's therefore give a second, rather more intuitive, presentation of the same proof idea: this should make things clearer.

Another proof Consider this time the set $\mathbb{B}$ of infinite binary strings, i.e. the set of unending strings like '0110001010011 . . .'. There's an infinite number of different such strings. Suppose, for reductio, that there is an enumerating function f which maps the natural numbers onto the strings, for example like this:

$$
\begin{aligned}
0 &\rightarrow b_0 : \underline{0}110001010011\ldots \\
1 &\rightarrow b_1 : 1\underline{1}00101001101\ldots \\
2 &\rightarrow b_2 : 11\underline{0}0101100001\ldots \\
3 &\rightarrow b_3 : 000\underline{1}111010101\ldots \\
4 &\rightarrow b_4 : 1101\underline{1}11011101\ldots
\end{aligned}
$$

. . . . . .

Go down the diagonal, taking the n-th digit of the n-th string b_n (in our example, this produces 01011 . . .). Now flip each digit, swapping 0s and 1s (in our example, yielding 10100 . . .). By construction, this 'flipped diagonal' string d differs from

---

[2]Georg Cantor first established this key result in Cantor (1874), using the completeness of the reals. The neater 'diagonal argument' first appears in Cantor (1891).

b_0 in the first place; it differs from the next string b_1 in the second place; and so on. So our diagonal construction defines a new string d that differs from each of the b_j, contradicting the assumption that our map f is 'onto' and enumerates *all* the binary strings.

Now, our supposed enumerating function f was just one example; but the same 'flipped diagonal' construction will plainly work to show that any other candidate map must also fail to enumerate all the strings. So $\mathbb{B}$ is not enumerable. ⊠

A moment's reflection shows that our two proofs use essentially the same idea. For take an infinite binary string $b = \beta_0\beta_1\beta_2\ldots$ where each bit β_i is either zero or one. Now this string characterizes a corresponding set B of natural numbers, where $n \in B$ if $\beta_n = 0$ and $n \notin B$ if $\beta_n = 1$. So, for example, the first string b_0 above corresponds to the set of numbers $\{0, 3, 4, 5, 7, 9, 10, \ldots\}$.

Suppose, then, that we enumerate some binary strings $b_0, b_1, b_2, \ldots$, which respectively represent the sets of numbers B_0, B_1, B_2 .... 'Going down the diagonal' gives us a string representing the set K such that $n \in K$ iff $n \in B_n$. And 'flipping' gives us the string d representing the diagonal set $D = \overline{K}$, the complement of K (i.e. the set of numbers not in K). $n \in D$ iff $n \notin B_n$; hence D differs from each B_n, so can't appear anywhere in the enumeration.[3] Our second proof, about $\mathbb{B}$, therefore also shows that any enumeration of sets B_n leaves out some set of numbers. In other words, as we showed in the first proof, the set of sets of natural numbers can't be enumerated.

(b)   Let's just add two more quick comments about the second version of our simple but profound proof.

First, an infinite binary string $b = \beta_0\beta_1\beta_2\ldots$ can be thought of as characterizing a corresponding function f_b, i.e. the function that maps each natural number to one of the numbers $\{0, 1\}$, where $f_b(n) = \beta_n$. So our proof idea also shows that the set of *functions* $f\colon \mathbb{N} \to \{0, 1\}$ can't be enumerated. Put in terms of functions, the trick is to suppose that these functions *can* be enumerated $f_0, f_1, f_2, \ldots$, define another function by 'going down the diagonal and flipping digits', i.e. define $\delta(n) = f_n(n) + 1 \mod 2$, and then note that this diagonal function δ can't be 'on the list' after all.

Second, infinite strings $b = \beta_0\beta_1\beta_2\ldots$ not ending in an infinite string of 1s correspond one-to-one to the real numbers $0 \leq b \leq 1$. The same proof idea then shows that the real numbers in the interval $[0, 1]$ can't be enumerated (and hence that we can't enumerate *all* the reals either).

(c)   We'll meet another version of the 'diagonalization' trick again at the end of the very next chapter. And then, as we'll see, there's a lot more diagonalization to come later. In fact, it is the pivotal proof idea that we'll be repeatedly using in this book.

---

[3]Maybe we should really call K the diagonal set, and its complement D the *anti-diagonal* set. But we won't fuss about that.

3 Effective computability

The previous chapter talked about functions rather generally. We now narrow the focus and concentrate more specifically on *effectively computable* functions. Later in the book, we will want to return to some of the ideas we introduce here and give sharper, technical, treatments of them. But for present purposes, informal intuitive presentations are enough.

We also introduce the crucial related notion of an *effectively enumerable* set, i.e. a set that can be enumerated by an effectively computable function.

3.1 Effectively computable functions

(a) Familiar school-room arithmetic routines – e.g. for squaring a number or finding the highest common factor of two numbers – give us ways of *effectively computing* the value of some function for a given input: the routines are, we might say, entirely mechanical.

Later, in the logic classroom, we learn new computational routines. For example, there's a quite trivial syntactic computation which takes two well-formed formulae (wffs) and forms their conjunction, and there's an only slightly less trivial procedure for effectively computing the truth value of a propositional calculus wff as a function of the values of its atoms.

What is meant by talking of an *effective* computational procedure? The core idea is that an effective computation involves (1) executing an *algorithm* which (2) *successfully terminates*.

1. An algorithm is a set of step-by-step instructions (instructions which are pinned down in advance of their execution), with each small step clearly specified in every detail (leaving no room for doubt as to what does and what doesn't count as executing the step, and leaving no room for chance).

   The idea, then, is that executing an algorithm (i) involves an entirely determinate sequence of discrete step-by-small-step procedures (where each small step is readily executable by a very limited calculating agent or machine). (ii) There isn't any room left for the exercise of imagination or intuition or fallible human judgement. Further, in order to execute the algorithm, (iii) we don't have to resort to outside 'oracles' (i.e. independent sources of information), and (iv) we don't have to resort to random methods (coin tosses).

   In sum, we might say that executing an algorithm is something that can be done by a suitable deterministic computing machine.

2. But there's more: plainly, if executing an algorithm is actually to compute a total function – i.e. a function which outputs a value for any relevant input(s) – then the procedure must *terminate* in a finite number of steps for every input, and produce the right sort of output. Note, then, it isn't part of the very idea of an algorithm that its execution always terminates; so in general, an algorithm might only compute a partial function.

Putting these two thoughts together, then, we can give an initial definition:

A one-place total function $f: \Delta \to \Gamma$ is *effectively computable* iff there is an algorithm which can be used to calculate, in a finite number of steps, the value of the function for any given input from the domain Δ.

The generalization to many-place functions is obvious.

(b) Informally, then, the idea is that an effectively computable function is one that a computer could compute! But what kind of computer do we have in mind here? We need to say something more about the relevant sort of computer's *size and speed*, and about its *architecture*.

A real-life computer is limited in size and speed. There will be some upper bound on the size of the inputs it can handle; there will be an upper bound on the size of the set of instructions it can store; there will be an upper bound on the size of its working memory. And even if we feed in inputs and instructions which our computer can handle, it is of little use to us if it won't finish executing its algorithmic procedure for centuries.

Still, we are cheerfully going to abstract from all these 'merely practical' considerations of size and speed – which is why we said nothing about them in explaining what we mean by an effectively computable function. In other words, we will count a function as being effectively computable if there is a finite set of step-by-step instructions which a computer could in principle use to calculate the function's value for any particular arguments, given time and memory enough. Let's be very clear, then: 'effective' here does *not* mean that the computation must be feasible for us, on existing computers, in real time. So, for example, we still count a function as effectively computable in this broad sense even if on existing computers it might take longer to compute a value for a given input than we have time left before the heat death of the universe. It is enough that there's an algorithm that works in theory and would deliver a result in the end, if only we had the computational resources to use it and could wait long enough.

'But then,' you might well ask, 'why bother with such a radically idealized notion of computability? If we allow procedures that may not deliver an output in the lifetime of the universe, what good is that? If we are interested in issues of computability, shouldn't we really be concerned not with idealized-computability-in-principle but with some notion of *practicable* computability?'

That's a fair challenge. And modern computer science has much to say about grades of computational complexity and levels of feasibility. However, we will stick to our ultra-idealized notion of computability. Why? Because later we will be proving a range of limitative theorems, e.g. about what *can't* be algorithmically computed. By working with a very weak 'in principle' notion of what is required for being computable, our impossibility results will be correspondingly very strong – for a start, they won't depend on any mere contingencies about what is practicable, given the current state of our software and hardware, and given real-world limitations of time or resources. They show that some numerical functions can't be computed algorithmically, even on the most generous understanding of that idea.

(c)   We've said that we are going to be abstracting from limitations on storage, etc. But you might suspect that this still leaves much to be settled. Doesn't the 'architecture' of a computing device affect what it can compute?

The short answer is that it doesn't (at least, once we are dealing with devices of a certain degree of complexity, which can act as 'general purpose' computers). And intriguingly, some of the central theoretical questions here were the subject of intensive investigation even before the first electronic computers were built. Thus, in the mid 1930s, Alan Turing famously analysed what it is for a numerical function to be step-by-step computable in terms of the capacities of a *Turing machine* (a computer implementing an algorithm built up from particularly simple basic steps: for explanations and examples, see Chapter 41). Now, a standard Mark I Turing machine has just a single 'tape' or workspace to be used for both storing and manipulating data. But we can readily describe a rather more efficient Mark II machine which has two tapes – one to be used as a main workspace, and a separate one for storing data. Or, more radically, we can consider a computer with unlimited 'Random Access Memory' – that is to say, an idealized version of a modern computer, with an unlimited set of registers in which it can store various items of working data ready to be retrieved into its workspace when needed.[1] The details don't matter here and now. What does matter is that exactly the same functions are computable by algorithms written for Mark I Turing machines, by algorithms written for Mark II machines (or its variants), and by algorithms written for register machines, despite their different architectures.

Indeed, *all* the detailed definitions of algorithmic computability by idealized computers with different architectures that have ever been seriously proposed turn out to be equivalent. In a slogan, *algorithmic computability is architecture independent.*

(d)   It's worth pausing over that last major claim, and taking it a bit more slowly, in two stages.

First stage: there's a technical result about the mutual equivalence of various different proposed ways of refining the ideas of effectively computing *numerical* functions (i.e. where the domain and codomain are natural numbers) – we end

---

[1] The theoretical treatment of unlimited register machines was first given in Shepherdson and Sturgis (1963); there is a very accessible presentation in the excellent Cutland (1980).

up, every time, with the same class of Turing-computable numerical functions. That's a formal mathematical theorem (or rather cluster of theorems, that we have to prove case by case). But it supports the conjecture which Turing famously makes in his classic paper published in 1936:

> *Turing's Thesis* The numerical functions that are effectively computable in the informal sense are just those functions that are in fact computable by a suitable Turing machine.

As we'll see, however, Turing machines are horrible to work with (you have to program them at the level of 'machine code'). So you might want to think instead in terms of the numerical functions which – abstracting from limitations of time and memory space – are computable on a modern general purpose computer, using programs written in your favourite general purpose language, C++ perhaps. Then Turing's Thesis is provably equivalent to this: the numerical functions that are effectively computable in the informal sense are just those functions that are in principle computable using algorithms written in C++.

Turing's Thesis[2] – we'll further explore its content in Chapter 44 – correlates an informal notion with a sharp technical analysis. So you might think it isn't the sort of thing which we can strictly speaking *prove* (though see Chapter 45). But be that as it may. Certainly, after more than seventy five years, no successful challenge to Turing's Thesis has ever been mounted. Which means that we can continue to talk informally about effectively computable numerical functions, and yet be very confident that we are referring to a fully determinate class.

(e) Second stage: what about extending the idea of being effectively computable to *non-numerical* functions? Does the idea remain determinate?

Here's a natural suggestion: any computation dealing with sufficiently discrete and distinguishable finite objects – the Xs – can be turned into an equivalent numerical computation via the trick of using simple numerical codes for the different Xs. In other words, by a relatively trivial algorithm, we can map Xs to numbers; we can then do the appropriate core computation on the numbers; and then another trivial algorithm translates the result back into a claim about Xs.[3] So if it is determinate what numerical functions are effectively computable, the same goes for other functions over suitable finite objects.

Fortunately, however, we don't need to pause to assess this line of argument in its fullest generality. For the purposes of this book, the non-numerical computations we are most interested in are cases where the Xs are *expressions* from standard formal languages, or *sequences of expressions*, etc. And in these cases, there's no doubt at all that we can – in a quite uncontentious sense – algorithmically map claims about such things to corresponding claims about numbers

---

[2]Turing's Thesis has a twin you might have heard of, namely Church's Thesis: we'll explore that as well, and the interrelation between the two, in Chapters 38 and 44.

[3]After all, this is how real-world digital computers work: they code up non-numerical data into numbers using representations in binary form, operate on them, and then decode to present the results in some nice human-consumable form.

(see Sections 4.5, 19.1, 19.3). So we can indeed assume, given Turing's Thesis, that it is also determinate what counts as an effectively computable operation over expressions, etc.

3.2   Effectively decidable properties and sets

(a)   We often use algorithmic routines not only to compute *functions* but also to decide whether a *property* holds. For example, there are familiar school-room routines for algorithmically testing whether a given number is divisible by nine, or whether it is prime. Later, in the logic room, we learn computational routines for deciding whether a given string of symbols is a wff of the propositional calculus, and for deciding whether such a wff is a tautology.

Inspired by such cases, here is another definition:

> A property/relation is *effectively decidable* iff there is an algorithmic procedure that a suitably programmed computer could use to decide, in a finite number of steps, whether the property/relation applies to any appropriate given item(s).

Note that again 'effective' here doesn't mean 'practicable' or 'efficient'; a property can be effectively decidable even if it would take a vast amount of time and/or computational resources to settle whether a given object has it. All that is required is that there is some algorithm that in principle would do the job, '[h]ad we but world enough, and time'.[4]

(b)   In what follows, we are crucially going to be interested in decidable properties of numbers and decidable relations holding between numbers. And now we can deploy the notion of a characteristic function which we introduced in Section 2.2, and alternatively say

> A numerical property or relation is effectively decidable iff its characteristic function is effectively computable.

Let's just check that our two definitions do indeed tally in the case of numerical properties. So suppose there is an algorithm which decides whether n is P, delivering a 'yes'/'no' verdict: just add the instruction 'if the answer is *yes*, output 0: otherwise output 1' and you get an algorithm for computing $c_P(n)$.[5] Conversely, suppose you have an algorithm for computing $c_P(n)$. Then you can add the line 'if the value is 0 output *yes*: otherwise output *no*' to get an algorithm to decide whether n is P.

Since, by Turing's Thesis, the notion of an effectively computable numerical function is determinate, so too is the notion of an effectively decidable numerical property or relation.

---

[4] The phrase is Andrew Marvell's, the opening line of 'To his Coy Mistress' (c. 1650).
[5] If you think that gets the '0' and '1' the wrong way around, check Section 2.2 again.

So far, so good. What, though, about the idea of a non-numerical effectively decidable property? Can we be sure that *that* notion is determinate?

Again yes, at least in the cases that interest us. For as we said, we can code up finite objects like expressions or sequences of expressions using numbers in simple algorithmic ways. So the question whether e.g. a certain property of formulae is a decidable one can be uncontentiously translated into the question whether a corresponding property of numbers is a decidable one. Since it is quite determinate what counts as a decidable property of numbers, it then follows that it is quite determinate what counts as a decidable property of formal expressions or sequences of expressions.

(c)   Moving from talk of properties to talk of the sets which are their extensions,[6] we'll also now say that

> A set Σ is effectively decidable iff there is an algorithmic procedure that a suitably programmed computer could use to decide, in a finite number of steps, whether any relevant given object is a member of Σ.

For the particular case where Σ is a set of numbers, we can sharpen up this definition in a way parallel to our sharpened account of decidable properties:

> A set of natural numbers $\Sigma \subseteq \mathbb{N}$ is effectively decidable iff the characteristic function of the property of belonging to Σ (i.e. the function c_Σ such that $c_\Sigma(n) = 0$ if $n \in \Sigma$ and $c_\Sigma(n) = 1$ otherwise) is effectively computable.

Given Turing's Thesis, this makes the idea of an effectively decidable set of numbers another sharply determinate one. And again, at least when we are dealing with finite objects that can be nicely coded up by numbers, the idea of effectively decidable sets of those objects can be made equally determinate.

(d)   As a reality check, let's just pause to note a pair of mini-theorems:

> **Theorem 3.1** *Any finite set of natural numbers is effectively decidable.*

> **Theorem 3.2** *If Σ is an effectively decidable set of numbers, so is its complement $\overline{\Sigma}$.*

Proofs (if needed!)   For the first, note that if $\Sigma \subseteq \mathbb{N}$ is finite, the characteristic function c_Σ always takes the value 1 except for a finite number of arguments when it goes to 0. Such a function is effectively computable by a brute-force algorithm which checks the input against the finite list of exceptional arguments, outputs 0 if it gets a match and otherwise defaults to 1.

---

[6]Recall: the extension of a property is the set of things that have the property. The extension of a relation is the set of ordered pairs of things that stand in the relation. We also – by extension! – talk of the extension of a predicate which expresses the property or relation.

For the second, note that if the characteristic function c_Σ is effectively computable, so (trivially) is the function $\bar{c}$ defined by $\bar{c}(n) = 1 - c_\Sigma(n)$. But $\bar{c}$ is evidently the characteristic function of $\overline{\Sigma}$. $\boxtimes$

Since the complement of a finite set is infinite, it follows that there are infinite sets of numbers which are decidable. There are also infinite sets with infinite complements that are decidable, e.g. the set of primes. But we'll soon see that there are undecidable sets of numbers too.

3.3 Effective enumerability

(a)   We said: a non-empty set Σ is enumerable so long as there is *some* surjective function $f \colon \mathbb{N} \to \Sigma$ which enumerates it.

As we noted before, this definition does not require that the enumerating function be a nice computable one; in other words, the enumerating function f here can be any arbitrary correlation of numbers with elements of Σ (so long as it is 'onto'). It need not even be finitely specifiable. By contrast, then, we'll say

> The set Σ is *effectively enumerable* (e.e.) iff either Σ is empty or there is an effectively computable function that enumerates it.[7]

By the same reasoning as in Section 2.3, you can equivalently think of it like this (when the members of Σ are the sort of things you can put on a list): Σ is e.e. just if an (idealized) computer could be programmed to generate a list of Σ's members such that any member will eventually be mentioned – the list may be empty, or have no end, and may contain repetitions, so long as every item in the set eventually makes an appearance.

(b)   Let's have some examples, concentrating for now on numerical ones. First, any finite set of numbers is e.e.: any listing will do, and – being finite – can be stored in an idealized computer and spat out on demand.

For a trivial example of an e.e. infinite set, the computable function $f(n) = n^2$ effectively enumerates the natural numbers which are perfect squares.

And for an only slightly more complex case imagine an algorithm that takes the natural numbers one at a time in order and applies the well-known test for deciding whether a number is prime, and lists the successes: this procedure generates a never-ending list on which every prime will eventually appear – so the primes are effectively enumerable.

Evidently, that argument about primes generalizes:

> **Theorem 3.3** *If Σ is an effectively decidable set of numbers, it is effectively enumerable.*

---

[7]NB: whether a set is effectively enumerable, enumerable but not effectively so, or neither, depends on what functions there *are*, not on which functions we *know* about. Also note that terminology hereabouts isn't entirely stable: some writers use 'enumerable' to mean *effectively* enumerable, and use e.g. 'denumerable' for the wider notion of enumerability.

Proof The case where Σ is empty is trivial. So assume $s \in \Sigma$, and consider the algorithm which, for input n, effectively tests whether n is in Σ (by hypothesis, that can be done), and if it gets a 'yes' outputs n, and otherwise outputs s. This algorithm computes a total surjective function $f \colon \mathbb{N} \to \Sigma$, so Σ is effectively enumerable. $\boxtimes$

(c) Does the result reverse? If a set Σ is effectively enumerable must it be effectively decidable?

Intuitively, that *shouldn't* hold in general. Suppose we want to decide whether $s \in \Sigma$. If we are only given that there is a computable enumerating function $f \colon \mathbb{N} \to \Sigma$, then it seems that all we can do is start evaluating $f(0), f(1), f(2), \ldots$. If eventually we find some n such that $f(n) = s$, that settles it: $s \in \Sigma$. But if we haven't (yet) found such an n, everything is still to play for; perhaps we've just not looked long enough and it will still turn out that $s \in \Sigma$, or perhaps $s \notin \Sigma$. A finite search along the $f(n)$, however long, may not settle anything.

However, we do have the following important little theorem:

> **Theorem 3.4** *If Σ and also its complement $\overline{\Sigma}$ are both effectively enumerable sets of numbers, then Σ is effectively decidable.*

Proof Suppose Σ is enumerated by the effectively computable function f, and $\overline{\Sigma}$ by g. Here's how to effectively determine, of any given number s, whether it is in Σ or not.

Compute in turn $f(0), g(0), f(1), g(1), f(2), g(2), \ldots$. Eventually we must get the output value s (since either $s \in \Sigma$ or $s \in \overline{\Sigma}$). If we get to some m such that $f(m) = s$, then the algorithm tells us to return the verdict that $s \in \Sigma$; if we get to some n such that $g(n) = s$, then the algorithm tells us to return the verdict that $s \in \overline{\Sigma}$, i.e. $s \notin \Sigma$. $\boxtimes$

We are, of course, relying here on our ultra-generous notion of decidability-in-principle. We might have to twiddle our thumbs for an immense time to see whether s is in Σ or is in $\overline{\Sigma}$. Still, our 'wait and see' method is guaranteed by our assumptions to produce a result in finite time, in an entirely mechanical way – so this counts as an effective procedure in our official generous sense.

So, if every e.e. set of numbers had an effectively enumerable complement, then every e.e. set of numbers would be decidable after all, contrary to our intuitive line of thought above. But as we'll see in Section 3.5, some e.e. sets have complements which are not e.e., and there are indeed undecidable e.e. sets of numbers.

3.4 Another way of defining e.e. sets of numbers

To lead up to our key proof in the next section, this section introduces an alternative way of characterizing the e.e. sets of numbers.

(a)   A one-place total numerical function $f \colon \mathbb{N} \to \mathbb{N}$ is effectively computable, we said, iff there is an algorithmic procedure Π that can be used to compute its value for any given number as input. But of course, not every algorithm Π computes a total numerical one-place function. Many will just do nothing or get stuck in a loop when fed a number as input. Other algorithms deliver outputs for some numbers as inputs but not for others (i.e. compute partial functions). So let's introduce the following definition:

> The *numerical domain* of an algorithm Π is the set of natural numbers n such that, when the algorithm Π is applied to the number n as input, then the run of the algorithm will (in principle) eventually terminate and deliver some number as output.

Then we can prove a perhaps rather surprising result: whatever the algorithm Π, its numerical domain will *always* be an effectively enumerable set of numbers. Indeed, we can make that a biconditional:

> **Theorem 3.5** *W is an effectively enumerable set of numbers if and only if it is the numerical domain of some algorithm Π.*

Proof of the 'only if' direction   Suppose W *is* an e.e. set of numbers. Then by definition either (i) W is empty, or (ii) there is an effectively computable function f which enumerates W (so $n \in W$ iff for some i, $f(i) = n$).

In case (i), choose any algorithm that never produces any output and we are done.

In case (ii) there must be some algorithm Π which computes the total function f. We can use Π in constructing the following more complex algorithm Π^+: given number n as input, loop around using Π to compute the values of $f(0), f(1), f(2), \ldots$ in turn, and keep going on and on, unless and until for some i, $f(i) = n$ – in which case stop, and output the number i. Then the numerical domain of Π^+ is obviously W (for Π^+ will terminate just when fed an $n \in W$).

Proof of the 'if' direction   Suppose that W is the numerical domain of some algorithm Π. Then basically what we want to do is to interleave runs of Π on inputs $0, 1, 2 \ldots$, and spit out again those inputs for which Π terminates with some output, giving us an effective enumeration of W.[8] Here is a way of implementing this idea.

If W is empty, then trivially it is effectively enumerable. So suppose W isn't empty and o is some member of it. And now recall the pairing functions we introduced in Section 2.4. Each possible pair of numbers $\langle i, j \rangle$ gets effectively correlated one-to-one with a number n, and there are computable functions $fst(n)$ and $snd(n)$ which return, respectively, the first member i and the second member j of the n-th pair. Using these, define the new algorithm Π' as follows:

---

[8]Why do we interleave rather than do the runs on different inputs sequentially? Well, consider what would happen if Π runs forever on some inputs.

Given input n, compute $i = fst(n)$, and $j = snd(n)$. Then run Π on input i for j steps (we defined algorithms as involving discrete step-by-step procedures, so we can number the steps). If Π on input i has halted with some output by step j, then Π' outputs i. Otherwise Π' outputs the default value o.

As n increases, this procedure Π' evaluates Π for every input i, for any number of steps j; it outputs just those arguments i for which Π eventually delivers some output. So Π' computes a total function whose range is the whole of Π's numerical domain W. Hence W is indeed effectively enumerable. $\boxtimes$

(b)   Now let's fix ideas, and suppose we are working within some particular general-purpose programming language like C++. If there's an algorithm for computing a numerical function f at all, then we can implement it in this language. Of course, it is a non-trivial claim that there *are* general purpose languages like this, apt for regimenting instructions for all kinds of numerical algorithms; still, even a passing familiarity with modern computing practice should make this assumption seem entirely reasonable.[9] Our last theorem then comes to this: W is an e.e. set of numbers iff it is the numerical domain of some algorithm regimented in our favourite general-purpose programming language.

Now start listing off all the possible strings of symbols of our chosen programming language (all the length 1 strings in some 'alphabetical order', then all the length 2 strings in order, then all the length 3 strings, . . .), and retain just those that obey the rules for being a series of syntactically well-formed program instructions in that language. This gives us an effectively generated list of all the possible algorithms $\Pi_0, \Pi_1, \Pi_2, \ldots$ as stated in our language (most of them will be garbage of course, useless algorithms which 'crash'). Let the numerical domain of Π_e be W_e. Then our last theorem tells us that every effectively enumerable set of numbers is W_e for some index e. Which implies

Theorem 3.6 *The set $\mathcal{W}$ of all effectively enumerable sets of natural numbers is itself enumerable.*

For take the function $f : \mathbb{N} \to \mathcal{W}$ defined by $f(e) = W_e$: this function enumerates $\mathcal{W}$.[10]

We can immediately deduce an important corollary:

Theorem 3.7 *Some sets of numbers are not effectively enumerable, and hence not effectively decidable.*

Proof   We already know that the powerset $\mathcal{P}$ of $\mathbb{N}$ – i.e. the collection of *all* sets of numbers *can't* be enumerated (see the first proof of Theorem 2.2). But we've just seen that $\mathcal{W}$, the set of effectively enumerable sets of numbers, *can*

---

[9]Much later, when we return to the discussion of Turing's Thesis, we'll say more in defence of the assumption.

[10]Question: is f an *effective* enumeration?

be enumerated. So $\mathcal{W} \neq \mathcal{P}$. But trivially, $\mathcal{W} \subseteq \mathcal{P}$. So there must be members of $\mathcal{P}$ which aren't in $\mathcal{W}$, i.e. sets of numbers which aren't effectively enumerable. Theorem 3.3 entails that these sets, a fortiori, aren't effectively decidable.   ⊠

3.5   The Basic Theorem about e.e. sets

With the simple but crucial Theorem 3.5 to hand, we can now prove what arguably deserves to be called the *Basic Theorem* about effectively enumerable sets of numbers (this is our first deep result – savour it!):

> **Theorem 3.8** *There is an effectively enumerable set of numbers K such that its complement $\overline{K}$ is not effectively enumerable.*

Proof   We use another diagonal construction. Put $K =_{\text{def}} \{e \mid e \in W_e\}$ and the result follows quickly:[11]

$\overline{K}$ *is not effectively enumerable.*   By definition, for any e, $e \in \overline{K}$ if and only if $e \notin W_e$. Hence, $\overline{K}$ cannot be identical with any given W_e (since e is in one but not the other). Therefore $\overline{K}$ isn't one of the effectively enumerable sets (since the W_e are all of them).

K *is effectively enumerable.*   We use a little variation of the argument idea that we used in proving Theorem 3.5.

Since $\overline{K}$ is not effectively enumerable, $\overline{K}$ isn't the whole of $\mathbb{N}$ (for that is trivially effectively enumerable!), so K isn't empty. So let o be some member of K. Now consider the effective procedure Π'' defined as follows:

> Given input n, compute $i = \mathit{fst}(n)$, and $j = \mathit{snd}(n)$. Then find the algorithm Π_i, and run it on input i for j steps. If Π_i on input i has halted with some output by step j, then Π'' outputs i. Otherwise Π'' outputs the default value o.

As n increases, this procedure runs through all pairs of values i, j; so the output of Π'' is the set of *all* numbers i such that i is in the numerical domain of Π_i, i.e. is the set of i such that $i \in W_i$, i.e. is K. So K is effectively enumerable.   ⊠

As an immediate corollary we can strengthen Theorem 3.7 which told us that some sets of numbers are not decidable. Now, more particularly, we have

> **Theorem 3.9** *Some effectively enumerable sets of numbers are not decidable*

Proof   Take K again, an e.e. set with a non-e.e. complement. If K were decidable, its complement $\overline{K}$ would be decidable too, by mini-Theorem 3.2. But then $\overline{K}$ would be e.e., by mini-Theorem 3.3. Contradiction.   ⊠

---

[11] Compare the analogous construction in our comments on Cantor's Theorem 2.2.

4 Effectively axiomatized theories

Gödel's Incompleteness Theorems tell us about the limits of theories of arithmetic. More precisely, they tell us about the limits of *effectively axiomatized formal theories* of arithmetic. But what exactly does that mean?

4.1 Formalization as an ideal

Rather than just dive into a series of definitions, it is well worth pausing to remind ourselves of why we might *care* about formalizing theories.

So let's get back to basics. In elementary logic classes, beginners are drilled in translating arguments into an appropriate formal language and then constructing formal deductions of the stated conclusions from given premises.

Why bother with formal languages? Because everyday language is replete with redundancies and ambiguities, not to mention sentences which simply lack clear truth-conditions. So, in assessing complex arguments, it helps to regiment them into a suitable artificial language which is expressly designed to be free from obscurities, and where surface form reveals logical structure.[1]

Why bother with formal deductions? Because everyday arguments often involve suppressed premises and inferential fallacies. It is only too easy to cheat. Setting out arguments as formal deductions in one style or another enforces honesty: we have to keep a tally of the premises we invoke, and of exactly what inferential moves we are using. And honesty is the best policy. For suppose things go well with a particular formal deduction. Suppose we get from the given premises to some target conclusion by small inference steps each one of which is obviously valid (no suppressed premises are smuggled in, and there are no suspect inferential moves). Our honest toil then buys us the right to confidence that our premises really do entail the desired conclusion.

Granted, outside the logic classroom we almost never set out deductive arguments in fully formalized versions. No matter. We have glimpsed a first ideal – arguments presented in an entirely perspicuous formal language with maximal clarity and with everything entirely open and above board, leaving no room for misunderstanding, and with all the arguments' commitments systematically and frankly acknowledged.[2]

---

[1] Of course the benefits may come at a price; for it is not always clear how faithful the regimentation is to the intentions of the original. Still, investigating *that* can itself be a useful clarificatory exercise.

[2] For an early and very clear statement of this ideal, see Frege (1882), where he explains the point of the first modern formal system of logic – albeit with a horrible notation – presented in his *Begriffsschrift* (i.e. *Conceptual Notation*) of 1879.

Traditional presentations of Euclidean geometry illustrate the pursuit of a related second ideal – the axiomatized theory. The aim is to discipline a body of knowledge by showing how everything follows from a handful of basic assumptions. So, like beginning logic students, school students used to be drilled in providing deductions, though these deductions were framed in ordinary geometric language. We thereby establish a whole body of theorems about (say) triangles inscribed in circles, by deriving them from simpler results, that in turn can ultimately be established by appeal to some small stock of fundamental principles or axioms. And what is gained by doing this? Suppose the derivations of our various theorems are set out in a laborious step-by-step style – where each small move is warranted by simple inferences from propositions that have already been proved.[3] We thereby develop a unified body of results that we can be confident must hold if the initial Euclidean axioms indeed are true.[4]

Old-style axiomatized geometry was presented very informally. These days, many mathematical theories are presented axiomatically in a much more formal style from the outset. For example, set theory is typically presented by laying down some axioms expressed in a partially formalized language and exploring their deductive consequences. The aim, again, is to discover exactly what is guaranteed by the fundamental principles embodied in the axioms. However, even the most tough-minded mathematics texts which explore axiomatized theories continue to be written in an informal mix of ordinary language and mathematical symbolism. Proofs are very rarely spelt out in every formal detail, and so their presentation still falls short of the logical ideal of full formalization.

But we will hope that nothing stands in the way of our more informally presented mathematical proofs being sharpened up into fully formalized ones – i.e. we hope that they *could* be set out in a strictly regimented formal language of the kind that logicians describe, with absolutely every inferential move made totally explicit and checked as being in accord with some overtly acknowledged rules of inference, with all the proofs ultimately starting from our stated axioms. True, the extra effort of laying out everything in this kind of detail will almost never be worth the cost in time and ink. In mathematical practice we use enough formalization to convince ourselves that our results don't depend on illicit smuggled premisses or on dubious inference moves, and leave it at that – our motto is 'sufficient unto the day is the rigour thereof'.[5] Still, it *is* absolutely essential for good mathematics to achieve precision and to avoid the use of unexamined

---

[3] We will also want to allow inferences from temporary assumptions that are later discharged, as in arguments by reductio ad absurdum. But let's not worry about rounding out the story here; we are just sketching the Big Picture.

[4] Or at least, that's what is supposed to happen (on the surface, school geometry perhaps doesn't seem very deep: yet making all its fundamental assumptions fully explicit in fact turns out to be surprisingly difficult). For a classic defence, extolling the axiomatic method in mathematics, see Hilbert (1918).

[5] 'Most mathematical investigation is concerned not with the analysis of the complete process of reasoning, but with the presentation of such an abstract of the proof as is sufficient to convince a properly instructed mind.' (Russell and Whitehead, 1910–13, vol. 1, p. 3)

inference rules or unacknowledged assumptions. So, putting together the logician's aim of perfect clarity and honest inference with the mathematician's project of regimenting a theory into a tidily axiomatized form, we can see the point of the notion of an *axiomatized formal theory* as a composite ideal.

Note, we are not saying that mathematicians somehow fall short in not routinely working inside fully formalized theories. Mathematics is hard enough even when done using the usual strategy of employing just as much rigour as seems appropriate to the case in hand. And, as mathematicians (and some philosophical commentators) are apt to stress, there is a lot more to mathematical practice than striving towards the logical ideal.[6] For a start, we typically aim for proofs which are not merely correct but *explanatory* – i.e. proofs which not only show that some proposition must be true, but in some sense make it clear *why* it is true. However, such observations don't affect our present point, which is that the business of formalization just takes to the limit features that we expect to find in good proofs anyway, i.e. precise clarity and lack of inferential gaps.

4.2 Formalized languages

Putting together the ideal of formal precision and the ideal of regimentation into an axiomatic system, we have arrived then at the concept of an *axiomatized formal theory* – i.e. a theory built in a formalized *language*, with a set of formulae from the language which are treated as *axioms* for the theory, and a *deductive system* for proof-building, so that we can derive theorems from the axioms.

In this section, we'll say just a bit more about the idea of a properly formalized language: our concern now is to highlight points about effective decidability.

(a) But first let's stress that we are normally interested in *interpreted* languages – i.e. we are concerned with formal expressions which have some intended significance, which can be true or false. After all, our formalized proofs are ideally supposed to be just that, i.e. *proofs* with content, which show things to be true.

Agreed, we'll often be very interested in certain features of proofs that can be assessed independently of their significance (for example, we will want to know whether a putative proof does obey the formal syntactic rules of a given deductive system). But it is one thing to set aside their semantics for some purposes; it is another thing entirely to drain formal proofs of all semantic significance.

We will think of a formal language L, therefore, as being a pair $\langle \mathcal{L}, \mathcal{I} \rangle$, where $\mathcal{L}$ is a syntactically defined system of expressions and $\mathcal{I}$ gives the interpretation of these expressions. In the next chapter, we'll give an account of the syntax and semantics of the particular language L_A, a formal counterpart of what we called

---

[6]See Lakatos (1976) for a wonderful exploration of how mathematics evolves. This gives some real sense of how regimenting proofs in order to clarify their assumptions – the process which formalization idealizes – is just one phase in the complex process that leads to the growth of mathematical knowledge.

'the language of basic arithmetic' in Section 1.1. But for the moment, we'll stick to generalities.

(b)   Start with L's syntactic component $\mathcal{L}$. We can, for our purposes, assume that this is based on a finite alphabet of symbols (for we can always construct e.g. an unending supply of variables from a finite base by standard tricks like using repeated primes to yield 'x', 'x′', 'x″', etc.).[7] Then:

1. We first need to specify which symbols or finite strings of symbols make up $\mathcal{L}$'s *non-logical vocabulary*, e.g. the individual constants (names), predicates, and function-signs of L.

2. We also need to settle which symbols or strings of symbols make up $\mathcal{L}$'s *logical vocabulary*: typically this will comprise at least variables (perhaps of more than one kind), symbols for connectives and quantifiers, the identity sign, and bracketing devices.

3. Now we turn to syntactic constructions for building up more complex expressions from the logical and non-logical vocabulary. The *terms* of L, for example, are what you can construct by applying and perhaps re-applying function expressions to constants and/or variables. Then we need rules to determine which sequences of symbols constitute $\mathcal{L}$'s *wffs* (its well-formed formulae). For technical purposes it can be useful to allow wffs with free variables; but of course, our main interest will be in the *$\mathcal{L}$-sentences*, i.e. the closed wffs without variables dangling free.

This general pattern should be very familiar: but now we need to make something explicit which is often left unsaid in introductory presentations. Given that the whole point of using a formalized language is to make everything as clear and determinate as possible, we plainly do not want it to be a disputable matter whether a given symbol or cluster of symbols is e.g. a constant or one-place predicate or two-place function of a system $\mathcal{L}$. Nor, crucially, do we want disputes about whether a given string of symbols is an $\mathcal{L}$-wff or, more specifically, is an $\mathcal{L}$-sentence. So, whatever the fine details, for a properly formalized syntax $\mathcal{L}$, there should be clear and objective procedures, agreed on all sides, for *effectively deciding* whether a putative constant-symbol really is a constant, a putative one-place predicate is indeed one, etc. Likewise we need to be able to effectively decide whether a string of symbols is an $\mathcal{L}$-wff/$\mathcal{L}$-sentence. And, if we are to be able to give such wffs unique interpretations, it must also be effectively decidable how the wffs parse uniquely into their constituent parts. It goes almost without saying that the formal languages familiar from elementary logic standardly have these features.

---

[7] If you have done some model theory, then you'll have encountered an exception: for there it is a useful dodge to construct 'languages' in an extended sense which have one primitive constant for every element of a domain, even when the domain has a very large cardinality (e.g. by using the element as a symbol for itself). But these are languages-as-formal-objects-we-can-theorize-about, not languages which we could actually master and use to express theories.

(c) Let's move on, then, to the interpretation $\mathcal{I}$. The prime aim is, of course, to fix the content of each closed $\mathcal{L}$-wff, i.e. each $\mathcal{L}$-sentence. And standardly, we fix the content of formal sentences by giving truth-conditions, i.e. by saying what it would take for a given sentence to be true.

However, we can't, in the general case, do this in a manageable way just by giving a list associating $\mathcal{L}$-sentences with truth-conditions (for the simple reason that there will be an unlimited number of sentences). We therefore have to aim for a 'compositional semantics', which tells us how to systematically work out the truth-condition of any $\mathcal{L}$-sentence in terms of the semantic significance of the expressions which it contains.

What does such a compositional semantics $\mathcal{I}$ look like? The basic pattern should again be very familiar. First we fix a domain for our language's variables to range over (or domains, if there are different sorts of variables). We then fix the semantic significance of the non-logical vocabulary – i.e. we assign values to the individual constants, give satisfaction conditions for predicates, and assign functions to the function symbols. Then we explain how to assign truth-conditions to the atomic wffs. For example, in the simplest case, the wff 'φa' is true iff the value of a satisfies the predicate φ. Finally, we have to explain the truth-conditions of the more complex wffs built up from atomic ones using connectives and quantifiers running over the assigned domain(s).

As promised, we will spell out this kind of story for the particular case of the language L_A in the next chapter, so we need not go into more detail here. For now, the point we want to highlight is just this: given the aims of formalization, a compositional semantics needs to yield an *unambiguous* truth-condition for each sentence, and moreover to do this *in an effective way*.

The usual accounts of the semantics of the standard formal languages of logic do indeed have this feature of effectively generating unique readings for sentences. Careful, though! The claim is only that the semantics effectively tells us the conditions under which a given sentence is true; it doesn't at all follow that there is an effective way of telling whether those conditions actually hold and whether the sentence really *is* true.

4.3 Formalized theories

Now for the idea of an axiomatized formal theory, built in a formalized language (normally, of course, an interpreted one). Once more, it is issues about decidability which need to be emphasized: we do this in (a) and (b) below, summarize in (c), and add an important remark in (d).

(a) First, some sentences of our theory's language are to be selected as *(non-logical) axioms*, i.e. as the fundamental non-logical assumptions of our theory. Of course, we will normally want these axioms to be sentences which are true on interpretation; but note that this is not built into the very notion of an axiomatized theory.

Since the fundamental aim of the axiomatization game is to see what follows
from a bunch of axioms, we certainly don't want it to be a matter for dispute
whether a given proof does or doesn't appeal only to axioms in the chosen set.
Given a purported derivation of some result, there should be an absolutely clear
procedure for settling whether the input premises are genuinely to be found
among the official axioms. In other words, for a usefully axiomatized formal
theory, *we must be able to effectively decide whether a given sentence is an axiom
or not.*

That doesn't, by the way, rule out theories with infinitely many axioms. We
might want to say 'every wff of such-and-such a form is an axiom' (where there
is an unlimited number of instances): that's permissible so long as it is still
effectively decidable what counts as an instance of that form.

(b)   Second, an axiomatized formal theory needs some deductive apparatus, i.e.
some sort of formal *proof system.* And we'll take proof derivations always to be
finite arrays of wffs, arrays which are built up in ways that conform to the rules
of the relevant proof system.[8]

We will take it that the core idea of a proof system is once more very familiar from elementary logic. The differences between various equivalent systems of
proof presentation – e.g. old-style linear proof systems which use logical axioms
vs. different styles of natural deduction proofs vs. tableau (or 'tree') proofs –
don't essentially matter. What is crucial, of course, is the strength of the overall system we adopt. We will predominantly be working with some version of
standard first-order logic with identity. But whatever system we adopt, we need
to be able to specify it in a way which enables us to settle, without room for
dispute, what counts as a well-formed derivation.

In other words, *we require the property of being a well-formed proof from
premises $\varphi_1, \varphi_2, \ldots, \varphi_n$ to conclusion ψ in the theory's proof system to be an
effectively decidable one.* The whole point of formalizing proofs is to set out
the deductive structure of an argument with absolute determinacy; so we don't
want it to be a disputable or subjective question whether the inference moves
in a putative proof do or do not conform to the rules for proof-building for the
formal system in use. Hence there should be a clear and effective procedure for
deciding whether a given array of symbols counts as a well-constructed derivation
according to the relevant proof system.[9]

---

[8]We are not going to put any finite upper bound on the permissible length of proofs. So you
might well ask: why not allow infinite arrays to count as proofs too? And indeed, there is some
interest in theorizing about infinite proofs. For example, there are formal systems including
the so-called ω-rule, which says that from the infinite array of premises $\varphi(0)$, $\varphi(1)$, $\varphi(2)$, $\ldots$,
$\varphi(n)$, $\ldots$ we can infer $\forall x \varphi(x)$ where the quantifier runs over all natural numbers.

But do note that finite minds can't really take in the infinite number of separate premises
in an application of the ω-rule: that's an impossible task. Hence, in so far as the business
of formalization is primarily concerned to regiment and formalize the practices of ordinary
mathematicians, albeit in an idealized way, it's natural at least to start by restricting ourselves
to finite proofs, even if we don't put any particular bound on the length of proofs.

[9]When did the idea clearly emerge that properties like being a wff or an axiom or a proof

Again be very careful here! The claim is only that it should be effectively decidable whether an array of wffs which is presented as a well-constructed derivation really *is* a derivation in the given proof-system. This is *not* to say that we can always decide in advance whether a derivation from given premises to a target conclusion exists to be discovered. Even in familiar first-order quantificational logic, for example, it is not in general decidable in advance whether there exists a proof from certain premises to a given conclusion (we will be proving this really rather remarkable undecidability result later, in Section 40.3).

(c)   For a nicely axiomatized formal theory T, then, we want it to be effectively decidable which wffs are its logical or non-logical axioms and also want it be be effectively decidable which arrays of wffs conform to the derivation rules of T's proof system. It will therefore also be decidable which arrays of wffs are axiomatic T-proofs – i.e. which arrays are properly constructed proofs, all of whose premises are indeed T-axioms.

So, to summarize these desiderata on decidability,

> T is an (interpreted) *effectively axiomatized theory* just if (i) T is couched in an (interpreted) formalized language $\langle \mathcal{L}, \mathcal{I} \rangle$, such that it is effectively decidable what is a wff/sentence of $\mathcal{L}$, and what the unique truth-condition of any sentence is, etc., (ii) it is effectively decidable which $\mathcal{L}$-wffs are axioms of T, (iii) T has a proof system such that it is effectively decidable whether an array of $\mathcal{L}$-wffs conforms to the proof-building rules, and hence (iv) it is effectively decidable whether an array of $\mathcal{L}$-wffs constitutes a proof from T's axioms.

(d)   It needs to be remarked that logic texts often define an axiomatized theory more generously, to be a set of sentences which are the consequences of some set of axioms Σ, without any constraints on Σ. *In this book, however, when we talk of axiomatized theories, we are always going to mean formal theories where it is effectively decidable what is in the relevant set of axioms Σ and effectively decidable what is a well-formed proof*: we will usually explicitly write *effectively axiomatized theory* to highlight this key point. It is only theories in this preferred sense which Gödel's theorems apply to.

4.4   More definitions

Here are six standard definitions. They do in fact apply to theories more generally, but our interest will of course be in cases where we are dealing with effectively axiomatized theories of the kind we have just been discussing. We have already met some of these defined notions.

---

ought to be decidable? It was arguably already implicit in Hilbert's conception of rigorous proof. But Richard Zach has suggested that an early source for the *explicit* deployment of the idea is von Neumann (1927); see (Mancosu et al., 2008, pp. 447–48, endnote 76).

1. Given a derivation of the *sentence* φ from the axioms of the theory T using the background logical proof system, we will say that φ is a *theorem* of T. Using the standard abbreviatory symbol, we write: $T \vdash \varphi$.

2. A theory T is *sound* iff every theorem of T is true (i.e. true on the interpretation built into T's language). Soundness is, of course, normally a matter of having true axioms and a truth-preserving proof system.

3. A theory T is *effectively decidable* iff the property of being a theorem of T is an effectively decidable property – i.e. iff there is an algorithmic procedure for determining, for any given sentence φ of T's language, whether or not $T \vdash \varphi$.

4. Assume now that T has a standard negation connective '$\neg$'. A theory T *decides* the sentence φ iff either $T \vdash \varphi$ or $T \vdash \neg\varphi$. A theory T *correctly decides* φ just when, if φ is true (on the interpretation built into T's language), $T \vdash \varphi$, and if φ is false, $T \vdash \neg\varphi$.

5. A theory T is *negation-complete* iff T decides every sentence φ of its language (i.e. for every sentence φ, either $T \vdash \varphi$ or $T \vdash \neg\varphi$).

6. T is *inconsistent* iff for some sentence φ, we have both $T \vdash \varphi$ and $T \vdash \neg\varphi$.

Note our decision to restrict the theorems, properly so-called, to the derivable *sentences*: so wffs with free variables derived as we go along through a proof don't count. This decision is for convenience as much as anything, and nothing deep hangs on it.

Here's a very elementary toy example to illustrate some of these definitions. Consider a trivial pair of theories, T_1 and T_2, whose shared language consists of the (interpreted) propositional atoms 'p', 'q', 'r' together with all the wffs that can be constructed from them using the familiar propositional connectives, whose shared underlying logic is a standard natural deduction system for propositional logic, and whose axioms are respectively

$T_1 : \neg$p,

$T_2 : \neg$p, q, $\neg$r.

T_1 and T_2 are then both axiomatized formal theories. For it is effectively decidable what is a wff of the theory, and whether a purported proof is a proof from the given axioms. Both theories are consistent. Moreover, both are decidable theories; just use the truth-table test to determine whether a candidate theorem really follows from the axioms.

However, note that although T_1 is a *decidable theory* that doesn't mean T_1 *decides every wff*; it doesn't decide e.g. the wff '(q $\wedge$ r)', since T_1's sole axiom doesn't entail either '(q $\wedge$ r)' or '$\neg$(q $\wedge$ r)'. To stress the point: it is one thing to have an algorithm for deciding what is a theorem; it is another thing for a theory to be negation-complete, i.e. to have the resources to prove or disprove every wff.

By contrast, T_2 *is* negation-complete: any wff constructed from the three atoms can either be proved or refuted using propositional logic, given the three axioms. (Why?)

Our toy example illustrates another crucial terminological point. Recall the familiar idea of a deductive system being 'semantically complete' or 'complete with respect to its standard semantics'. For example, a natural deduction system for propositional logic is said to be semantically complete when every inference which is semantically valid (in effect, is valid by the truth-table test) can be shown to be valid by a proof in the deductive system. But now note that a theory's having a semantically complete logic is one thing, being a negation-complete theory is something else entirely. For example, T_1 by hypothesis has a complete truth-functional *logic*, but is not a complete *theory*. For a more interesting example, we'll soon meet a formal arithmetic which we label 'Q'. This theory uses a standard quantificational deductive logic, which again is a semantically complete *logic*: but we can easily show that Q is not a negation-complete *theory*.

Putting it symbolically may help. To say that a theory T with the set of axioms Σ is negation-complete is to say that, for any sentence φ,

$$\text{either } \Sigma \vdash \varphi \text{ or } \Sigma \vdash \neg\varphi;$$

while to say that a logic is semantically complete is to say that for any set of wffs Σ and any sentence φ,

$$\text{if } \Sigma \vDash \varphi \text{ then } \Sigma \vdash \varphi,$$

where as usual '$\vdash$' signifies the relation of formal deducibility, and '$\vDash$' signifies the relation of semantic consequence.[10]

Do watch out for this potentially dangerous double use of the term 'complete';[11] beware too of the use of '(effectively) decidable' and 'decides' for two not unconnected but significantly different ideas. These dual usages are now entirely entrenched; you just have to learn to live with them.

---

[10] As it happens, the first proof of the semantic completeness of a proof system for quantificational logic was also due to Gödel, and the result is often referred to as 'Gödel's Completeness Theorem' (Gödel, 1929). The topic of *that* theorem is therefore evidently not to be confused with the topic of his (First) Incompleteness Theorem: the semantic completeness of a proof system for quantificational logic is one thing, the negation incompleteness of certain theories of arithmetic quite a different thing.

[11] The double use isn't a case of terminological perversity – even though logicians can be guilty of that! For there's the following parallel. A negation-complete theory is one such that, if you add as a new axiom some proposition that can't already be derived in the theory, then the theory becomes useless by virtue of becoming inconsistent. Likewise a semantically complete deductive system is one such that, if you add a new logical axiom that can't already be derived (or a new rule of inference that can't be established as a derived rule of the system) then the logic becomes useless by virtue of warranting arguments that aren't semantically valid.

4.5   The effective enumerability of theorems

Deploying our notion of effective enumerability from Section 3.3, we can now state and prove the following portmanteau theorem (the last claim is the crucial part):

> **Theorem 4.1** *If T is an effectively axiomatized theory then (i) the set of wffs of T, (i′) the set of sentences of T, (ii) the set of proofs constructible in T, and (iii) the set of theorems of T, can each be effectively enumerated.*

Proof sketch for (i)   By hypothesis, T has a formalized language with a finite basic alphabet.

But that implies we can give an algorithm for mechanically enumerating all the possible finite strings of symbols formed from a finite alphabet. For example, put the symbols of the finite alphabet into some order. Then start by listing all the strings of length 1, followed by the length 2 strings in 'alphabetical order', followed by the length 3 strings in 'alphabetical order', and so on and so forth.

By the definition of a formalized language, however, there is an algorithmic procedure for deciding as we go along which of these symbol strings count as wffs. So, putting these algorithmic procedures together, as we ploddingly enumerate all the possible strings we can throw away the non-wffs that turn up, leaving us with an effective enumeration of all the wffs.   ☒

Proof sketch for (i′)   As for (i), replacing 'wff' by 'sentence'.   ☒

Proof sketch for (ii)   Assume first that T-proofs are linear sequences of wffs. Just as we can effectively enumerate all the possible wffs, so we can effectively enumerate all the possible finite sequences of wffs in some 'alphabetical order'. One brute-force way is to start effectively enumerating all possible strings of symbols, and throw away any that isn't a sequence of wffs. By the definition of an (effectively) axiomatized theory, there is then an algorithmic recipe for deciding which of these sequences of wffs are well-formed derivations from axioms of the theory. So as we go along we can mechanically select out these proof sequences from the other sequences of wffs, to give us an effective enumeration of all the possible proofs.

If T-proofs are more complex arrays of wffs – as in tree systems – then the construction of an effective enumeration of the arrays needs to be correspondingly more complex; but the core proof-idea remains the same.   ☒

Proof sketch for (iii)   Start effectively enumerating the well-constructed proofs again. But this time, just record their conclusions when they pass the algorithmic test for being closed sentences. This effectively generated list now contains all and only the theorems of the theory.   ☒

One comment on all this (compare Section 3.3(c)). Be very clear that to say that the theorems of an effectively axiomatized theory can be effectively *enumerated* is not to say that the theory is *decidable*. It is one thing to have a mechanical method which is bound to generate any theorem eventually; it is quite another thing to have a mechanical method which, given an arbitrary wff φ, can determine – without going on forever – whether φ will ever turn up on the list of theorems. Most interesting axiomatized theories are, as we'll see, *not* decidable.

4.6 Negation-complete theories are decidable

Despite that last point, however, we do have the following important result in the rather special case of negation-complete theories (compare Theorem 3.4):

> **Theorem 4.2** *Any consistent, effectively axiomatized, negation-complete theory T is effectively decidable.*[12]

Proof We know from Theorem 4.1 that there is an algorithm for effectively enumerating the theorems of T. So to decide whether the sentence φ of T's language is a T-theorem, start effectively listing the theorems, and do this until either φ or $\neg\varphi$ turns up and then stop. If φ turns up, declare it to be a theorem. If $\neg\varphi$ turns up, declare that φ is *not* a theorem.

Why does this work as an effective decision procedure? Well first, by hypothesis, T is negation-complete, so either φ is a T-theorem or $\neg\varphi$ is. So it is guaranteed that in a finite number of steps either φ or $\neg\varphi$ will be produced in our enumeration of the theorems, and hence our 'do until' procedure terminates. And second, if φ is produced, φ is a theorem of course, while if $\neg\varphi$ is produced, we can conclude that φ is not a theorem, since the theory is assumed to be consistent.

Hence, in this special case, there *is* a dumbly mechanical procedure for deciding whether φ is a theorem. ◻

---

[12]By the way, it is trivial that an *inconsistent* effectively axiomatized theory with a classical logic is decidable. For if T is inconsistent, every wff of T's language is a theorem by the classical principle *ex contradictione quodlibet* (a contradiction entails whatever proposition you choose). So all we have to do to determine whether φ is a T-theorem is to decide whether φ is a sentence of T's language, which by hypothesis you can if T is an effectively axiomatized formal theory.

5 Capturing numerical properties

The previous chapter concerned effectively axiomatized theories in general. This chapter introduces some key concepts we need in describing formal arithmetics in particular, notably the concepts of *expressing* and *capturing* numerical properties and functions. But we need to start with ...

5.1 Three remarks on notation

(a)   Gödel's First Incompleteness Theorem is about the limitations of axiomatized formal theories of arithmetic: if T satisfies some minimal constraints, we can find arithmetical truths that can't be derived in T. Evidently, in discussing Gödel's result, it will be *very* important to be clear about when we are working 'inside' a given formal theory T and when we are talking informally 'outside' that particular theory (e.g. in order to establish truths that T can't prove).

However, we do want our informal talk to be compact and perspicuous. Hence we will want to borrow some standard logical notation from our formal languages for use in augmenting mathematical English (so, for example, we might write '$\forall x \forall y (x + y = y + x)$' as a compact way of expressing the 'ordinary' arithmetic truth that the order in which you sum numbers doesn't matter).

Equally, we will want our formal wffs to be readable. Hence we will tend to use notation in building our formal languages that is already familiar from informal mathematics (so, for example, if we want to express the addition function in a formalized theory of arithmetic, we will use the usual sign '+', rather than some unhelpfully anonymous two-place function symbol like 'f_3^2').

This two-way borrowing of notation inevitably makes expressions of informal everyday arithmetic and their formal counterparts look very similar. And while context alone should make it pretty clear which is which, it is best to have a way of explicitly marking the distinction. To that end, *we will adopt the convention of using our ordinary type-face (mostly in italics) for informal mathematics, and using a* sans-serif *font for expressions in our formal languages.* Thus compare

$$1 + 2 = 3 \qquad\qquad 1 + 2 = 3$$
$$\forall x \forall y (x + y = y + x) \quad \forall \mathsf{x} \forall \mathsf{y}(\mathsf{x} + \mathsf{y} = \mathsf{y} + \mathsf{x})$$
$$\forall x \exists y\, y = Sx \qquad\quad \forall \mathsf{x} \exists \mathsf{y}\, \mathsf{y} = \mathsf{Sx}$$

The expressions on the left will belong to our mathematicians'/logicians' augmented English (borrowing 'S' to mean 'the successor of'); the expressions on the right are wffs – or abbreviations for wffs – of one of our formal languages, with the symbols chosen to be reminiscent of their intended interpretations.

(b)   In addition to *italic symbols* for informal mathematics and sans-serif symbols for formal wffs, we also need another layer of symbols. For example, we need a compact way of generalizing about formal expressions, as when we defined negation-completeness in Section 4.4 by saying that for any sentence φ, the theory T entails either φ or its negation $\neg\varphi$. We'll standardly use Greek letters for this kind of 'metalinguistic' duty. So note that Greek letters will never belong to our formal languages themselves: these symbols belong to logicians' augmented English.

What exactly is going on, then, when we are talking about a formal language L and say e.g. that the negation of φ is $\neg\varphi$, when we are apparently mixing a symbol from augmented English with a symbol from L? Answer: there are hidden quotation marks, and '$\neg\varphi$' is to be read as meaning 'the expression that consists of the negation sign "$\neg$" followed by φ'.

(c)   Sometimes, when being *very* pedantic, logicians use so-called Quine-quotes when writing mixed expressions which contain both formal and metalinguistic symbols (thus: $\ulcorner\neg\varphi\urcorner$). But this is excessive. We are not going to bother, and no one will get confused by our more casual (and entirely normal) practice. In any case, we'll soon want to use corner-quotes for a quite different purpose.

We will be pretty relaxed about ordinary quotation marks too. We've so far been rather punctilious about using them when mentioning, as opposed to using, wffs and other formal expressions. But from now on, we will normally drop them other than around single symbols. Again, no confusion should ensue.

Finally, we will also be very relaxed about dropping unnecessary brackets in formal expressions (and we'll cheerfully change the shape of pairs of brackets, and even occasionally insert redundant ones, when that aids readability).

5.2   The language L_A

Now to business. There is no single language which could reasonably be called *the* language for formal arithmetic: rather, there is quite a variety of different languages, apt for framing theories of different strengths.

However, the core theories of arithmetic which we'll be discussing most are framed in the interpreted language $L_A = \langle \mathcal{L}_A, \mathcal{I}_A \rangle$, which is a formalized version of what we called 'the language of basic arithmetic' in Section 1.1. So let's concentrate for the moment on characterizing this simple language.

(a) *Syntax* $\mathcal{L}_A$ has a standard first-order[1] syntax, with one built-in two-place predicate and three built-in function expressions.

1. The *logical* vocabulary of $\mathcal{L}_A$ comprises the usual connectives (say, '$\neg$', '$\wedge$', '$\vee$', '$\rightarrow$', '$\leftrightarrow$'), an inexhaustible but enumerable supply of variables

---

[1] Recall: 'first-order' means that the quantified variables occupy positions held by constants; compare 'second-order' logic which also allows another sort of quantifier whose variables occupy predicate position. See Sections 9.4(b) and 29.2.

(including, let's suppose, 'a' to 'e', 'u' to 'z'), the usual quantifier symbols ('∀' and '∃'), and brackets. It also has the identity symbol '=' (which will be the sole atomic predicate).

2. The *non-logical* vocabulary of $\mathcal{L}_A$ is $\{0, S, +, \times\}$, where

   i. '0' is a constant,

   ii. 'S' is a one-place function-expression (read 'the successor of'),

   iii. '+' and '×' are two-place function-expressions.

   For readability, we'll allow ourselves to write e.g. $(a + b)$ and $(a \times b)$ rather than $+(a, b)$ and $\times(a, b)$.

3. A *term* of $\mathcal{L}_A$ is an expression that you can build up from '0' and/or variables using the successor function 'S', addition and multiplication – as in SSS0, $(S0 + x)$, $(SSS0 \times (Sx + y))$, and so on. Putting it more carefully,

   i. '0' is a term, as is any variable,

   ii. if σ and τ are terms, so are $S\sigma$, $(\sigma + \tau)$, $(\sigma \times \tau)$,

   iii. nothing else is a term.

   The *closed* terms are the variable-free terms, such as the numerals we introduce next.

4. A *(standard) numeral* of $\mathcal{L}_A$ is a term built up from our single constant '0' using just the successor function, i.e. they are expressions of the form SS...S0 with zero or more occurrences of 'S'.[2] We'll abbreviate the numerals S0, SS0, SSS0, etc. by '1', '2', '3', etc. Further, when we want to generalize, we'll write e.g. '$\overline{n}$' to indicate the standard numeral SS...S0 with n occurrences of 'S'. (Overlining is a common convention, and it helpfully distinguishes numerals from variables of $\mathcal{L}_A$.)

5. Since the only predicate built into $\mathcal{L}_A$ is the identity sign, the *atomic wffs* all have the form $\sigma = \tau$, where again σ and τ are terms. Then the *wffs* are the atomic wffs plus those formed from them by, perhaps repeatedly, using connectives and quantifiers in some standard way (the fine details don't matter though for convenience we will allow wffs with free variables).

With details completed in a standard way, it will be effectively decidable what is a term, what is a wff, and what is a sentence of $\mathcal{L}_A$.

(b) *Semantics*   The story about the interpretation $\mathcal{I}_A$ begins as you would expect:

---

[2]In using 'S' rather than 's', we depart from the normal logical practice which we follow elsewhere of using upper-case letters for predicates and lower-case letters for functions; but this particular departure is sanctioned by aesthetics and common usage.

A very common alternative convention you should know about even though we won't employ it here is to use a postfixed prime as the symbol for the successor function. In that variant notation the standard numerals are then 0, 0′, 0″, 0‴, ...

1. $\mathcal{I}_A$ assigns values to closed terms as follows:

   i. The value of '0' is zero. Or in an obvious shorthand, $val[0] = 0$.

   ii. If τ is a closed term, then $val[S\tau] = val[\tau] + 1$.

   iii. If σ and τ are closed terms, then $val[(\sigma + \tau)] = val[\sigma] + val[\tau]$, and $val[(\sigma \times \tau)] = val[\sigma] \times val[\tau]$.

   It immediately follows from (i) and (ii), by the way, that numerals have the values that they should have, i.e. for all n, $val[\bar{n}] = n$.

2. The atomic sentences of $\mathcal{L}_A$ must all have the form $\sigma = \tau$, where σ and τ are closed terms. So given the standard reading of the identity relation, this is immediate:

   A sentence of the form $\sigma = \tau$ is true iff $val[\sigma] = val[\tau]$.

3. Molecular sentences built up using the truth-functional connectives are treated in the familiar ways.

4. So that just leaves the quantifiers to deal with. Now, every natural number n has a numeral $\bar{n}$ to pick it out (there are no stray numbers, outside the sequence of successors of zero named by expressions of the form 'SSSS...S0'). Hence, in this special case, we can put the rule for the existential quantifier very simply like this:[3]

   A sentence of the form $\exists \xi \varphi(\xi)$ (where 'ξ' can be any variable) is true iff, for some number n, $\varphi(\bar{n})$ is true.

Similarly,

   A sentence of the form $\forall \xi \varphi(\xi)$ is true iff, for any n, $\varphi(\bar{n})$ is true.

Again it is easy to see that $\mathcal{I}_A$ will, as we want, effectively assign a unique interpretation to every $\mathcal{L}_A$ sentence (we needn't worry about interpreting wffs with free variables). So L_A is an effectively formalized language.

5.3 A quick remark about truth

The semantics $\mathcal{I}_A$ entails that the sentence $(1 + 2) = 3$, i.e. $(S0 + SS0) = SSS0$, is true just so long as one plus two is three. Likewise the sentence $\exists v\, 4 = (v \times 2)$, i.e. $\exists v\, SSSS0 = (v \times SS0)$, is true just so long as there is some number such that four is twice that number (i.e. so long as four is even). But, by any normal arithmetical standards, one plus two *is* three, and four *is* even. So by the same workaday standards, those two L_A-sentences are indeed true.

---

[3]We here don't need to fuss about using the Tarskian notion of satisfaction to explain the truth-conditions of sentences involving quantifiers, precisely because we are in the special position where every element of the domain has a term which denotes it.

Later, when we come to present Gödel's Theorems, we will describe how to take an arithmetical theory T built in the language L_A, and construct a sentence G_T which turns out to be true but unprovable-in-T. And while the sentence in question is a bit exotic, there is nothing in the least exotic about the notion of truth being applied to it here either: it is the very same workaday notion we've just so simply explained. $\mathcal{I}_A$ explicitly defines what it takes for *any* L_A-sentence, however complex, to be true in this humdrum sense.

Now there are, to be sure, philosophers who will say that strictly speaking no L_A-sentence *is* genuinely true in the humdrum sense – because they are equally prepared to say that, speaking really strictly, one plus two *isn't* three and four *isn't* even.[4] Such common-or-garden arithmetic claims, they aver, presuppose the existence of numbers as mysterious kinds of objects in some Platonic heaven, and they are sceptical about the literal existence of such things. In the view of many of these philosophers, arithmetical entities should be thought of as useful *fictions*; and, at least when we are on our very best behaviour, we really ought not to claim that one plus two equals three but only that *in the arithmetical fiction* one plus two equals three.

We can't, however, tangle with this surprisingly popular philosophical view here: and fortunately we needn't do so, for the issues it raises are quite orthogonal to our main concerns in this book. Fictionalists about arithmetic can systematically read our talk of various L_A sentences being true in their favoured way – i.e. as talk 'within the arithmetical fiction'. It won't significantly affect the proofs and arguments that follow.

5.4 Expressing numerical properties and functions

A competent formal theory of arithmetic should surely be able to talk about a lot more than just the successor function, addition and multiplication. But 'talk about' *how*?

(a) Let's assume that we are dealing with a theory built in the rather minimal language L_A. So, for a first example, consider L_A-sentences of the type

1. $\exists v(2 \times v = \bar{n})$.

For $n = 4$, for example, this unpacks into '$\exists v(SS0 \times v = SSSS0)$'. Abbreviate such a wff by $\psi(\bar{n})$. Then it is obvious that, for any n,

if n is even, then $\psi(\bar{n})$ is true,
if n isn't even, then $\neg\psi(\bar{n})$ is true,

where we mean, of course, true on the arithmetic interpretation built into L_A.

So consider the corresponding open wff[5] $\psi(x)$ with one free variable, i.e.

---

[4]See e.g. Field (1989, ch. 1) and Balaguer (1998) for discussion.

[5]Usage varies: in this book, an open wff is one which isn't closed, i.e. which has at least one free variable, though it might have other bound variables.

1'. $\exists v(2 \times v = x)$.

This is, as the logicians say, satisfied by the number n just when $\psi(\bar{n})$ is true, i.e. just when n is even. Or to put it another way, $\psi(x)$ has the set of even numbers as its extension. Which means that our open wff expresses the property *even*, at least in the sense of being true of the right objects, i.e. having the right extension.

Another example: n has the property of being prime iff it is greater than 1, and its only factors are 1 and itself. Or equivalently, n is prime just in case it is not 1, and of any two numbers that multiply to give n, one of them must be 1. So consider wffs of the type

2. $(\bar{n} \neq 1 \land \forall u \forall v(u \times v = \bar{n} \to (u = 1 \lor v = 1)))$

(where we use $\alpha \neq \beta$ for $\neg \alpha = \beta$). Abbreviate such a wff by $\chi(\bar{n})$. Then $\chi(\bar{n})$ holds just in case n is prime, i.e. for every n,

> if n is prime, then $\chi(\bar{n})$ is true,
> if n isn't prime, then $\neg\chi(\bar{n})$ is true.

The corresponding open wff $\chi(x)$, i.e.

2'. $(x \neq 1 \land \forall u \forall v(u \times v = x \to (u = 1 \lor v = 1)))$,

is therefore satisfied by exactly the prime numbers. In other words, $\chi(x)$ expresses the property *prime*, again in the sense of having the right extension.

In this sort of way, a formal language like L_A with minimal basic resources can in fact come to express a whole variety of arithmetical properties by means of complex open wffs with the right extensions. And our examples motivate the following official definition that in fact applies to *any* language L in which we can form the standard numerals (and allows free variables):

> A numerical property P is *expressed* by the open wff $\varphi(x)$ with one free variable in a language L iff, for every n,
> > if n has the property P, then $\varphi(\bar{n})$ is true,[6]
> > if n does not have the property P, then $\neg\varphi(\bar{n})$ is true.

'True' of course continues to mean true on the given interpretation built into L.

(b) We can now extend our definition in the obvious way to cover relations. Note, for example, that in a language like L_A (allowing redundant brackets)

3. $\psi(\bar{m}, \bar{n}) =_{\text{def}} \exists v(v + \bar{m} = \bar{n})$

is true just in case $m \leq n$. And so it is natural to say that

3'. $\psi(x, y) =_{\text{def}} \exists v(v + x = y)$

---

[6]Do we need to spell it out? $\varphi(\bar{n})$ is the result of substituting the numeral for n for each free occurrence of 'x' in $\varphi(x)$.

expresses the relation *less-than-or-equal-to*, in the sense of getting the extension right. Generalizing again:

> A two-place numerical relation R is expressed by the open wff $\varphi(\mathsf{x}, \mathsf{y})$ with two free variables in a language L iff, for any m, n,
>> if m has the relation R to n, then $\varphi(\overline{\mathsf{m}}, \overline{\mathsf{n}})$ is true,
>> if m does not have the relation R to n, then $\neg\varphi(\overline{\mathsf{m}}, \overline{\mathsf{n}})$ is true.

Likewise for many-place relations.[7]

(c)   Let's highlight again that 'expressing' in our sense is just a matter of getting the extension right. Suppose $\varphi(\mathsf{x})$ expresses the property P in L, and let θ be any true L-sentence. Then whenever $\varphi(\overline{\mathsf{n}})$ is true so is $\varphi(\overline{\mathsf{n}}) \wedge \theta$. And whenever $\varphi(\overline{\mathsf{n}})$ is false so is $\varphi(\overline{\mathsf{n}}) \wedge \theta$. Which means that $\varphi'(\mathsf{x}) =_{\text{def}} \varphi(\mathsf{x}) \wedge \theta$ also expresses P – irrespective of what θ means.

Hence, we might say, $\varphi(\mathsf{x})$'s expressing P in our sense can really only be a necessary condition for its expressing that property in the more intuitive sense of having the right meaning. However, the intuitive notion is murky and notoriously difficult to analyse (even if we can usually recognize wffs which 'express the right meaning' when we meet them). By contrast, our notion is sharply defined. And, rather blunt instrument though it is, it will serve us perfectly well for most purposes.

(d)   Let's say that a two-place numerical relation R is *functional* iff for every m there is one and only one n such that Rmn. Then evidently any one-place numerical function f has an associated two-place functional relation R_f such that $f(m) = n$ just in case $R_f mn$. And if, as is standard, you say the extension of a one-place function f is the set of pairs $\langle m, n \rangle$ such that $f(m) = n$, then f has the *same* extension as its corresponding functional relation R_f.

If we are interested, then, in expressing functions in the same sense of getting their extensions right, then expressing a one-place function f and expressing the corresponding two-place functional relation R_f should be equivalent.[8]   Which

---

[7] A pernickety footnote for very-well-brought-up logicians. We could have taken the canonical way of expressing a monadic property to be not a complete wff $\varphi(\mathsf{x})$ with a free variable but a predicative expression $\varphi(\xi)$ – where 'ξ' here isn't a variable but a metalinguistic *place-holder*, marking a *gap* to be filled by a term (i.e. by a name or variable). Similarly, we could have taken the canonical way of expressing a two-place relation to be a doubly gappy predicative expression $\varphi(\xi, \zeta)$, etc.

Now, there are technical and philosophical reasons for rather liking the latter notation to express properties and relations. However, it is certainly the default informal mathematical practice to prefer to use complete expressions with free variables rather than expressions with place-holders which mark gaps. Sticking to this practice, as we do in this book, therefore makes for a more familiar-looking notation and hence aids readability. (Trust me! – I did at one stage try writing this book systematically using Greek letters as place-holders, and some passages would look quite unnecessarily repellent to the ordinary mathematician's eye.)

[8] Some identify a function with its extension, and also identify a relation with *its* extension. If you take that line, then you will say that a function f and its corresponding functional relation R_f are in fact the very *same* thing. But we aren't committing ourselves to that line here, nor for our purposes do we need to deny it (though it is arguably a mistake).

motivates the following definition:

A one-place numerical function f is expressed by the open wff $\varphi(\mathsf{x}, \mathsf{y})$ in a language L iff, for any m, n,
       if $f(m) = n$, then $\varphi(\overline{\mathsf{m}}, \overline{\mathsf{n}})$ is true,
       if $f(m) \neq n$, then $\neg\varphi(\overline{\mathsf{m}}, \overline{\mathsf{n}})$ is true.

Likewise for many-place functions.

(e) We have a simple theorem:

Theorem 5.1 *L can express a property P iff it can express P's characteristic function c_P.*

Proof We just need to check that

1. if $\varphi(\mathsf{x})$ expresses P, then $(\varphi(\mathsf{x}) \wedge \mathsf{y} = 0) \vee (\neg\varphi(\mathsf{x}) \wedge \mathsf{y} = 1)$ expresses c_P;

2. if $\psi(\mathsf{x}, \mathsf{y})$ expresses c_P, then $\psi(\mathsf{x}, 0)$ expresses P.

For (1), suppose $\varphi(\mathsf{x})$ expresses P. If $c_P(m) = 0$, then m is P, so by our supposition $\varphi(\overline{\mathsf{m}})$ must be true, so $(\varphi(\overline{\mathsf{m}}) \wedge 0 = 0) \vee (\neg\varphi(\overline{\mathsf{m}}) \wedge 0 = 1)$ is true.

And if $c_P(m) \neq 0$, then m is not P, so $\neg\varphi(\overline{\mathsf{m}})$ must be true, so both disjuncts of $(\varphi(\overline{\mathsf{m}}) \wedge 0 = 0) \vee (\neg\varphi(\overline{\mathsf{m}}) \wedge 0 = 1)$ are false, so the negation of the whole is true.

Similarly for the remaining cases $c_P(m) = 1$, $c_P(m) \neq 1$. And the four cases together verify (1).

(2) is trivial, so we are done.   ⊠

5.5 Capturing numerical properties and functions

(a) Of course, we don't merely want various numerical properties, relations and functions to be *expressible* in the language of a formal theory of arithmetic. We also want to be able to use the theory to *prove* facts about which numbers have which properties or stand in which relations (more carefully: we want formal derivations which will be proofs in the intuitive sense if we can take it that the axioms are indeed secure truths). We likewise want to be able to use the theory to *prove* facts about the values of various functions for given inputs.

Now, it is a banal observation that to establish facts about *individual* numbers typically requires less sophisticated proof-techniques than proving general truths about *all* numbers. So let's focus here on the relatively unambitious task of case-by-case proving that particular numbers have or lack a certain property. This level of task is reflected in the following general definition:

The theory T *captures* the property P by the open wff $\varphi(\mathsf{x})$ iff, for any n,
       if n has the property P, then $T \vdash \varphi(\overline{\mathsf{n}})$,
       if n does not have the property P, then $T \vdash \neg\varphi(\overline{\mathsf{n}})$.

43

For example, even in theories of arithmetic T with very modest axioms, the wff $\psi(\mathsf{x}) =_{\mathrm{def}} \exists\mathsf{v}(2 \times \mathsf{v} = \mathsf{x})$ not only expresses but captures the property *even*. In other words, for each even n, T can prove $\psi(\overline{\mathsf{n}})$, and for each odd n, T can prove $\neg\psi(\overline{\mathsf{n}})$. Likewise, in the same theories, the wff $\chi(\mathsf{x})$ from the previous section not only expresses but captures the property *prime*.

We can extend the notion of 'capturing' to the case of relations in the entirely predictable way:

> The theory T captures the two-place relation R by the open wff $\varphi(\mathsf{x}, \mathsf{y})$ iff, for any m, n,
>> if m has the relation R to n, then $T \vdash \varphi(\overline{\mathsf{m}}, \overline{\mathsf{n}})$,
>> if m does not have the relation R to n, then $T \vdash \neg\varphi(\overline{\mathsf{m}}, \overline{\mathsf{n}})$.

Likewise for many-place relations.

(b)   We should add an important comment which parallels the point we made about 'expressing'. Suppose $\varphi(\mathsf{x})$ captures the property P in T, and let θ be any T-theorem. Then whenever $T \vdash \varphi(\overline{\mathsf{n}})$, then $T \vdash \varphi(\overline{\mathsf{n}}) \wedge \theta$. And whenever $T \vdash \neg\varphi(\overline{\mathsf{n}})$, then $T \vdash \neg(\varphi(\overline{\mathsf{n}}) \wedge \theta)$. Which means that $\varphi'(\mathsf{x}) =_{\mathrm{def}} \varphi(\mathsf{x}) \wedge \theta$ also captures P – irrespective of θ's content.

Hence, we might say, $\varphi(\mathsf{x})$'s capturing P in our initial sense is just a necessary condition for its capturing that property in the more intuitive sense of proving wffs with the right meaning. But again the intuitive notion is murky, and our sharply defined notion will serve our purposes.

(c)   What about the related idea of 'capturing' a function?

Given the tightness of the link between a function and the corresponding functional relation, one obvious first-shot thing to say is that a theory T captures a function iff it captures the corresponding functional relation. It turns out, though, that it is useful to require a little more than that, and the story gets a bit complicated. However, as we won't actually *need* an official story about capturing functions until Chapter 16, we can usefully shelve further discussion until then.

5.6   Expressing vs. capturing: keeping the distinction clear

To be frank, the jargon we have used here in talking of a theory's 'capturing' a numerical property is deviant. But terminology here varies very widely anyway. Perhaps most commonly these days, logicians talk of P being 'represented' by a wff $\varphi(\mathsf{x})$ satisfying our conditions for capture. But I'm rather unapologetic: '*cap*ture' is very helpfully mnemonic for '*ca*se-by-case *p*rove'.

But whatever your favoured jargon, the key thing is to be absolutely clear about the distinction we need to mark – so let's highlight it again. Sticking to the case of properties,

1. whether P is *expressible* in a given theory just depends on the richness of that theory's *language*;

2. whether a property P can be *captured* by the theory depends on the richness of its *axioms* and *proof system*.[9]

Note that expressibility does not imply capturability: indeed, we will prove later that – for any respectable theory of arithmetic T – there are numerical properties that are expressible in T's language but not capturable by T (see e.g. Section 24.7). However, there *is* a link in the other direction. Suppose T is a *sound* theory of arithmetic, i.e. one whose theorems are all true on the given arithmetic interpretation of its language. Hence if $T \vdash \varphi(\bar{n})$, then $\varphi(\bar{n})$ is true. And if $T \vdash \neg\varphi(\bar{n})$, then $\neg\varphi(\bar{n})$ is true. Which immediately entails that *if $\varphi(x)$ captures P in the sound theory T, then $\varphi(x)$ expresses P.*

---

[9]'Expresses' for properties and relations is used in our way by e.g. Smullyan (1992, p. 19). As alternatives, we find e.g. 'arithmetically defines' (Boolos et al., 2002, p. 199), or simply 'defines' (Leary 2000, p. 130; Enderton 2002, p. 205).

Gödel originally talked of a numerical relation being *entscheidungsdefinit* – translated in his *Collected Works* as 'decidable' – when it is captured by an arithmetical wff (Gödel, 1931, p. 176). As later alternatives to our 'captures' we find 'numeralwise expresses' (Kleene 1952, p. 195; Fisher 1982, p. 112), and also simply 'expresses'(!) again (Mendelson, 1997, p. 170), 'formally defines' (Tourlakis, 2003, p. 180) and plain 'defines' (Boolos et al., 2002, p. 207). At least 'binumerate' – (Smoryński 1977, p. 838; Lindström 2003, p. 9) – won't cause confusion. But as noted, 'represents' (although it is perhaps too close for comfort to 'expresses') seems the most common choice in recent texts: see e.g. Leary (2000, p. 129), Enderton (2002, p. 205), Cooper (2004, p. 56).

As we will see, when it comes to talking of expressing vs. capturing for functions, there's the same terminological confusion, though this time compounded by the fact that there is more than one notion of capturing functions which can be in play.

The moral is plain: *when reading other discussions, always very carefully check the local definitions of the jargon!*

6 The truths of arithmetic

In Chapter 4, we proved that the theorems of any effectively axiomatized theory *can* be effectively enumerated. In this chapter, we prove by contrast that the truths of any language which is sufficiently expressive of arithmetic *can't* be effectively enumerated (we will explain in just a moment what 'sufficiently expressive' means). As we'll see, it immediately follows that a sound axiomatized theory with a sufficiently expressive language can't be negation-complete.

6.1 Sufficiently expressive languages

Recall: a one-place numerical function f can be expressed in language L just when there is an open L-wff φ such that $\varphi(\overline{m}, \overline{n})$ is true iff $f(m) = n$ (Section 5.4). We will now say:

> An interpreted formal language L is *sufficiently expressive* iff (i) it can express every effectively computable one-place numerical function, and (ii) it can form wffs which quantify over numbers.

Clause (ii) means, of course, that L must have quantifiers. But – unless the domain of L's built-in interpretation is already just the natural numbers – L will also need to be able to form a predicate we'll abbreviate $\mathsf{Nat}(\mathsf{x})$ which picks out (whatever plays the role of) the natural numbers in L's domain. E.g. we need '$\exists \mathsf{x}(\mathsf{Nat}(\mathsf{x}) \wedge \varphi(\mathsf{x}))$' to be available to say that some number satisfies the condition expressed by φ.

As we've just announced, we are going to show that sound effectively axiomatized theories with sufficiently expressive languages can't be negation-complete. But of course, that wouldn't be an interesting result if a theory's having a sufficiently expressive language were a peculiarly tough condition to meet. But it isn't. Much later in this book, in Section 39.2, we'll show that even L_A, the language of basic arithmetic, is sufficiently expressive.

However, we can't yet establish this claim about L_A: doing that would obviously require having a general theory of computable functions, and we so far haven't got one. For the moment, then, we'll just *assume* that theories with sufficiently expressive languages are worth thinking about, and see what follows.[1]

---

[1] Though perhaps we can do rather better than mere assumption even at this early point, for consider the following line of argument. Suppose we have a suitably programmed general purpose computer M which implements a program computing f. Now imagine that we use numerical coding to associate M's programs, its memory states, its outputs, etc., with numbers. Then we can encode claims about M's performance as it evaluates $f(m)$. There will,

6.2 The truths of a sufficiently expressive language

It now quickly follows, as we initially announced, that

> **Theorem 6.1** *The set of truths of a sufficiently expressive language L is not effectively enumerable.*

Proof Take the argument in stages. (i) The Basic Theorem about e.e. sets of numbers (Theorem 3.8) tells us that there is a set K which is e.e. but whose complement $\overline{K}$ isn't. Suppose the effectively computable function k enumerates it, so $n \in K$ iff $\exists x\, k(x) = n$, with the variable running over numbers.

(ii) Since k is effectively computable, in any given sufficiently expressive arithmetical language L there will be some wff of L which expresses k: let's abbreviate that wff $\mathsf{K(x, y)}$. Then $k(m) = n$ just when $\mathsf{K(\overline{m}, \overline{n})}$ is true.

(iii) By definition, a sufficiently expressive language can form wffs which quantify over numbers. So $\exists x\, k(x) = n$ just when $\exists \mathsf{x}(\mathsf{Nat(x)} \wedge \mathsf{K(x, \overline{n})})$ is true (where $\mathsf{Nat(x)}$ stands in for whatever L-predicate might be needed to explicitly restrict L's quantifiers to numbers).

(iv) So from (i) and (iii) we have

$n \in K$ if and only if $\exists \mathsf{x}(\mathsf{Nat(x)} \wedge \mathsf{K(x, \overline{n})})$ is true; therefore,
$n \in \overline{K}$ if and only if $\neg \exists \mathsf{x}(\mathsf{Nat(x)} \wedge \mathsf{K(x, \overline{n})})$ is true.

(v) Now suppose for a moment that the set $\mathcal{T}$ of true sentences of L *is* effectively enumerable. Then, given a description of the expression K, we could run through the supposed effective enumeration of $\mathcal{T}$, and whenever we come across a truth of the type $\neg \exists \mathsf{x}(\mathsf{Nat(x)} \wedge \mathsf{K(x, \overline{n})})$ for some n – and it will be effectively decidable if a wff has that particular syntactic form – list the number n. That procedure would give us an effectively generated list of all the members of $\overline{K}$.

(vi) But by hypothesis $\overline{K}$ is *not* effectively enumerable. So $\mathcal{T}$ can't be effectively enumerable after all. Which is what we wanted to show.[2] ⊠

6.3 Unaxiomatizability

Here's a first easy corollary. We need a simple definition:

> A set of wffs Σ is *effectively axiomatizable* iff there is an effectively axiomatized formal theory T such that, for any wff φ, $\varphi \in \Sigma$ if and only if $T \vdash \varphi$ (i.e. Σ is the set of T-theorems).

in particular, be a statement $\varphi(\overline{m}, \overline{n})$ in an arithmetical language L which encodes the claim that M, running the appropriate program, gives output n on input m, so $\varphi(\overline{m}, \overline{n})$ is true just when $f(m) = n$. Hence φ expresses f, which makes L sufficiently expressive, given that L is rich enough to have the resources to code up descriptions of the behaviour of programs in general purpose computers. However, it should seem plausible that a fairly simple arithmetical language should suffice for such coding – as we'll indeed confirm in later chapters.

[2] We can now see that, to get the argument to fly, we don't really need the full assumption that our language can express *all* effectively computable functions, so long as it can express k.

Then it is immediate that

Theorem 6.2 *The set $\mathcal{T}$ of true sentences of a sufficiently expressive language L is not effectively axiomatizable.*

Proof Suppose otherwise, i.e. suppose that T is an effectively axiomatized theory, framed in a sufficiently expressive language L, such that the T-theorems are just the truths $\mathcal{T}$ expressible in that language. Then, because it is effectively axiomatized, T's theorems could be effectively enumerated (by the last part of Theorem 4.1). That is to say, the truths $\mathcal{T}$ of L could be effectively enumerated, contrary to Theorem 6.1. Hence there can be no such theory as T.　　　　☒

6.4 An incompleteness theorem

Suppose we build an effectively axiomatized theory T in a sufficiently expressive language L. Then because T is effectively axiomatized, its theorems *can* be effectively enumerated. On the other hand, because T's language is sufficiently expressive, the truths expressible in its language *cannot* be effectively enumerated. There is therefore a mismatch between the truths and the T-theorems here.

Now suppose that T is also a *sound* theory, i.e. its theorems are all true. The mismatch between the truths and the T-provable sentences must then be due to there being truths which T can't prove. Suppose φ is one of these. Then T doesn't prove φ. And since $\neg\varphi$ is false, the sound theory T doesn't prove that either. Which entails our first version of an incompleteness theorem:

Theorem 6.3 *If T is a sound effectively axiomatized theory whose language is sufficiently expressive, then T is not negation-complete.*

But we announced – though of course at this stage we can't yet prove – that even the minimal language L_A is sufficiently expressive, so any more inclusive language will be too: *hence a vast number of theories which can express some arithmetic must be incomplete if sound.*

Astonishing! We have reached an arithmetical incompleteness theorem already. And note that we can't patch up T by adding more true axioms and/or a richer truth-preserving logic and make it complete while continuing to have an effectively axiomatized theory. For the augmented theory T' will still be sound, still have a sufficiently expressive language, and hence (if it remains effectively axiomatized) it still can't be complete. So we could equally well call Theorem 6.3 an *incompletability* theorem.

The great mathematician Paul Erdős had the fantasy of *The Book* in which God keeps the neatest and most elegant proofs of mathematical theorems. Our sequence of proofs of Theorems 3.8 and 6.1 and now of our first incompletability result Theorem 6.3 surely belongs in *The Book*.[3]

---

[3]For more on Erdős's conceit of proofs from *The Book*, see Aigner and Ziegler (2004).

7 Sufficiently strong arithmetics

Theorem 6.3, our first shot at an incompleteness theorem, applies to sound theories. But we have already remarked in Section 1.2 that Gödel's arguments show that we don't need to assume soundness to prove incompleteness. In this chapter we see how to argue from *consistency* to incompleteness.

But if we are going to weaken one assumption (from soundness to mere consistency) we'll need to strengthen another assumption: we'll now consider theories that don't just *express* enough but which can *capture*, i.e. *prove*, enough.

Starting in Chapter 10, we'll begin examining various formal theories of arithmetic 'from the bottom up', in the sense of first setting down the axioms of the theories and then exploring what the different theories are capable of proving. For the moment, however, we are continuing to proceed the other way about. In the previous chapter, we considered theories that have sufficiently expressive languages, and so can express what we'd like any arithmetic to be able to express. Now we introduce the companion concept of a *sufficiently strong* theory, which is one that by definition can prove what we'd like any moderately competent theory of arithmetic to be able to prove about decidable properties of numbers. We then establish some easy but deep results about such theories.

7.1 The idea of a 'sufficiently strong' theory

Suppose that P is some effectively decidable property of numbers, i.e. one for which there is an algorithmic procedure for deciding, given a natural number n, whether n has property P or not.

Now, when we construct a formal theory of the arithmetic of the natural numbers, we will surely want deductions inside our theory to be able to track, case by case, any mechanical calculation that we can already perform informally. We don't want going formal to *diminish* our ability to determine whether n has this property P. As we stressed in Section 4.1, formalization aims at regimenting what we can already do; it isn't supposed to hobble our efforts. So while we might have some passing interest in more limited theories, we will naturally aim for a formal theory T which at least (a) is able to frame some open wff $\varphi(\mathsf{x})$ which expresses the decidable property P, and (b) is such that if n has property P, $T \vdash \varphi(\bar{\mathsf{n}})$, and if n does not have property P, $T \vdash \neg\varphi(\bar{\mathsf{n}})$. In short, we want T to capture P (in the sense of Section 5.5).

The suggestion therefore is that, if P is any effectively decidable property of numbers, we ideally want a competent theory of arithmetic T to be able to capture P. Which motivates the following definition:

> A formal theory of arithmetic T is *sufficiently strong* iff it captures all effectively decidable numerical properties.

And it seems a reasonable and desirable condition on a formal theory of the arithmetic of the natural numbers that it be sufficiently strong.[1]

Much later (in Section 39.2), when we've done some more investigation into the general idea of effective decidability, we'll finally be in a position to warrant the claim that some simple, intuitively sound, and (by then) very familiar theories built in L_A do indeed meet this condition. We will thereby show that the condition of being 'sufficiently strong' is actually easily met. But we can't establish that now: this chapter just supposes that there *are* such theories and derives some consequences.

7.2   An undecidability theorem

A trivial way for a theory T to be sufficiently strong (i.e. to prove lots of wffs about properties of individual numbers) is by being inconsistent (i.e. by proving *every* wff about individual numbers). It goes without saying, however, that we are interested in *consistent* theories.

We also like to get *decidable* theories when we can, i.e. theories for which there is an algorithm for determining whether or not a given wff is a theorem (see Section 4.4).

But, sadly, we have the following key result:[2]

> **Theorem 7.1** *No consistent, sufficiently strong, effectively axiomatized theory of arithmetic is decidable.*

Proof   We suppose T is a consistent and sufficiently strong axiomatized theory yet also decidable, and derive a contradiction.

By hypothesis, T's language can frame open wffs with 'x' free. These will be effectively enumerable: $\varphi_0(x), \varphi_1(x), \varphi_2(x), \ldots$. For by Theorem 4.1 we know that the complete set of wffs of T can be effectively enumerated. It will then be a mechanical business to select out the ones with just 'x' free (there are standard mechanical rules for determining whether a variable is free or bound).

Now let's fix on the following definition:

> n has the property D if and only if $T \vdash \neg\varphi_n(\overline{n})$.

---

[1] Why is being 'sufficiently expressive' defined in terms of expressing *functions*, and being 'sufficiently strong' defined in terms of capturing *properties*?

No deep reason at all; it's a superficial matter of convenience. Given that capturing properties goes with capturing their characteristic functions (Theorem 16.1), we could have defined being sufficiently strong by means of a condition on functions too.

[2] The undecidability of arithmetic was first shown by Church (1936b). For a neater proof, see Tarski et al. (1953, pp. 46–49). I learnt the informal proof as given here from Timothy Smiley, who was presenting it in Cambridge lectures in the 1960s. A version of this line of argument can be found in Hunter (1971, pp. 224–225).

Note that the construction here links the subscripted index with the standard numeral to be substituted for the variable in $\neg\varphi_n(\mathsf{x})$. So this is a cousin of the 'diagonal' constructions which we encountered in proving Theorem 2.2 and again in proving Theorem 3.8.

We next show that the supposition that T is a decidable theory entails that the 'diagonal' property D is an effectively decidable property of numbers. For given any number n, it will be a mechanical matter to enumerate the open wffs until the n-th one, $\varphi_n(\mathsf{x})$, is produced. Then it is a mechanical matter to form the numeral $\bar{n}$, substitute it for the variable and prefix a negation sign. Now we just apply the supposed algorithm for deciding whether a sentence is a T-theorem to test whether the wff $\neg\varphi_n(\bar{n})$ is a theorem. So, on our current assumptions, there is an algorithm for deciding whether n has the property D.

Since, by hypothesis, the theory T is sufficiently strong, it can capture all decidable numerical properties: so it follows, in particular, that D is capturable by some open wff. This wff must of course occur somewhere in our enumeration of the $\varphi(\mathsf{x})$. Let's suppose the d-th wff does the trick: that is to say, property D is captured by $\varphi_d(\mathsf{x})$.

It is now entirely routine to get out a contradiction. For, by definition, to say that $\varphi_d(\mathsf{x})$ captures D means that for any n,

> if n has the property D, $T \vdash \varphi_d(\bar{n})$,
> if n doesn't have the property D, $T \vdash \neg\varphi_d(\bar{n})$.

So taking in particular the case $n = d$,

  i. if d has the property D, $T \vdash \varphi_d(\bar{d})$,
  ii. if d doesn't have the property D, $T \vdash \neg\varphi_d(\bar{d})$.

But note that our initial definition of the property D implies

  iii. d has the property D if and only if $T \vdash \neg\varphi_d(\bar{d})$.

From (ii) and (iii), it follows that whether d has property D or not, the wff $\neg\varphi_d(\bar{d})$ is a theorem either way. So by (iii) again, d does have property D, hence by (i) the wff $\varphi_d(\bar{d})$ must be a theorem too. So a wff and its negation are both theorems of T. Therefore T is inconsistent, contradicting our initial assumption that T is consistent.

In sum, the supposition that T is a consistent and sufficiently strong axiomatized formal theory of arithmetic *and* decidable leads to contradiction. ⊠

There's an old hope (which goes back to Leibniz) that can be put in modern terms like this: we might one day be able to mechanize mathematical reasoning to the point that a suitably primed computer could solve all mathematical problems in a domain like arithmetic by deciding theoremhood in an appropriate formal theory. What we've just shown is that this is a false hope: as soon as a theory is strong enough to capture the results of boringly mechanical reasoning about decidable properties of individual numbers, it must itself cease to be decidable.

7.3   Another incompleteness theorem

Now let's put together Theorem 4.2, *Any consistent, effectively axiomatized, negation-complete theory is decidable*, and Theorem 7.1, *No consistent, sufficiently strong, effectively axiomatized theory of arithmetic is decidable*. These, of course, immediately entail

> **Theorem 7.2** *If T is a consistent, sufficiently strong, effectively axiomatized theory of arithmetic, then T is not negation-complete.*

Let us temporarily say that a consistent, sufficiently strong, effectively axiomatized theory is a *good* theory. Then we have shown that for any good theory of arithmetic, there will be a pair of sentences φ and $\neg\varphi$ in its language, neither of which is a theorem. But one of the pair must be true on the given interpretation of T's language. Therefore, for any good theory of arithmetic T, there are true sentences of its language which T cannot decide.

And adding in new axioms won't help. To re-play the sort of argument we gave in Section 1.2, suppose T is a good theory of arithmetic, and suppose φ is a true sentence of arithmetic that T can't prove or disprove. The theory T^+ which you get by adding φ as a new axiom to T will, of course, now trivially prove φ, so we've plugged that gap. But note that T^+ is consistent (for if T^+, i.e. $T + \varphi$, were inconsistent, then $T \vdash \neg\varphi$ contrary to hypothesis). And T^+ is sufficiently strong (since it can still prove everything T can prove). It is still decidable which wffs are axioms of T^+, so the theory still counts as an effectively axiomatized formal theory. So T^+ is another good theory and Theorem 7.2 applies: so there is a wff φ^+ (distinct from φ, of course) which is again true-on-interpretation but which T^+ cannot decide (and if T^+ can't prove either φ^+ or $\neg\varphi^+$, then neither can the weaker T). In sum, the good theory T is therefore not only incomplete but also in a good sense incompletable.[3]

Which is another proof for *The Book*.

---

[3] Perhaps we should note that, while the informal incompleteness argument of Chapter 6 depended on assuming that there are general-purpose programming languages in which we can specify (the equivalent of) any numerical algorithm, the argument of this chapter doesn't require that assumption. On the other hand, our previous incompleteness result didn't make play with the idea of theories strong enough to capture all decidable numerical properties, whereas our new incompleteness result does. What we gain on the roundabouts we lose on the swings.

8 Interlude: Taking stock

8.1 Comparing incompleteness arguments

Our informal incompletability results, Theorems 6.3 and 7.2, aren't the same as Gödel's own theorems. But they are close cousins, and they seem quite terrific results to arrive at so very quickly.

Or are they? Everything depends, for a start, on whether the ideas of a 'sufficiently expressive' arithmetic language and a 'sufficiently strong' theory of arithmetic are in good order. Still, as we've already briefly indicated in Section 3.1, there are a number of standard, well-understood, ways of formally refining the intuitive notions of effective computability and effective decidability, ways that turn out to specify the same entirely definite and well-defined class of numerical functions and properties. Hence the ideas of a 'sufficiently expressive' language (which expresses all computable one-place functions) and a 'sufficiently strong' theory (which captures all decidable properties of numbers) can in fact also be made perfectly determinate.

But, by itself, that claim doesn't take us very far. For it leaves wide open the possibility that a language expressing all computable functions or a theory that captures all decidable properties has to be very rich indeed. However, we announced right back in Section 1.2 that Gödel's own arguments rule out complete theories even of the truths of basic arithmetic. Hence, if our easy Theorems are to have the full reach of Gödel's work, we'll really have to show (for starters) that the language of basic arithmetic is already sufficiently expressive, and that a theory built in that language can be sufficiently strong.

In sum, if something like our argument for Theorem 6.3 is to be used to establish a variant of one of Gödel's own results, then it needs to be augmented with (i) a general treatment of the class of computable functions, *and* (ii) a proof that (as we claimed) even L_A can express at least the one-place computable functions. And if something like our argument for Theorem 7.2 is to be used, it needs to be augmented by (iii) a proof that (as we claimed) common-or-garden theories couched in L_A can be sufficiently strong.

But even with (i), (ii) and (iii) in play, there would still remain a significant difference between our easy theorems and Gödel's arguments. For our lines of argument don't yet give us any specific examples of unprovable truths. By contrast, Gödel's proof tells us how to take a consistent effectively axiomatized theory T and actually construct a true but unprovable-in-T sentence (the one that encodes 'I am unprovable in T'). *Moreover, Gödel does this without needing the fully general treatment in (i) and without needing all (ii) or (iii) either.*

There is a significant gap, then, between our two intriguing, quickly-derived, theorems and the industrial-strength results proved by Gödel. So, while what we have shown so far is highly suggestive, it is time to start turning to Gödel's own arguments. But before we press on, let's highlight two important points:

A. Our arguments for incompleteness came in two flavours. First, we combined the premiss (a) that we are dealing with a *sound* theory with the premiss (b) that our theory's language is *expressively* rich enough. Second, we weakened one assumption and beefed up the other: in other words, we used the weaker premiss (a′) that we are dealing with a *consistent* theory but added the stronger premiss (b′) that our theory can *prove* enough facts. We'll see that Gödel's arguments too come in these two flavours.

B. Arguments for incompleteness don't have to depend on the construction of Gödel sentences that somehow say of themselves that they are unprovable. Neither of our informal proofs do.

8.2   A road-map

So now we turn to Gödel's proofs. And to avoid getting lost in what follows, it will help to have in mind an overall road-map of the route we are taking:

1. After a preliminary chapter on the idea of induction, we begin by describing some standard effectively axiomatized systems of arithmetic, in particular the benchmark PA, so-called 'First-order Peano Arithmetic', and an important subsystem Q, 'Robinson Arithmetic'. (Chapters 10–13)

2. These systems are framed in L_A, the language of basic arithmetic. So they only have successor, addition and multiplication as 'built-in' functions. But we go on to describe the large family of 'primitive recursive' functions, properties and relations (which includes all familiar arithmetical functions like the factorial and exponential, and familiar arithmetical properties like being prime, and relations like one number being the square of another). And we then show that Q and PA can not only express but capture all the primitive recursive functions, properties and relations – a major theorem that was, in essence, first proved by Gödel. (Chapters 14–17)

3. We next turn to Gödel's simple but crucial innovation – the idea of systematically associating expressions of a formal arithmetic with numerical codes. Any sensibly systematic scheme of 'Gödel numbering' will do; but Gödel's original style of numbering has a certain naturalness, and makes it tolerably straightforward to prove arithmetical results about the codings. With a coding scheme in place, we can reflect properties and relations of strings of symbols of PA (to concentrate on that theory) by properties and relations of their Gödel numbers. For example, we can define the numerical properties *Term* and *Wff* which hold of a number just when it is

the code number for a symbol sequence which is, respectively, a term or a wff of PA. And we can, crucially, define the numerical relation $Prf(m, n)$ which holds when m codes for an array of wffs that is a PA proof, and n codes the closed wff that is thereby proved. This project of coding up various syntactic relationships is often referred to as *the arithmetization of syntax*. And what Gödel showed is that – given a sane system of Gödel numbering – these and a large family of related arithmetical properties and relations are primitive recursive. (Chapters 19, 20)

4. Next – the really exciting bit! – we use the fact that relations like Prf are expressible in PA to construct a 'Gödel sentence' G. This will be true when there is no number that is the Gödel number of a PA proof of the wff that results from a certain construction – where the wff that results is none other than G itself. So G is true just if it is unprovable in PA. Given PA is sound and only proves truths, G can't be provable; hence G is true; hence ¬G is false, and so is also unprovable in PA. In sum, given PA is sound, it cannot decide G. Further, it turns out that we can drop the *semantic* assumption that PA is sound. Using the fact that PA can capture relations like Prf (as well as merely express them), we can still show that G is undecidable while just making a *syntactic* assumption. (Chapter 21)

5. Finally, we note that the true-but-unprovable sentence G for PA is generated by a method that can be applied to any other arithmetic that satisfies some modest conditions. In particular, adding G as a new axiom to PA just gives us a revised theory for which we can generate a new true-but-unprovable wff G'. Throwing in G' as a further axiom then gives us another theory for which we can generate yet another true-but-unprovable wff. And so it goes. PA is therefore not only incomplete but incompletable. In fact, *any* properly axiomatized consistent theory that contains the weak theory Q is incompletable. (Chapter 22)

Just one comment. This summary makes Gödel's formal proofs in terms of *primitive recursive* properties and relations sound rather different from our informal proofs using the idea of theories which express/capture *decidable* properties or relations. But a link can be made when we note that the primitive recursive properties and relations are in fact a large subclass of the intuitively decidable properties and relations. Moreover, showing that Q and PA can express/capture all primitive recursive properties and relations takes us most of the way to showing that those theories are sufficiently expressive/sufficiently strong.

However, we'll leave exploring this link until much later. Only after we have travelled the original Gödelian route (which doesn't presuppose a *general* account of computability) will we return to consider how to formalize the arguments of Chapters 6 and 7 (a task which *does* presuppose such a general account).

9 Induction

Before we start looking at particular formalized arithmetics in detail, we need to remind ourselves of a standard method (from ordinary, informal mathematics) for establishing general truths about numbers. This chapter, then, discusses the idea of a proof by induction. We will later be using this kind of proof both *inside* some of our formal arithmetics and in establishing results *about* the formal theories.

Many readers will have encountered inductive proofs already and will be able to skim through the chapter quite speedily. But we do need to fix some terminology and draw some distinctions – so don't skip completely.

9.1 The basic principle

(a) Suppose we want to show that *all* natural numbers have some property P. We obviously can't give separate proofs, one for each n, that n has P, because that would be a never-ending task. So how can we proceed?

One route forward is to appeal to the *principle of arithmetical induction*. Suppose we can show that (i) 0 has some property P, and also that (ii) if any given number has the property P then so does the next: then we can infer that *all* numbers have property P.

Let's borrow some logical notation (in the spirit of Section 5.1), use φ for an expression attributing some property to numbers, and put the principle like this:

> **Induction** Given (i) $\varphi(0)$ and (ii) $\forall n(\varphi(n) \to \varphi(n+1))$, we can infer $\forall n\varphi(n)$.

Why are arguments which appeal to this principle good arguments? Well, suppose we establish both *the base case* (i) and the *induction step* (ii).[1] By (i) we have $\varphi(0)$. By (ii), $\varphi(0) \to \varphi(1)$. Hence we can infer $\varphi(1)$. By (ii) again, $\varphi(1) \to \varphi(2)$. Hence we can now infer $\varphi(2)$. Likewise, we can use another instance of (ii) to infer $\varphi(3)$. And so on and so forth, running as far as we like through the successors of 0 (i.e. through the numbers that can be reached by starting from zero and repeatedly adding one). But the successors of 0 are the only natural numbers. So for *every* natural number n, $\varphi(n)$.

The arithmetical induction principle is underwritten, then, by the basic structure of the number sequence, and in particular by the absence of 'stray' numbers

---

[1]Note: *arithmetical* induction takes us from a universal premiss, the induction step (ii), to a universal conclusion. That's why it can be deductively valid, and is of course not to be confused with an *empirical* induction from a restricted sample to a general conclusion!

that you can't get to step-by-step from zero by applying and reapplying the successor function.

(b)   Let's give a not-quite-trivial high-school illustration of an inductive argument at work. Consider then the equation

S. $0^2 + 1^2 + 2^2 + 3^2 + \ldots + n^2 = n(n+1)(2n+1)/6.$

It can be spot-checked that this holds for any particular given value of n you choose. But we want to prove that it holds for *all* n.

Let's stipulate, then, that $\varphi(n)$ is true iff equation (S) holds good for the given value of n. It is trivially true that (i) $\varphi(0)$.

Now suppose $\varphi(n)$, i.e. suppose that (S) holds, for some arbitrary n. Then, adding $(n+1)^2$ to each side, we get

$$(0^2+1^2+2^2+3^2+\ldots+n^2) + (n+1)^2 = n(n+1)(2n+1)/6 + (n+1)^2.$$

Elementary manipulation of the right hand side yields

$$0^2+1^2+2^2+3^2+\ldots+(n+1)^2 = (n+1)((n+1)+1)(2(n+1)+1)/6.$$

Hence (S) holds for $n+1$, i.e. $\varphi(n+1)$. Discharging the temporary supposition gives us $\varphi(n) \to \varphi(n+1)$. But n was arbitrary, so (ii), $\forall n(\varphi(n) \to \varphi(n+1))$.

Now we can invoke induction. From (i) and (ii) we can conclude by the induction rule that $\forall n\varphi(n)$. As we wanted, the equation (S) holds for every n.

(c)   Let's give another simple example, this time from logic. Again, we spell things out in very plodding detail.

It is an elementary observation that a wff of classical logic starting with a negation sign followed by a block of quantifiers is always logically equivalent to the result of pushing in the negation sign past all the quantifiers, 'flipping' universals into existentials and vice versa as you go.

But suppose you want a full-dress *proof* of that. Let's write Q for a quantifier ($\forall v$ or $\exists v$ for some variable v) and let $\overline{Q}$ be the dual quantifier. Then what we want to show is that, however many quantifiers are involved,

Q. $\neg Q_n Q_{n-1} \ldots Q_1 \psi \leftrightarrow \overline{Q}_n \overline{Q}_{n-1} \ldots \overline{Q}_1 \neg \psi$

is logically true, for any ψ. So this time stipulate that $\varphi(n)$ holds iff (Q) is indeed a logical truth for n quantifiers. The quantifier-free base case (i) $\varphi(0)$ is then trivially true.

Suppose $\varphi(n)$. Then consider (1) $\neg \exists v Q_n Q_{n-1} \ldots Q_1 \psi(v)$. Elementarily, we can derive $\neg Q_n Q_{n-1} \ldots Q_1 \psi(a)$ for arbitrary a. But by our supposition $\varphi(n)$, this implies $\overline{Q}_n \overline{Q}_{n-1} \ldots \overline{Q}_1 \neg \psi(a)$. Generalizing we get (2) $\forall v \overline{Q}_n \overline{Q}_{n-1} \ldots \overline{Q}_1 \neg \psi(v)$. Conversely, by a similar argument, we can show that (2) implies (1). So (1) is equivalent to (2).

Now consider (3) $\neg \forall v Q_n Q_{n-1} \ldots Q_1 \psi(v)$. Another pair of arguments will show it is equivalent to (4) $\exists v \overline{Q}_n \overline{Q}_{n-1} \ldots \overline{Q}_1 \neg \psi(v)$. (Check that!)

Putting everything together establishes $\varphi(n+1)$, still on the assumption $\varphi(n)$. Now, the argument continues as before. Infer (ii) $\forall n(\varphi(n) \to \varphi(n+1))$, and appeal to the induction rule again to conclude $\forall n\varphi(n)$, i.e. (Q) holds for all n.

9.2 Another version of the induction principle

(a)   We need next to note what is, on the surface, a slightly different version of the induction principle. This is *course-of-values induction*, also called *strong* or *complete* induction. The idea this time is that if we can show that (i) 0 has some property P, and also that (ii′) if all numbers up to and including n have P, so does $n+1$, then we can infer that *all* numbers have property P.

In symbols the new rule is this:

> **Course-of-values induction** Given the assumptions (i) $\varphi(0)$ and (ii′) $\forall n\{(\forall k \leq n)\varphi(k) \to \varphi(n+1)\}$, we can infer $\forall n\varphi(n)$.

A slight tweak of the argument we gave before shows that this too is a sound inferential principle. For suppose we establish both (i) and the new induction step (ii′). By (i) we have $\varphi(0)$. By (ii′), $\varphi(0) \to \varphi(1)$, hence we get $\varphi(1)$, and so we have $\varphi(0) \wedge \varphi(1)$. By (ii′) again, $(\varphi(0) \wedge \varphi(1)) \to \varphi(2)$. Hence we can infer $\varphi(2)$, so we now have $\varphi(0) \wedge \varphi(1) \wedge \varphi(2)$. Likewise, we can use another instance of (ii′) to infer $\varphi(3)$. And so on and so forth, through all the successors of 0, arbitrarily far. So for every natural number n, $\varphi(n)$.

(b)   Let's go straight to elementary logic for a simple illustration of our second version of an induction rule at work. It is another elementary observation that every wff of the propositional calculus is balanced, i.e. has the same number of left-hand and right-hand brackets. But suppose again you want to prove that.

We proceed by (course-of-values) induction on the number of connectives in the wff. This time, let $\varphi(n)$ hold iff an n-connective wff is balanced. Once more, the base case (i), i.e. $\varphi(0)$, holds trivially.

So now suppose that, for every number up to and including n, an n-connective wff is balanced. Now if ψ is a wff with $n+1$ connectives, it must be of one of the forms $\neg\theta$ or $(\theta \circ \chi)$ where $\circ$ is one of the dyadic connectives. In the first case, θ must have n connectives and is by assumption balanced, and so $\neg\theta$ is still balanced. In the second case, θ and χ both have no more than n connectives, so by assumption both are balanced, so $(\theta \circ \chi)$ is still balanced. Either way, ψ is balanced. Which establishes the induction step (ii′) that if, for every number k up to and including n, $\varphi(k)$, then $\varphi(n+1)$.

Given (i) and (ii′), a course-of-values induction proves that $\forall n\varphi(n)$; i.e. every wff, however complex, is balanced.

(c)   For a weightier (though probably still familiar) example from logic, let's consider the induction argument used in standard soundness proofs.

To minimize irrelevant detail, we will take the soundness proof for a Hilbert-style deductive system for the propositional calculus, showing that any wff derivable from the axioms of logic by repeated applications of modus ponens is a truth-table tautology.

So let $\varphi(n)$ hold when the result of an n line proof is a tautology. Trivially, (i) $\varphi(0)$ – since a zero line proof has no conclusion, so – vacuously – any conclusion is a tautology.

Now assume that any proof with no more than n lines results in a tautology, and consider an $n + 1$ line proof. This must extend an n line proof either (a) by adding an instance of an axiom, which will be a tautology, or (b) by applying modus ponens to two previous proved wffs φ and $(\varphi \to \psi)$, to derive ψ. But in case (b), by assumption, φ and $(\varphi \to \psi)$ are results of proofs of no more than n lines, so are tautologies, and therefore ψ is a tautology too. So either way, the $n + 1$ line proof also results in a tautology.

That establishes (ii$'$) which says that if, for every number k up to and including n, $\varphi(n)$, then $\varphi(n + 1)$. Course-of-values induction from (i) and (ii$'$) then shows us that for any n, an n line proof delivers a tautology: i.e. all derivable theorems of the system are tautologies.

Soundness proofs for fancier logical systems will work in the same way – by a course-of-values induction on the length of the proof.

(d)   Given that the intuitive justification of course-of-values induction is so very close to the justification of our original induction principle, you might well suspect that in fact the two principles must really be equivalent. You would be right.

Suppose we have both (i) $\varphi(0)$, and (ii$'$) $\forall n\{(\forall k \leq n)\varphi(k) \to \varphi(n + 1)\}$. The first evidently implies $(\forall k \leq 0)\varphi(k)$. The second equally evidently implies $\forall n\{(\forall k \leq n)\varphi(k) \to (\forall k \leq n + 1)\varphi(k)\}$.

Putting $\psi(n) =_{\text{def}} (\forall k \leq n)\varphi(k)$, we therefore have both $\psi(0)$ and also $\forall n(\psi(n) \to \psi(n+1))$. So by the *original* induction principle, we can infer $\forall n \psi(n)$.

But, quite trivially, $\forall n \psi(n)$ implies $\forall n \varphi(n)$. Which all goes to show that we can get from (i) and (ii$'$) to $\forall n \varphi(n)$ via our original induction principle.

Although we presented course-of-value induction as a separate principle, it therefore does not allow us to prove anything that we can't prove by ordinary induction. Hence, when we come later to adding a formal version of ordinary arithmetical induction to a formal theory of arithmetic, there would be little point in adding a separate formal version of course-of-values induction.

So we won't.

9.3   Induction and relations

Note next that we can use inductive arguments to prove general results about *relations* as well as about monadic properties.

Let's take a simple illustrative example from informal mathematics. We want to prove that $2^m 2^n = 2^{m+n}$ for any natural numbers m, n. Let $\varphi(m, n)$ express the relation that holds between m and n if we indeed have $2^m 2^n = 2^{m+n}$. Fix on an arbitrary number m. Then we can easily prove (i) $\varphi(m, 0)$, and (ii) for any n, if $\varphi(m, n)$ then $\varphi(m, n + 1)$. So we can conclude by induction that for all n, $\varphi(m, n)$. But m was arbitrary, so we can infer that for *any* m and n, $\varphi(m, n)$ which was to be proved.

Evidently, this sort of argument is in good order. But note that in the middle of the reasoning, at the point where we applied induction, 'm' was acting as a 'parameter', a temporary name not a bound variable. So when we come to implementing induction inside formal arithmetics we will need to set things up so that we can handle inductive arguments with parameters (or do something equivalent). More about this in due course.

9.4   Rule, schema, or axiom?

We have given enough illustrations of inductive arguments for the moment.[2] We will finish this chapter by thinking just a little more about how the basic induction principle should be framed.

(a)   Here again is the principle as we presented it at the outset, now explicitly marked as a *rule of inference*:

> **Induction Rule**   Given (i) $\varphi(0)$ and (ii) $\forall n(\varphi(n) \to \varphi(n + 1))$,
> we can infer $\forall n \varphi(n)$.

This rule can be brought to bear on any pair of premises (i) and (ii) involving a predicate φ which expresses a property that can coherently be attributed to numbers.

There is a corresponding *family of true conditionals*. These are the propositions obtained by filling out the following schema, one for every suitable way of instantiating φ:

> **Induction Schema** $\{\varphi(0) \land \forall n(\varphi(n) \to \varphi(n + 1))\} \to \forall n \varphi(n)$.

Now suppose (1) you have derived $\forall n \varphi(n)$ by applying the Rule to the relevant assumptions (i) $\varphi(0)$ and (ii) $\forall n(\varphi(n) \to \varphi(n + 1))$. Then you could have got to the same conclusion from the same assumptions by invoking the corresponding instance of the Schema and then applying modus ponens. Conversely (2) suppose you have got to $\forall n \varphi(n)$ by invoking the relevant Schema-instance and assumptions (i) and (ii). Then you could have got to the same conclusion by using the Rule on (i) and (ii).

Informally, therefore, there is little to choose between presenting the induction principle by means of a Rule or as a family of conditional truths specified by

---

[2] If you want more examples of induction in operation in elementary informal mathematics, see e.g. Velleman (1994), ch. 6.

the Schema. Similarly when we turn to adding induction to a formal theory of arithmetic. Typically, we can either add a rule of inference, or get the same effect by giving a formal template or *schema* and saying that any instance of that schematic form is an axiom.

(b) Isn't there a third way of presenting the intuitive principle of induction? Instead of giving one rule of inference, or giving many separate conditional axioms via a schema, can't we give a single universally quantified axiom? Putting it informally,

> **Induction Axiom** For all numerical properties X, if 0 has property X and also any number n is such that, if it has X, then its successor also has X, then *every* number has property X.

Borrowing symbols from logic, we could put it like this,

$$\forall X(\{X(0) \wedge \forall n(X(n) \rightarrow X(n+1))\} \rightarrow \forall n X(n)),$$

where 'X' is a second-order variable which can occupy predicate-position.[3] You might even think that this is the most natural way to frame the principle of induction.

But not too fast! What exactly does the second-order quantifier here quantify over? The notion of a 'numerical property' is, to say the least, a bit murky. Still, perhaps we don't need to distinguish such properties more finely than their extensions (see Section 14.3(c)), and so can treat the quantification here as tantamount to a quantification over sets of numbers. However that still leaves the question: *which* sets of numbers are we to quantify over? Well, the usual answer goes, over every arbitrary collection of numbers, finite or infinite, whether we can specify its members or not. But now we might wonder about *that* idea of arbitrary collections of numbers, including infinite ones: does an understanding of arithmetic and the ordinary use of inductive arguments really require us to conceive of such sets whose membership we can't specify?[4]

We won't pause now over such questions, however; we can leave them as a topic for Chapter 29. For even if we do resolve things in a way that can sustain the thought that the *second-order* Induction Axiom is, in some sense, the root principle of induction, we can't use that thought in formulating formal principles of induction apt for theories built in the *first-order* language L_A. And it is theories built in L_A which are going to be our concern for quite a few chapters.

---

[3]See Section 5.2, fn. 1, and also (for much more) Sections 29.1, 29.2.

[4]Recall our proof of Theorem 2.2, where we showed that you cannot enumerate the subsets of $\mathbb{N}$. So if we read the second-order quantifier as running over all arbitrary sets of numbers, the Induction Axiom tells us about an indenumerable collection of cases. By contrast, since the open wffs $\varphi(\mathsf{x})$ of a normal formal language are enumerable by Theorem 4.1(i), there are only enumerably many instances of the Induction Schema. So, with the suggested interpretation of its second-order quantifier, the Axiom is significantly stronger than the Schema. We will return to this point.

10 Two formalized arithmetics

We move on from the generalities of the previous chapters, and start looking at some particular formal arithmetics. Our main task in this chapter is to introduce Robinson Arithmetic Q, which will feature centrally in what follows. But we limber up by first looking at an even simpler theory, namely . . .

10.1 BA, Baby Arithmetic

We begin, then, with a theory which 'knows' about the addition and multiplication of particular numbers, but doesn't 'know' *any* arithmetical generalizations at all (for it lacks the whole apparatus of quantification). Hence our label 'Baby Arithmetic', or BA for short.

As with any formal theory, we need to characterize (a) its language, (b) its logical apparatus, and (c)–(e) its non-logical axioms.

(a)   BA's language is $L_B = \langle \mathcal{L}_B, \mathcal{I}_B \rangle$, where $\mathcal{L}_B$'s basic non-logical vocabulary is the same as that of $\mathcal{L}_A$ (see Section 5.2). In other words, $\mathcal{L}_B$ has a single individual constant '0', the one-place function symbol 'S', and the two-place function symbols '+' and '×'. Note, we can still construct the standard numerals in $\mathcal{L}_B$ just as before.

However, $\mathcal{L}_B$'s logical vocabulary is impoverished compared with $\mathcal{L}_A$. It lacks the whole apparatus of quantifiers and variables. Crucially, it still has the identity sign so that it can express equalities, and it has negation so that it can express inequalities. We will also give BA the other propositional connectives too (but which you chose as primitive is matter of taste).

The intended interpretation $\mathcal{I}_B$ built into L_B is the obvious one. '0' still has the value zero. 'S' still signifies the successor function S, and '+' and '×' of course continue to be interpreted as addition and multiplication.

(b)   BA needs some standard deductive apparatus to deal with identity and with the propositional connectives.

As far as identity is concerned, our logic needs to prove every instance of $\tau = \tau$, where τ is a term of our language $\mathcal{L}_B$. And we need a version of Leibniz's Law: given the premises (i) $\varphi(\sigma)$ and (ii) $\sigma = \tau$ (or $\tau = \sigma$), then Leibniz's Law will allow us to infer $\varphi(\tau)$.

The deductive system for the propositional connectives can then be your favourite version of classical logic. The details don't matter.

(c)   Now, crucially, for the (non-logical) axioms of BA. To start with, we want to pin down formally at least the following intuitive facts about the structure of the natural number sequence: (1) Zero is the *first* number, i.e. it isn't a successor; so for every n, $0 \neq Sn$. (2) The number sequence never circles back on itself; so different numbers have different successors. Contraposing, for any m, n, if $Sm = Sn$ then $m = n$.

We haven't got quantifiers in BA's formal language, however, so it can't express these general facts directly. Rather, we need to employ *schemata* (i.e. general templates) and say: *any sentence that you get from one of the following schemata by taking particular values for m and n is an axiom.*

Schema 1 $0 \neq S\overline{n}$

Schema 2 $S\overline{m} = S\overline{n} \rightarrow \overline{m} = \overline{n}$

Let's pause to show that instances of these initial schemata do indeed determine that different terms in the sequence 0, S0, SS0, SSS0, ..., pick out different numbers (for example $4 \neq 2$, i.e. SSSS0 $\neq$ SS0).[1]

Recall, we use '$\overline{n}$' to represent the numeral SS...S0 with n occurrences of 'S': so the general result we need is

Theorem 10.1 *For any m, n, if $m \neq n$, then* BA $\vdash \overline{m} \neq \overline{n}$.

Proof   Suppose $m < n$. Put $(n - m) - 1 = k$, so $k \geq 0$. Now consider the BA argument that begins

1. $0 \neq S\overline{k}$ Instance of Schema 1
2. $S0 = SS\overline{k} \rightarrow 0 = S\overline{k}$ Instance of Schema 2
3. $S0 \neq SS\overline{k}$ From 1, 2
4. $SS0 = SSS\overline{k} \rightarrow S0 = SS\overline{k}$ Instance of Schema 2
5. $SS0 \neq SSS\overline{k}$ From 3, 4
6. ...

And keep on going in the same way, adding an 'S' to each side of the inequation, until at line $(2m + 1)$ we reach $\overline{m} \neq \overline{n}$.

If $n < m$, we put $(m - n) - 1 = k$ and then in the same way derive $\overline{n} \neq \overline{m}$ and finally use the symmetry of identity to derive $\overline{m} \neq \overline{n}$ again.   ⊠

(d)   We next pin down the addition function by saying that any wffs of the following forms are axioms:

Schema 3 $\overline{m} + 0 = \overline{m}$

Schema 4 $\overline{m} + S\overline{n} = S(\overline{m} + \overline{n})$

(As usual, we allow ourselves to drop unnecessary brackets round additions and multiplications.) Instances of Schema 3 tell us the result of adding 0. Instances

---

[1]Reality check: why 'S' in the previous paragraph and 'S' in this one?

of Schema 4 with $n = 0$ tell us how to add 1 (i.e. add $S0$) by adding 0 and then applying the successor function to the result. Once we know about adding 1, we can use another instance of Schema 4 with $n = 1$ to explain how to add 2 (i.e. add $SS0$) in terms of adding $S0$. And so it goes.

Here, for example, is a BA derivation of $2 + 3 = 5$, or rather (putting that in unabbreviated form) of SS0 + SSS0 = SSSSS0:

1.	SS0 + 0 = SS0	Instance of Schema 3
2.	SS0 + S0 = S(SS0 + 0)	Instance of Schema 4
3.	SS0 + S0 = SSS0	From 1, 2 by LL
4.	SS0 + SS0 = S(SS0 + S0)	Instance of Schema 4
5.	SS0 + SS0 = SSSS0	From 3, 4 by LL
6.	SS0 + SSS0 = S(SS0 + SS0)	Instance of Schema 4
7.	SS0 + SSS0 = SSSSS0	From 5, 6 by LL

where 'LL' of course indicates the use of Leibniz's Law which allows us to inter-substitute identicals.

Evidently, this sort of proof is always available to give the correct result for any addition of two numerals. In other words, we have

Theorem 10.2 *For any m, n,* BA $\vdash \overline{m} + \overline{n} = \overline{m + n}$.[2]

(e)   We can similarly pin down the multiplication function by requiring every numeral instance of the following to be an axiom too:

Schema 5 $\overline{m} \times 0 = 0$

Schema 6 $\overline{m} \times S\overline{n} = (\overline{m} \times \overline{n}) + \overline{m}$

Instances of Schema 5 tell us the result of multiplying by zero. Instances of Schema 6 with $m = 0$ define how to multiply by 1 in terms of multiplying by 0 and then applying the already-defined addition function. Once we know about multiplying by 1, we can use another instance of Schema 6 with $m = 1$ to tell us how to multiply by 2 (multiply by 1 and do some addition). And so on and so forth.

More formally, here's a BA derivation of $2 \times 3 = 6$:

1.	SS0 × 0 = 0	Instance of Schema 5
2.	SS0 × S0 = (SS0 × 0) + SS0	Instance of Schema 6
3.	SS0 × S0 = 0 + SS0	From 1, 2 by LL
4.	0 + SS0 = SS0	By Theorem 10.2
5.	SS0 × S0 = SS0	From 3, 4 by LL

---

[2]Given our overlining convention, '$\overline{m} + \overline{n}$' is shorthand for the expression you get by writing the numeral for m followed by a plus sign followed by the numeral for n, while '$\overline{m + n}$' is shorthand for the numeral for $m + n$. (Likewise, for $n \geq m$, '$\overline{n - m}$' stands in for the numeral for $n - m$; while '$\overline{n} - \overline{m}$' would be ill-formed since our formal language lacks a subtraction sign.)

6.	$SS0 \times SS0 = (SS0 \times S0) + SS0$	Instance of Schema 6
7.	$SS0 \times SS0 = SS0 + SS0$	From 5, 6 by LL
8.	$SS0 + SS0 = SSSS0$	By Theorem 10.2
9.	$SS0 \times SS0 = SSSS0$	From 7, 8 by LL
10.	$SS0 \times SSS0 = (SS0 \times SS0) + SS0$	Instance of Schema 6
11.	$SS0 \times SSS0 = SSSS0 + SS0$	From 9, 10 by LL
12.	$SSSS0 + SS0 = SSSSSS0$	By Theorem 10.2
13.	$SS0 \times SSS0 = SSSSSS0$	From 11, 12 by LL.

Which is tedious, but makes it very plain that we will similarly be able to derive the value of the product of any two numerals. Hence we have

Theorem 10.3 *For any m, n, BA $\vdash \overline{m} \times \overline{n} = \overline{m \times n}$.*

10.2 BA is negation-complete

BA, then, is the theory whose non-logical axioms are all the instances of the Schemata 1–6, and whose rules are the classical rules for identity, negation, and the other propositional connectives. It is easily checked to be an effectively axiomatized theory.

We have just seen that BA can prove equations correctly evaluating any term of the form $\overline{m} + \overline{n}$ or $\overline{m} \times \overline{n}$. We now note that BA can in fact correctly evaluate *any* term at all, however complex. That is to say,

> **Theorem 10.4** *If τ is a term of $\mathcal{L}_B$, which takes the value t on $\mathcal{I}_B$, then BA $\vdash \tau = \overline{t}$.*

Proof sketch Three initial observations:

(i) If BA correctly evaluates τ, it correctly evaluates $S\tau$. For suppose τ takes the value t, and so $S\tau$ takes the value $t+1$. By assumption BA proves $\tau = \overline{t}$, and hence we can derive $S\tau = S\overline{t}$. But that means BA proves $S\tau = \overline{t+1}$ (since, unpacked, '$S\overline{t}$' is of course none other than '$\overline{t+1}$').

(ii) If BA correctly evaluates σ and τ, it correctly evaluates $\sigma + \tau$. For suppose, σ takes the value s, τ takes the value t and BA proves both $\sigma = \overline{s}$ and $\tau = \overline{t}$. Then $\sigma + \tau$ takes the value $s + t$. But from our assumptions, BA proves $\sigma + \tau = \overline{s} + \overline{t}$. And by Theorem 10.2, BA proves $\overline{s} + \overline{t} = \overline{s+t}$. So by the transitivity of identity BA proves $\sigma + \tau = \overline{s+t}$.

(iii) Similarly, if BA correctly evaluates σ and τ, it correctly evaluates $\sigma \times \tau$.

Now put those observations together with the point that every term, however complicated, is constructed from repeated applications of S, $+$, $\times$, starting from 0. Quite trivially, BA correctly evaluates 0, which is just to say that it proves $0 = 0$. But we've just seen that starting from terms it correctly evaluates, each time we apply S, + or × we get another term which BA correctly evaluates. So BA correctly evaluates every term it can construct. ⊠

(Exercise: turn that proof-sketch into a more careful proof by a course-of-values induction on the complexity of the term τ.)

Here is an easy corollary of that last result taken together with Theorem 10.1. Recall, to say that BA correctly decides φ is to say that if φ is true BA $\vdash \varphi$, and if φ is false BA $\vdash \neg\varphi$ (see Section 4.4, defn. 4). Then,

Theorem 10.5 BA *correctly decides every atomic wff of* $\mathcal{L}_B$.

Proof The only atomic wffs of BA are equations. Suppose that the equation $\sigma = \tau$ is true. Then some number n is the value of both σ and τ. So by the previous theorem, BA $\vdash \sigma = \bar{n}$ and BA $\vdash \tau = \bar{n}$. Hence, by the identity rules, BA $\vdash \sigma = \tau$.

Suppose the equation $\sigma = \tau$ is false. If the value of σ is s, and the value of τ is t, then $s \neq t$. By the previous theorem, BA $\vdash \sigma = \bar{s}$ and BA $\vdash \tau = \bar{t}$. By Theorem 10.1 BA $\vdash \bar{s} \neq \bar{t}$. Hence, by the identity rules again, BA $\vdash \sigma \neq \tau$. ⊠

We can now readily beef up that last result:

Theorem 10.6 BA *correctly decides* every *wff of* $\mathcal{L}_B$.

Proof Suppose φ is any truth-functional combination of atomic wffs: then the truth-values of those atoms together fix the truth-value of φ. In other words, taking Σ to comprise the true atoms and the negations of the false atoms in φ, Σ will semantically entail φ (if it is true) or $\neg\varphi$ (if φ is false). So any complete propositional deductive logic will be able to deduce either φ or $\neg\varphi$ (depending which is the true one) from Σ.

Now, given any BA-wff φ, we have just seen that BA can prove all the members of the corresponding Σ – i.e. can prove the true atoms (equations) in φ, and the negations of the false ones. Further, BA includes a complete propositional deductive logic. Hence BA can indeed go on either to prove φ outright (if it is true) or to prove $\neg\varphi$ (if φ is false). ⊠

However, any $\mathcal{L}_B$-wff φ is either true or false, so our last theorem shows that BA always proves one of φ or $\neg\varphi$. In other words,

Theorem 10.7 BA *is negation-complete.*

BA is evidently a sound theory (as far as it goes); it has true axioms and a truth-preserving logic. If its language were 'sufficiently expressive' in the sense of Section 6.1, then by Theorem 6.3 it couldn't be complete. But the quantifier-free language $\mathcal{L}_B$ is very restricted indeed; it is nowhere near sufficiently expressive. That is why BA can settle any proposition in the very limited class it *can* express, and we get Theorem 10.7.

In sum, BA's completeness comes at the high price of being expressively extremely impoverished.

10.3   Q, Robinson Arithmetic

The obvious way to start beefing up BA into something rather more exciting is to restore the familiar apparatus of quantifiers and variables. So while keeping the same non-logical vocabulary, we will now allow ourselves again the resources of first-order quantification, so that we are working with the full language $L_A = \langle \mathcal{L}_A, \mathcal{I}_A \rangle$ of basic arithmetic (see Section 5.2). Our theory's deductive apparatus will be some version of classical first-order logic with identity. In a moment, we'll fix on our official logic.

Since we now have the quantifiers available to express generality, we can replace each of the metalinguistic Schemata we used in specifying the axioms of BA with a corresponding quantified Axiom. For example, we can replace the first two Schemata governing the successor function by

> **Axiom 1** $\forall x(0 \neq Sx)$
>
> **Axiom 2** $\forall x \forall y(Sx = Sy \to x = y)$

In Axiom 1 (and more Axioms below) we informally add redundant brackets for readability. Obviously, each instance of our earlier Schemata 1 and 2 can be deduced from the corresponding Axiom by instantiating the quantifiers.

Note, however, that while these Axioms tell us that zero isn't a successor, they leave it open whether there are other objects that aren't successors cluttering up the domain of quantification (there could be 'pseudo-zeros'). We don't want our quantifiers – now that we've introduced them – running over such stray objects. So let's explicitly rule them out:

> **Axiom 3** $\forall x(x \neq 0 \to \exists y(x = Sy))$

Next, we can similarly replace our previous Schemata for addition and multiplication by universally quantified Axioms:

> **Axiom 4** $\forall x(x + 0 = x)$
>
> **Axiom 5** $\forall x \forall y(x + Sy = S(x + y))$
>
> **Axiom 6** $\forall x(x \times 0 = 0)$
>
> **Axiom 7** $\forall x \forall y(x \times Sy = (x \times y) + x)$

Again it is obvious that each instance of one of our earlier Schemata 3–6 can be deduced from the corresponding generalization from Axioms 4–7 by instantiating the quantifiers.

The theory with language L_A, Axioms 1 to 7, plus a standard first-order logic, is called *Robinson Arithmetic*, or (very often) simply Q.[3] It is again an effectively axiomatized theory.

---

[3]This formal system was first isolated by Robinson (1952) and immediately became well-known through the classic Tarski et al. (1953).

10.4   Which logic?

We haven't yet specified a particular implementation of classical logic for our two arithmetics. But we ought now to be a bit more specific at least about the *kind* of logical system we officially give Q.

As will be very familiar, there is a wide variety of formal deductive systems for first-order logic, systems which are equivalent in the sense of proving the same sentences as conclusions from given sentences as premises. Let's contrast, in particular, Hilbert-style axiomatic systems with natural deduction systems.

A Hilbert-style system defines a class of logical axioms, usually by giving schemata such as $\varphi \to (\psi \to \varphi)$ and $\forall \xi \varphi(\xi) \to \varphi(\tau)$ and then stipulating – perhaps with some restrictions – that any instance of a schema is an axiom. Having a rich set of axioms, such a deductive system can operate with just one or two rules of inference. And a proof in a theory using an axiomatic logic is then simply a linear sequence of wffs, each one of which is either (i) a logical axiom, or (ii) an axiom belonging to the specific theory, or (iii) follows from previous wffs in the sequence by one of the rules of inference.[4]

A natural deduction system, on the other hand, will have no logical axioms but many rules of inference. And, in contrast to axiomatic logics, it will allow temporary assumptions to be made for the sake of argument and then later discharged. We will need some way, therefore, of keeping track of when temporary assumptions are in play. One common option is to use tree structures of the type Gerhard Gentzen introduced, with a system for labelling inferences to indicate when suppositions are discharged. Another option is to use Frederic Fitch's device of indenting a column of argument to the right each time a new assumption is made, and shifting back to the left when the assumption is discharged.[5]

So which style of logical system should we adopt in developing Q and other arithmetics with a first-order logic? Well, that will depend on whether we are more concerned with the ease of proving certain metalogical results *about* formal arithmetics or with the ease of proving results *inside* the theories. Hilbertian systems are very amenable to metalogical treatment but are horrible to use in practice. Natural deduction systems are indeed natural in use; but it takes more effort to theorize about arboriform proof structures.

I propose that in this book we cheerfully have our cake and eat it. So when we consider arithmetics like Q, *officially* we'll take their logic to be a Hilbertian, axiomatic one, so that proofs are simple linear sequences of wffs. This way, when we come to theorize *about* arithmetic proofs, and e.g. use the Gödelian trick of using numbers to code proofs, everything goes as simply as it possibly can. However, when we want to outline sample proofs *inside* formal arithmetics, we

---

[4]The *locus classicus* is Hilbert and Ackermann (1928). For a modern logic text which uses a Hilbert-style system, see e.g. Mendelson (1997).

[5]The *locus classicus* for natural deduction systems is, of course, Gentzen (1935). For a modern text which uses a natural deduction system set out in tree form, see e.g. van Dalen (1994). Frederic Fitch introduces his elegant way of setting out proofs in Fitch (1952).

will continue to give arguments framed in a more manageable natural deduction style. The familiar equivalences between the different logical systems will then warrant the conclusion that an official Hilbert-style proof of the same result will be available.

10.5 Q is not complete

Like BA, Q is evidently a sound theory (and so consistent). Its axioms are all true; its logic is truth-preserving; its derivations are therefore proper proofs in the intuitive sense of demonstrations of truth. In sum, every Q-theorem is a true L_A sentence. But just which L_A-truths are theorems?

Since, as we noted, any BA axiom – i.e. any instance of one of our previous Schemata – can be derived from one of our new Q Axioms, every $\mathcal{L}_B$-sentence that can be proved in BA is equally a quantifier-free $\mathcal{L}_A$-sentence which can be proved in Q. Hence, Q again correctly decides every quantifier-free sentence.

However, there are *very* simple true quantified sentences that Q can't prove. For example, Q can prove any particular wff of the form $0 + \bar{n} = \bar{n}$. But it can't prove its universal generalization. In other words,

Theorem 10.8 $Q \nvdash \forall x(0 + x = x)$.[6]

Proof sketch We use an elementary 'model-theoretic' argument. To show a wff χ is *not* a theorem of a given theory T, it is enough to find a model or interpretation (often a deviant, unintended, re-interpretation) for the T-wffs which makes the axioms of T true and hence makes all its theorems true, but which makes χ false.

So take $\mathcal{L}_A$, the syntax of Q. We want to find a deviant re-interpretation $\mathcal{I}_D$ of the same $\mathcal{L}_A$-wffs, where $\mathcal{I}_D$ still makes Q's Axioms true but allows cases where 'adding' a 'number' to the 'zero' yields a different 'number'. Here's an artificial – but still legitimate – example.

Take the domain of our deviant, unintended, interpretation $\mathcal{I}_D$ to be the set N^* comprising the natural numbers but with two other 'rogue' elements a and b added (these could be e.g. Kurt Gödel and his friend Albert Einstein). Let '0' still refer to zero. And take 'S' now to pick out the successor* function S^* which is defined as follows: $S^*n = Sn$ for any natural number in the domain, while for our rogue elements $S^*a = a$, and $S^*b = b$. It is immediate that Axioms 1 to 3 are still true on this deviant interpretation.

We now have to re-interpret Q's function '+'. Suppose we take this to pick out addition*, where $m +^* n = m + n$ for any natural numbers m, n in the domain, while $a +^* n = a$ and $b +^* n = b$. Further, for any x (whether number or rogue element), $x +^* a = b$ and $x +^* b = a$. It is easily checked that interpreting '+' as addition* still makes Axioms 4 and 5 true. But by construction, $0 +^* a \neq a$, so this interpretation makes $\chi =_{\text{def}} \forall x(0 + x = x)$ false.

---

[6]The notational shorthand here is to be read in the obvious way: i.e we write $T \nvdash \varphi$ as short for not-$(T \vdash \varphi)$, i.e. φ is unprovable in T.

We are not quite done, however, as we still need to show that we can give a co-ordinate re-interpretation of '×' in Q by some deviant multiplication* function. We can leave it as an exercise to fill in suitable details. Then, with the details filled in, we will have an overall interpretation $\mathcal{I}_D$ which makes the axioms of Q true and χ false. So Q can't prove χ (given that Q's logic is sound).   ⊠

Obviously, Q can't prove $\neg\chi$ either. Just revert to the standard interpretation $\mathcal{I}_A$. Q certainly has true axioms on this interpretation: so all theorems are true on $\mathcal{I}_A$. But $\neg\chi$ is false on $\mathcal{I}_A$, so it can't be a theorem.

Hence, in sum, Q $\nvdash \chi$ and Q $\nvdash \neg\chi$. Which gives us the utterly unsurprising

Theorem 10.9 Q *is not negation-complete,*

(still assuming, of course, that Q is sound and so consistent). We have already announced that Gödel's incompleteness theorem is going to prove that *no* sound effectively axiomatized theory in the language of basic arithmetic can be negation-complete. But what we've just shown is that we don't need to invoke anything as elaborate as Gödel's arguments to see that Q is incomplete: Robinson Arithmetic is, so to speak, *boringly* incomplete.

10.6   Why Q is interesting

Given it can't even prove $\forall x(0 + x = x)$, Q is evidently a *very* weak theory of arithmetic. You might suspect, then, that like BA this is another toy theory which is not really worth pausing long over.

But not so. Despite its great shortcomings, and perhaps rather unexpectedly, we'll later be able to show that Q *is 'sufficiently strong' in the sense of Chapter 7.* For 'sufficient strength' is a matter of being able to *case-by-case* prove enough wffs about decidable properties of individual numbers. And it turns out that Q's hopeless weakness at proving generalizations doesn't stop it from doing that.

So that's why Q is interesting. Suppose a theory of arithmetic is effectively axiomatized, consistent and can prove everything Q can prove (those do seem *very* modest requirements). Then what we've just announced and promised to prove is that any such theory will be sufficiently strong. And therefore Theorem 7.2 will apply – such a theory will be incomplete.

However, proving the crucial claim that Q *does* have sufficient strength to capture all decidable properties has to be business for much later; plainly, we can only establish it when we have a quite general theory of decidability to hand.

What we *will* prove quite soon is a somewhat weaker claim about Q: in Chapter 17, we show that it can capture all 'primitive recursive' properties, where these form an important subclass of the decidable properties. This pivotal theorem will be the crucial load-bearing part of various proofs of Gödel-style incompleteness theorems. The next chapter goes through some necessary preliminaries.

11 What Q can prove

As we saw, Robinson's Q is a very weak theory of arithmetic. But in this chapter, we will explore what it *can* establish. Quite unavoidably, some of the detailed proofs of our claims do get tedious; so you are very welcome to skip as many of them as you like (you won't miss anything exciting). However, you will need to carry forward to later chapters an understanding of the key *concepts* we'll be introducing. Note in particular the notions of Δ_0, Σ_1 and Π_1 wffs, and the important result that Q is 'Σ_1-complete'.

11.1 Capturing *less-than-or-equal-to* in Q

We know from Section 5.4(b) that the *less-than-or-equal-to* relation is *expressed* in L_A by the wff $\exists v(v + x = y)$. In this section, we show that the relation is *captured* by the same wff in Q. That is to say, for any particular pair of numbers, m, n, if $m \leq n$, then $Q \vdash \exists v(v + \overline{m} = \overline{n})$, and otherwise $Q \vdash \neg\exists v(v + \overline{m} = \overline{n})$.

Proof sketch Suppose $m \leq n$, so for some $k \geq 0$, $k + m = n$. Q can prove everything BA proves and therefore can correctly evaluate every addition of numerals. So we have $Q \vdash \overline{k} + \overline{m} = \overline{n}$. Hence by existential quantifier introduction $Q \vdash \exists v(v + \overline{m} = \overline{n})$, as was to be shown.

Suppose alternatively $m > n$. We need to show $Q \vdash \neg\exists v(v + \overline{m} = \overline{n})$. We'll first demonstrate this in the case where $m = 4$, $n = 2$, and prove $\neg\exists v(v + 4 = 2)$ inside Q. Then it will be easy to see that the proof strategy will work more generally.

Officially, Q's logic is a Hilbert-type system; but to keep things manageable, we'll pretend that it uses an equivalent natural deduction system instead (see Section 10.4). And because it is well known – but also simple to follow if you don't know it – we will use a Fitch-style system where we indent sub-proofs while a new temporary assumption is in force. We use symbols recruited from our stock of unused variables to act as temporary names ('parameters').

So consider the following argument (for brevity we will omit statements of Q's axioms and some other trivial steps):[1]

1.	$a + SSSS0 = SS0$	Supposition
2.	$a + SSSS0 = S(a + SSS0)$	From Axiom 5
3.	$S(a + SSS0) = SS0$	From 1, 2 by LL

---

[1] Does it need saying? 'LL' again indicates the use of Leibniz's Law, 'MP' stands for modus ponens, 'RAA' for reductio ad absurdum, 'UI' for universal quantifier introduction, and '∃/∀' indicates whatever steps we need to move between the quantifiers in the standard way.

4.	$S(a + SSS0) = SS0 \to a + SSS0 = S0$	From Axiom 2
5.	$a + SSS0 = S0$	From 3, 4 by MP
6.	$a + SSS0 = S(a + SS0)$	From Axiom 5
7.	$S(a + SS0) = S0$	From 5, 6 by LL
8.	$S(a + SS0) = S0 \to a + SS0 = 0$	From Axiom 2
9.	$a + SS0 = 0$	From 7, 8 by MP
10.	$a + SS0 = S(a + S0)$	From Axiom 5
11.	$S(a + S0) = 0$	From 9, 10 by MP
12.	$0 \neq S(a + S0)$	From Axiom 1
13.	$0 \neq 0$	From 11, 12 by LL
14.	$0 = 0$	From identity rule
15.	Contradiction!	From 13, 14
16.	$\neg(a + SSSS0 = SS0)$	From 1–15 by RAA
17.	$\forall v \neg(v + SSSS0 = SS0)$	From 16 by UI
18.	$\neg\exists v(v + SSSS0 = SS0)$	From 17 by $\exists/\forall$.

And having done the proof for the case $m = 4$, $n = 2$, inspection reveals that we can use the same general pattern of argument to show $Q \vdash \neg\exists v(v + \overline{m} = \overline{n})$ whenever $m > n$.

Thus, we suppose $a + \overline{m} = \overline{n}$. Then, just as in steps 2–9, we chip away initial occurrences of S from the numerals $\overline{m}$ and $\overline{n}$ until we derive $a + \overline{m - n} = 0$. A little routine like that in steps 10–15 will then quickly generate contradiction, so by reductio we get $\neg(a + \overline{m} = \overline{n})$ on no assumptions about a. We then invoke the quantifier rules to derive $\neg\exists v(v + \overline{m} = \overline{n})$. ⊠

11.2 '$\leq$' and bounded quantifiers

Given the result we've just proved, it is very natural to introduce '$\xi \leq \zeta$' as an abbreviation – at a first shot – for '$\exists v(v + \xi = \zeta)$'.[2]

But obviously that initial definition could get us into trouble. Suppose it happens that what we put into the slot marked by 'ξ' already has a free occurrence of 'v': we then get a nasty 'clash of variables'. So here is an improved proposal:

> Use '$\xi \leq \zeta$' as an abbreviation for '$\exists\nu(\nu + \xi = \zeta)$', where '$\nu$' holds the place of the alphabetically first variable which isn't in either of the terms which replace 'ξ' and 'ζ'.

Since it so greatly helps readability, we'll henceforth make very free use of this abbreviatory symbol for writing formal wffs of arithmetic.

---

[2]Here the metalinguistic Greek letters are serving as 'place-holders'. Some presentations treat '$\leq$' as a primitive symbol built into our formal theories from the start, governed by its own additional axioms. Nothing important hangs on the difference between that approach and our policy of introducing the symbol by definition.

Nothing hangs either on our policy of introducing '$\leq$' as our basic symbol rather than '$<$', which could have been defined by $\xi < \zeta =_{\mathrm{def}} \exists\nu(S\nu + \xi = \zeta)$.

In informal mathematics we often want to say that all/some numbers less than or equal to a given number k have some particular property. We can now express such claims in formal arithmetics by wffs of the shape '$\forall \xi(\xi \leq \kappa \to \varphi(\xi))$' and '$\exists \xi(\xi \leq \kappa \land \varphi(\xi))$'. And it is standard to introduce a further abbreviatory device to handle such wffs, thus:

'$(\forall \xi \leq \kappa)\varphi(\xi)$' is an abbreviation for '$\forall \xi(\xi \leq \kappa \to \varphi(\xi))$',

'$(\exists \xi \leq \kappa)\varphi(\xi)$' is an abbreviation for '$\exists \xi(\xi \leq \kappa \land \varphi(\xi))$',

where 'κ' is a term and '$\leq$' is to be unpacked as we've just explained. We'll say that such expressions '$(\forall \xi \leq \kappa)$', '$(\exists \xi \leq \kappa)$' are *bounded quantifiers*.

11.3   Q is order-adequate

Q can case-by-case capture the less-than-or-equal-to relation. We now note that Q can also prove a bunch of general facts about this relation. Let's say, for brevity, that a theory T is *order-adequate* if the following nine propositions hold:

O1. $T \vdash \forall x(0 \leq x)$.

O2. For any n, $T \vdash \forall x(\{x = 0 \lor x = 1 \lor \ldots \lor x = \overline{n}\} \to x \leq \overline{n})$.

O3. For any n, $T \vdash \forall x(x \leq \overline{n} \to \{x = 0 \lor x = 1 \lor \ldots \lor x = \overline{n}\})$.

O4. For any n, if $T \vdash \varphi(0)$, $T \vdash \varphi(1)$, ..., $T \vdash \varphi(\overline{n})$,
$$\text{then } T \vdash (\forall x \leq \overline{n})\varphi(x).$$

O5. For any n, if $T \vdash \varphi(0)$, or $T \vdash \varphi(1)$, ..., or $T \vdash \varphi(\overline{n})$,
$$\text{then } T \vdash (\exists x \leq \overline{n})\varphi(x).$$

O6. For any n, $T \vdash \forall x(x \leq \overline{n} \to x \leq S\overline{n})$.

O7. For any n, $T \vdash \forall x(\overline{n} \leq x \to (\overline{n} = x \lor S\overline{n} \leq x))$.

O8. For any n, $T \vdash \forall x(x \leq \overline{n} \lor \overline{n} \leq x)$.

O9. For any $n > 0$, $T \vdash (\forall x \leq \overline{n-1})\varphi(x) \to (\forall x \leq \overline{n})(x \neq \overline{n} \to \varphi(x))$.

Then we have the following summary result:

Theorem 11.1 Q *is order-adequate.*

This theorem is pivotal for what follows later. It does, however, belong squarely to the class of results which are rather trivial but are a bit tiresome to prove: it is indeed fiddly to check that Q satisfies the nine conditions.

If you have a taste for elementary logical brain-teasers, then by all means see how many of the nine conditions you can verify: but none of the proofs involves anything really interesting, as you will see when we give a few examples in the final section of this chapter. Otherwise you are very welcome to skip and take the results on trust.[3]

11.4 Q can correctly decide all Δ_0 sentences

(a)   Q can correctly decide every sentence of the form $\sigma = \tau$ which you can construct in L_A. That's because Q includes BA, and Theorem 10.5 tells us that BA can correctly decide every such equation. Q can also correctly decide every sentence of the form $\sigma \leq \tau$: that's because Q can capture the less-than-or-equals relation. Q also contains classical propositional logic. So it can correctly decide sentences built up from equations and/or order claims using propositional connectives.

But we have just seen in effect that, in Q, a wff of the form $(\forall x \leq \bar{n})\varphi(x)$ is equivalent to the conjunction $\varphi(0) \wedge \varphi(1) \wedge \ldots \wedge \varphi(\bar{n})$. Likewise $(\exists x \leq \bar{n})\varphi(x)$ is equivalent to $\varphi(0) \vee \varphi(1) \vee \ldots \vee \varphi(\bar{n})$. So this means that Q must also be able to correctly decide any sentence built up from equations and/or statements of order by using bounded quantifiers along with the propositional connectives.

(b)   That is the headline news. In the rest of this section we put this all rather more carefully. We start with a key definition:

1. If σ and τ are terms of L_A, then $\sigma = \tau$ and $\sigma \leq \tau$ are Δ_0 wffs.

2. If φ and ψ are Δ_0 wffs, so are $\neg\varphi$, $(\varphi \wedge \psi)$, $(\varphi \vee \psi)$, $(\varphi \rightarrow \psi)$ and $(\varphi \leftrightarrow \psi)$.

3. If φ is a Δ_0 wff, so are $(\forall \xi \leq \kappa)\varphi$ and $(\exists \xi \leq \kappa)\varphi$, where ξ is any variable free in φ, and κ is a numeral or a variable distinct from ξ.[4]

4. Nothing else is a Δ_0 wff.

Here are some simple examples of Δ_0 wffs (allowing abbreviations for numerals and the contraction '$\neq$'): (i) $1 \leq 3$, (ii) $(y + 1) = x$ (iii) $SS(y \times 2) \leq (y \times SSSy)$, (iv) $(x \leq y \wedge SSx \neq Sy)$, (v) $(\exists x \leq 2)x \neq 1$, (vi) $\neg(\exists z \leq 5)(\forall y \leq x)\, y \leq z + x$, and (vii) $(1 \leq x \rightarrow (\exists y \leq x)Sy = x)$.

(c)   We now note the following result:

Theorem 11.2 *We can effectively calculate the truth-value of any Δ_0 sentence.*

(Meaning, of course, the truth-value according to the interpretation $\mathcal{I}_A$ built into the language L_A.) This should indeed seem obvious because we only have

---

[3]'He has only half learned the art of reading who has not added to it the more refined art of skipping and skimming.' (A. J. Balfour, one-time British Prime Minister.) This remark applies in spades to the art of reading mathematical texts.

[4]Why the distinctness requirement? Because $(\forall x \leq x)\varphi(x)$, for example, unpacks as $\forall x(x \leq x \rightarrow \varphi(x))$ which is equivalent to the *unbounded* $\forall x\varphi(x)$. Not what we want!

to look through a finite number of cases in order to deal with any *bounded* quantifications.

Still, it is worth demonstrating the result properly, to illustrate a very important proof-technique. So let us say that a Δ_0 sentence has degree k iff it is built up from wffs of the form $\sigma = \tau$ or $\sigma \leq \tau$ by k applications of connectives and/or bounded quantifiers. Then we can use a course-of-values induction on the degree of complexity k.

Proof For the base case (i): a degree 0 sentence is just a simple sentence of the form $\sigma = \tau$ or $\sigma \leq \tau$, and we can evidently calculate the truth-value of any such sentence (why?).

Now for the induction step (ii): if we can calculate the truth-value of any Δ_0 sentence of degree no more than k by some effective algorithmic procedure, then we can calculate the truth-value of any degree $k + 1$ sentence χ too. To show this, there are three cases to consider:

a. χ is of the form $\neg\varphi$, $(\varphi \wedge \psi)$, $(\varphi \vee \psi)$, $(\varphi \rightarrow \psi)$ or $(\varphi \leftrightarrow \psi)$, where φ and ψ are Δ_0 sentences of degree no greater than k. The truth-value of the relevant φ and ψ is by hypothesis calculable, and hence (using truth-tables) so is the truth value of χ.

b. χ is of the form $(\forall \xi \leq \overline{n})\varphi(\xi)$, where $\varphi(\xi)$ is a Δ_0 wff of degree k.[5] This has the same truth-value as $\varphi(0) \wedge \varphi(1) \wedge \ldots \wedge \varphi(\overline{n})$. Each conjunct is a closed Δ_0 wff of degree k, i.e. it is a sentence whose truth-value is by hypothesis calculable. Hence the truth-value of χ is calculable too.

c. χ is of the form $(\exists \xi \leq \overline{n})\varphi(\xi)$. The argument is similar.

Putting (i) and (ii) together, an informal course-of-values induction yields the desired conclusion that we can effectively calculate the truth-value of *any* Δ_0 sentence, whatever its degree k. ⊠

(d) We now show that there is a sense in which Q can do what *we* can do – i.e. prove the true Δ_0 wffs and refute the false ones.

Theorem 11.3 Q *correctly decides every* Δ_0 *sentence.*

Proof We again proceed by an informal course-of-values induction on the complexity of sentences.

For the base case (i) of degree 0 sentences, we remark again that such a sentence is either (a) an equation $\sigma = \tau$ or else (b) a wff of the form $\sigma \leq \tau$, where σ and τ are closed terms denoting some numbers s and t respectively. In case (a), we are done, because we already know that Q correctly decides every such equation. For case (b), Q like BA correctly evaluates terms, i.e. can prove

---

[5] Because χ is a sentence, $\varphi(\xi)$ can only have the variable ξ free. And because it is a sentence, χ can't be of the form $(\forall \xi \leq \nu)\varphi(\xi)$ with ν a free variable.

$\sigma = \bar{s}$ and $\tau = \bar{t}$ with numerals on the right. But since '$\leq$' captures the less-than-or-equal-to relation, Q correctly decides $\bar{s} \leq \bar{t}$. Hence, plugging in the identities, Q correctly decides $\sigma \leq \tau$.

For the induction step (ii) let's assume Q correctly decides all Δ_0 sentences of degree up to k. We'll show that it correctly decides χ, an arbitrary degree $k + 1$ sentence. As in the last proof, there are three cases to consider.

a. χ is built using a propositional connective from φ and perhaps ψ, sentences of lower degree which by assumption Q correctly decides. But it is an elementary fact about classical propositional logic that, if a theory with a classical logic correctly decides φ and ψ, it correctly decides $\neg\varphi$, $(\varphi \wedge \psi)$, $(\varphi \vee \psi)$, $(\varphi \rightarrow \psi)$ and $(\varphi \leftrightarrow \psi)$. And so Q correctly decides χ.

b. χ is of the form $(\forall\xi \leq \bar{n})\varphi(\xi)$. If χ is a *true* sentence, then $\varphi(0)$, $\varphi(1)$, ..., $\varphi(\bar{n})$ must all be true sentences. Being of lower degree, these are – by hypothesis – all correctly decided by Q; so Q proves $\varphi(0)$, $\varphi(1)$, ..., $\varphi(\bar{n})$. Hence, by (O4) of Section 11.3, Q also proves $(\forall\xi \leq \bar{n})\varphi(\xi)$. On the other hand, if χ is *false*, $\varphi(\bar{m})$ is false for some $m \leq n$, and – being of lower degree – this is correctly decided by Q, so Q proves $\neg\varphi(\bar{m})$. Hence, by (O5), Q proves $(\exists\xi \leq \bar{n})\neg\varphi(\xi)$, which easily entails $\neg(\forall\xi \leq \bar{n})\varphi(\xi)$. In sum, Q correctly decides χ, i.e. $(\forall\xi \leq \bar{n})\varphi(\xi)$.

c. χ is of the form $(\exists\xi \leq \bar{n})\varphi(\xi)$. Dealt with similarly to case (ii).

Given (i) and (ii), we use course-of-values induction to deduce that Q decides all Δ_0 sentences, whatever their degree.   ☒

Be very clear. There is no induction rule *inside* Q: we don't start considering formal arithmetics with an induction principle until the next chapter. But that doesn't prevent us from using induction informally, *outside* Q, to establish meta-theoretical results about what that theory can prove.

11.5 Σ_1 and Π_1 wffs

(a) Δ_0 sentences can be arbitrarily long and messy, involving lots of bounded quantifications. But we have now seen that even Q – which is otherwise so hopelessly weak at proving generalities – can prove all the true Δ_0 sentences and refute all the false ones, because everything is bounded.

At the next step up in *quantifier complexity*, we find the so-called Σ_1 and Π_1 wffs. Basically, the Σ_1 wffs involve unbounded existential quantifications of Δ_0 wffs. And the Π_1 wffs involve unbounded universal quantifications of Δ_0 wffs.

To help remember the terminology here, note that the 'Σ' in the standard label 'Σ_1' comes from an old alternative symbol for the existential quantifier, as in ΣxFx – that's a Greek 'S' for '(logical) sum'. Likewise the 'Π' in 'Π_1' comes from the corresponding symbol for the universal quantifier, as in ΠxFx – that's a Greek 'P' for '(logical) product'. The subscript '1' indicates the first level of a hierarchy which we will say more about in a moment.

(b) Why are these notions of Σ_1 and Π_1 wffs of interest? Because – looking ahead – it turns out that Σ_1 wffs are just what we need to express and capture computable numerical functions and effectively decidable properties of numbers. While the sentences that Gödel constructs in proving his incompleteness theorem are equivalent to Π_1 sentences.

Given their significance, then, let's pause to pin down the notions more carefully. Here then is our definition for the class of Σ_1 wffs:[6]

1. Any Δ_0 wff is a Σ_1 wff.

2. If φ and ψ are Σ_1 wffs, so are $(\varphi \wedge \psi)$ and $(\varphi \vee \psi)$.

3. If φ is a Σ_1 wff, so are $(\forall \xi \leq \kappa)\varphi$ and $(\exists \xi \leq \kappa)\varphi$, where ξ is any variable free in φ, and κ is a numeral or a variable distinct from ξ.

4. If φ is a Σ_1 wff, so is $(\exists \xi)\varphi$ where ξ is any variable free in φ.

5. Nothing else is a Σ_1 wff.

Two comments. First, note the absence of the negation $\neg\varphi$ from clause (2) (as well as the related absences of the conditional $\varphi \to \psi$, i.e. $\neg\varphi \vee \psi$, and the biconditional $\varphi \leftrightarrow \psi$). But of course, if we allowed first applying an (unbounded) existential quantifier and then negating the result, that would be tantamount to allowing (unbounded) universal quantifications, which isn't what we want. Second, there is a certain redundancy in this way of putting it, because it is immediate that clause (4) covers the existential half of clause (3).

The companion definition of the class of Π_1 wffs goes as you would expect:

1. Any Δ_0 wff is a Π_1 wff.

2. If φ and ψ are Π_1 wffs, so are $(\varphi \wedge \psi)$ and $(\varphi \vee \psi)$.

3. If φ is a Π_1 wff, so are $(\forall \xi \leq \kappa)\varphi$ and $(\exists \xi \leq \kappa)\varphi$, where ξ is any variable free in φ, and κ is a numeral or a variable distinct from ξ.

4. If φ is a Π_1 wff, so is $(\forall \xi)\varphi$ where ξ is any variable free in φ.

5. Nothing else is a Π_1 wff.

(c) A couple of quick examples, to enable us to introduce another bit of terminology. First, put

---

[6]The general concepts of Σ_1 and Π_1 wffs are indeed standard. However, the official definitions do vary in various ways across different presentations. Don't be fazed by this. Given enough background setting, the various versions are equivalent, and the choice of definition is largely a matter of convenience. Compare Section 12.5(b).

$\psi(x) =_{\text{def}} (\exists v \leq x)(2 \times v = x),$

$\chi(x) =_{\text{def}} \{x \neq 1 \wedge (\forall u \leq x)(\forall v \leq x)(u \times v = x \rightarrow (u = 1 \vee v = 1))\}.$

Then $\psi(x)$ evidently expresses the property of being an *even* number. And $\chi(x)$ expresses the property of being *prime* (where we rely on the trivial fact that a number's factors can be no greater than it). Both these wffs are Δ_0.

The sentence

$\exists x(\psi(x) \wedge \chi(x))$

says that there is an even prime and is Σ_1. And Goldbach's conjecture that every even number greater than two is the sum of two primes can be expressed by the sentence

$\forall x\{(\psi(x) \wedge 4 \leq x) \rightarrow (\exists y \leq x)(\exists z \leq x)(\chi(y) \wedge \chi(z) \wedge y + z = x)\}$

which is Π_1, since what is after the initial quantifier is built out of Δ_0 wffs using bounded quantifiers and connectives and so is still Δ_0.

Because (the formal expression of) Goldbach's conjecture is Π_1, such Π_1 sentences are often said to be of *Goldbach type*. True, Goldbach's conjecture is very simply expressed while other Π_1 sentences can be arbitrarily long and complicated. But the thought is that all Π_1 sentences are like Goldbach's conjecture at least in involving just universal quantification(s) of bounded stuff.

(d)   Note the following simple result, to help fix ideas:

> **Theorem 11.4** *(i) The negation of a Δ_0 wff is Δ_0. (ii) The negation of a Σ_1 wff is equivalent to a Π_1 wff, (iii) the negation of a Π_1 wff is equivalent to a Σ_1 wff.*

Proof   (i) holds by definition.

(ii) Suppose $\neg\varphi$ is a negated Σ_1 wff. Use the equivalence of '$\neg\exists\xi$' with '$\forall\xi\neg$' (and so of '$\neg(\exists\xi \leq \kappa)$' with '$(\forall\xi \leq \kappa)\neg$'), the equivalence of '$\neg(\forall\xi \leq \kappa)$' with '$(\exists\xi \leq \kappa)\neg$', plus de Morgan's laws, to 'drive negations inwards', until the negation signs all attach to the Δ_0 wffs that φ is ultimately built from. You will end up with an equivalent wff ψ built up using universal quantifiers, bounded quantifiers, conjunction and disjunction applied to those now negated Δ_0 wffs. But negated Δ_0 wffs are themselves still Δ_0 by (i). So the equivalent wff ψ is indeed Π_1.[7]

(iii) is argued similarly.   ☒

Our proof here in fact delivers more than advertised. For in the general case, it can be highly non-trivial that a given pair of first-order wffs are logically equivalent to each other. But not here. We haven't just shown that a negated

---

[7]Exercise: make that into a proper proof by course-of-values induction over the degree of complexity of Σ_1 wffs, where complexity is measured by the number of applications of quantifiers and/or connectives to Δ_0 wffs.

Σ_1 wff is equivalent to some Π_1 wff (and likewise a negated wff Π_1 is equivalent to a Σ_1 wff): we've shown that the equivalences are essentially *elementary* ones, revealed by simply shifting negation signs around.

Some useful shorthand: we will call a wff which is equivalent to a Π_1 (Σ_1) wff by elementary logical manipulation Π_1-*equivalent* (Σ_1-*equivalent*).

(e)   Finally, just for information, we note that the Σ_1 and Π_1 wffs sit at the first level up from the Δ_0 wffs in a hierarchy which then continues on ever upwards. Thus at the next level we have the Σ_2 wffs (essentially got from Π_1 wffs by existentially quantifying them) and the Π_2 wffs (got from the Σ_1 wffs by universally quantifying them). Then there are the Σ_3 wffs (got from Π_2 wffs by existentially quantifying them). And so it goes.[8]

11.6   Q is Σ_1-complete

(a)   Suppose T is a theory such that, syntactically, any L_A sentence is also a sentence of T's language. Then we will say

i. T is Σ_1-*sound* iff, for any Σ_1-sentence φ, if $T \vdash \varphi$, then φ is true,

ii. T is Σ_1-*complete* iff, for any Σ_1-sentence φ, if φ is true, then $T \vdash \varphi$,

where 'true' here means *true on the interpretation built into L_A*. We also define Π_1-*sound* and Π_1-*complete* similarly.

Theorem 10.8 shows that Q is not Π_1-complete (why?). By contrast, we do have the following important theorem:

Theorem 11.5 Q *is Σ_1-complete.*

Proof   We use again the same proof idea as for the last two theorems. So define the degree k of a Σ_1 sentence to be the number of applications of connectives and (bounded or unbounded) quantifiers needed to build it up from Δ_0 wffs. We argue by an informal course-of-values induction on k.

For the base case (i), $k = 0$, we are dealing with Δ_0 sentences, and we already know that Q can prove any true Δ_0 sentence, by Theorem 11.3.

For the induction step (ii), we assume that Q can prove any true Σ_1 sentence of degree up to k, and show that it can therefore prove any true Σ_1 sentence χ of degree $k + 1$.

We have three cases to consider:

a. The sentence χ is of the form $(\varphi \wedge \psi)$ or $(\varphi \vee \psi)$ with φ and ψ both Σ_1 sentences. In the first case, given χ is true, φ and ψ must both be true sentences, and being of degree no more that k must by assumption be

---

[8]By the same convention of using the subscript to indicate how many times we are applying alternating kinds of (unbounded) quantifiers to bounded wffs, a Σ_0 wff will be one in which *no* unbounded quantifier gets applied – and therefore the Σ_0 wffs are just the Δ_0 wffs again. In fact both labels are widely current.

provable in Q; and hence their conjunction is provable in Q. In the second case, one of φ and ψ must be true, and being of degree no more than k must be provable in Q; so the disjunction of that true sentence with anything else is provable in Q.

b. The sentence χ is of the form $(\forall \xi \leq \bar{n})\varphi(\xi)$ with φ a Σ_1 wff of degree k. Since χ is true, so are $\varphi(0)$, $\varphi(1)$, ..., $\varphi(\bar{n})$. And hence by assumption Q proves each of them, so by (O4) of Section 11.3, Q also proves χ. The case where χ is of the form $(\exists \xi \leq \bar{n})\varphi(\xi)$ is dealt with similarly.

c. The sentence χ is of the form $\exists \xi \varphi(\xi)$ with φ a Σ_1 wff of degree k. Since χ is true, some number n must satisfy φ, so $\varphi(\bar{n})$ is true. But then Q can prove this by assumption, and hence can prove χ.

Given (i) and (ii), a course-of-values induction gives us the desired result. $\boxtimes$

11.7 Intriguing corollaries

(a) We will appeal to the result that Q is Σ_1-complete in the course of working up to our Gödelian proof of the incompleteness theorem (see the proof of Theorem 17.1). But that's for later. For now, we will derive a couple of immediate corollaries of our last theorem.

Consider Goldbach's conjecture G again. As we remarked back in Section 1.1, no proof of Goldbach's conjecture G is currently known. But note the following:

1. Suppose that G is *true*. Then $\neg G$ will be false and hence *not provable* in Q (given Q has true axioms and only proves truths).

2. Suppose that G is *false*. Then $\neg G$ will be (i) true, and (ii) logically equivalent to a Σ_1 sentence χ (by Theorem 11.4). But then χ, being a true Σ_1 sentence, will be provable in Q by Theorem 11.5. Hence $\neg G$ is *provable* in Q too.

Therefore G is true if and only if $\neg G$ is not provable in Q. That's equivalent to saying G *is true if and only if it is consistent with* Q.

This argument evidently generalizes to any Π_1 arithmetical claim – and there are lots of interesting ones (Fermat's Last Theorem turns out to be another example). So we have, still assuming Q's soundness,

> **Theorem 11.6** *A Π_1 sentence φ is true if and only if φ is consistent with* Q.

Hence, if only we had a method which we could use to decide whether a wff φ was deducible in first order logic from a given assumption (and so, in particular, from the conjunction of Q's axioms), we'd have a method for deciding whether Q is consistent (take φ to be a contradiction), and hence a method for deciding the truth or falsity of any Π_1 statement of arithmetic. Unfortunately, as we'll later show, there can be no such method. Questions of deducibility in first-order logic are not, in general, effectively decidable: that's Theorem 40.3.

(b) Let's say that the theory T *extends* Q if T proves every Q-theorem.[9]

It is immediate that, since Q is Σ_1-complete, so is any theory T which extends Q. It then follows that

Theorem 11.7 *If T extends Q, T is consistent iff it is Π_1-sound.*

Proof First, suppose T proves a *false* Π_1 sentence φ. $\neg\varphi$ will then be equivalent to a *true* Σ_1 sentence. But in that case, since T extends Q and so is Σ_1-complete, T will also prove $\neg\varphi$, making T inconsistent. Contraposing, if T is consistent, it proves no false Π_1 sentence, so by definition is Π_1-sound.

The converse is trivial, since if T is inconsistent, we can derive anything in T, including false Π_1 sentences and so T isn't Π_1-sound. ◻

Though easily proved, this is, in its way, a really rather remarkable observation. It means that we don't have to fully *believe* a theory T – i.e. don't have to accept *all* its theorems are true – in order to use it to establish that some Π_1 arithmetic generalization (such as Goldbach's conjecture or Fermat's Last Theorem) is true. We just have to believe that T is a *consistent* theory which extends Q and show that we can derive the Π_1 sentence in T: since T must be Π_1-sound that guarantees that the sentence is indeed true.

(c) One further observation. Suppose Q_G is the theory you get by adding Goldbach's conjecture as an additional axiom to Q. Then, by Theorem 11.7, Q_G is consistent only if the conjecture is true, since the conjecture is (trivially) a Π_1 theorem of Q_G. But no one really knows whether Goldbach's conjecture *is* true; *so no one knows whether Q_G is consistent.*

Which is an emphatic illustration that even really rather simple theories may not wear their consistency on their face.

11.8   Proving Q is order-adequate

Finally in this chapter, we will (partially) verify Theorem 11.1 since so much depends on it. But you really must not get bogged down in the tiresome details: so, even enthusiasts should feel free to skim or skip this section!

Proof for (O1) For arbitrary a, Q proves $a + 0 = a$, hence $\exists v(v + 0 = a)$, i.e. $0 \leq a$. Generalize to get the desired result. ◻

Proof for (O2) Arguing inside Q, suppose that $a = 0 \lor a = 1 \lor \ldots \lor a = \overline{n}$. We showed in Section 11.1 that if $k \leq m$, then Q proves $\overline{k} \leq \overline{m}$. Which means

---

[9]This is a *syntactic* notion: we mean, more carefully, that for any wff which is a Q theorem, T can prove the same string of symbols. We don't require that T natively interprets the symbols the same way.

that from each disjunct we can derive $a \leq \bar{n}$. Hence, arguing by cases, $a \leq \bar{n}$. So, discharging the supposition, Q proves $(a = 0 \vee a = 1 \vee \ldots \vee a = \bar{n}) \rightarrow a \leq \bar{n}$. The desired result is immediate since a was arbitrary. ⊠

Proof for (O3)   We use induction. For the base case (i), i.e. to show that the target wff is provable for $n = 0$, we need a proof of $\forall x (x \leq 0 \rightarrow x = 0)$. It is enough to suppose, inside a Q proof, that $a \leq 0$, for arbitrary a, and deduce $a = 0$. So suppose $a \leq 0$, i.e. $\exists v (v + a = 0)$. Then for some b, $b + a = 0$. Now Axiom 3 is equivalent to $\forall x (x = 0 \vee \exists y (x = Sy))$, so it tells us that either (1) $a = 0$, or (2) $a = Sa'$ for some a'. But (2) implies $b + Sa' = 0$, so by Axiom 5 $S(b + a') = 0$, contradicting Axiom 1. That rules out case (2). Therefore $a = 0$. Which establishes (i).

   (ii) We assume that Q proves $\forall x (x \leq \bar{n} \rightarrow \{x = 0 \vee x = 1 \vee \ldots \vee x = \bar{n}\})$. We want to show that Q proves $\forall x (x \leq \overline{n+1} \rightarrow \{x = 0 \vee x = 1 \vee \ldots \vee x = \overline{n+1}\})$. It is enough to suppose, inside a Q proof, that $a \leq \overline{n+1}$, for arbitrary a, and then deduce $a = 0 \vee a = 1 \vee \ldots \vee a = \overline{n+1}$.

   Our supposition, unpacked, is $\exists v (v + a = \overline{n+1})$. And by Axiom 3, either (1) $a = 0$ or (2) $a = Sa'$, for some a'.

   Exploring case (2), we then have $\exists v (v + Sa' = S\bar{n})$: hence, by Axiom 5 and Axiom 2, we can derive $\exists v (v + a' = \bar{n})$; i.e. $a' \leq \bar{n}$. So, using our assumption that the result holds for n, $a' = 0 \vee a' = 1 \vee \ldots \vee a' = \bar{n}$. Since $a = Sa'$, that implies $a = 1 \vee a = 2 \vee \ldots \vee a = \overline{n+1}$.

   Hence, since either (1) or (2) holds, $a = 0 \vee a = 1 \vee a = 2 \vee \ldots \vee a = \overline{n+1}$. Discharging the original assumption establishes the decided induction step (ii).

   Hence, from (i) and (ii) we get the desired conclusion (O3). ⊠

Proofs for (O4) to (O7)   Exercises!

Proof for (O8)   We proceed by induction again.

   (i) We saw in proving (O1) that for any a, $0 \leq a$. A fortiori, $0 \leq a \vee a \leq 0$. Generalizing gives us the desired result for the base case $n = 0$.

   (ii) We'll suppose that the result holds for $n = k$, and show that it holds for $n = k + 1$. Hence, for arbitrary a,

i.	$a \leq \bar{k} \vee \bar{k} \leq a$	By our supposition
ii.	$a \leq \bar{k} \rightarrow a \leq \overline{k+1}$	By (O6)
iii.	$\bar{k} \leq a \rightarrow (\bar{k} = a \vee \overline{k+1} \leq a)$	By (O7).

And since Q captures *less-than-or-equal-to*, we know it proves $\bar{k} \leq \overline{k+1}$, hence

iv.	$a = \bar{k} \rightarrow a \leq \overline{k+1}$	
v.	$a \leq \overline{k+1} \vee \overline{k+1} \leq a$	From (i) to (iv).

Since a is arbitrary, generalizing gives us what we needed to show to establish the induction step. ⊠

Proof for (O9)   Another exercise!

12 IΔ_0, an arithmetic with induction

When it comes to proving universal generalizations about numbers, Robinson Arithmetic Q is very weak. As we saw, it can't even prove $\forall x(0 + x = x)$. So what do we need to add to get a more competent formal theory? An induction principle, of course.

This chapter and the next considers how to implement induction inside formalized theories framed in L_A, the language of basic arithmetic. We start by discussing IΔ_0, which adds a minimal amount of induction to Q.

12.1 The formal Induction Schema

(a) Suppose, then, that we want to formulate an induction principle inside some theory T which is built in L_A. Since L_A lacks second-order quantifiers, we don't have the option of going for a single second-order induction axiom, even if we wanted to. But in Section 9.4 we saw that we have two other initial options, which are normally equivalent to each other. We can add a rule of inference: or we can get the same effect by giving an axiom schema, and saying every instance of it is an axiom. For convenience, we'll take the second line.

So, as a first shot, let's try saying that any instance of the formal

Induction Schema $\{\varphi(0) \wedge \forall x(\varphi(x) \to \varphi(Sx))\} \to \forall x\varphi(x)$

is to count as a T-axiom, where $\varphi(x)$ stands in for some suitable open wff of T's language with just 'x' free – and $\varphi(0)$ and $\varphi(Sx)$ are, of course, the results of systematically substituting '0' and 'Sx' for 'x'. We'll think about what might count as a 'suitable' wff in just a moment: but basically we want expressions $\varphi(x)$ which express what theory T regards as genuine properties, i.e. properties which really do fall under the intuitive induction principle.

(b) First, though, we need to recall that we can use inductive arguments to prove general results about *relations* as well as about monadic properties: see Section 9.3. For the reasons noted there, we need to be able to formally mirror in T those informal arguments where induction is applied to a predicate involving parameters. Exactly how we do this will depend on the details of T's proof system. But suppose we are using a Hilbert-style system (which we said was going to be our official choice, see Section 10.4). Then we will need to appeal to an instance of the Induction Schema with $\varphi(x)$ replaced by the likes of $R(v, x)$, with 'v' behaving as a free variable, so we get conditionals such as

(R) $\{R(v, 0) \wedge \forall x(R(v, x) \to R(v, Sx))\} \to \forall x R(v, x)$.

Because of the free variable, however, (R) is not a closed sentence with a fixed truth-value on interpretation. So treating this wff with its dangling variable as an axiom would clash with the desideratum that an arithmetic theory's axioms should be true-on-the-natural-arithmetic-interpretation.

An obvious technical dodge enables us to get around this wrinkle. So let us indeed allow any suitable open wff φ to be substituted into the Induction Schema, even if this leaves variables dangling free. But we will say – as a second shot – that it is *the universal closure* of any such instance of the Induction Schema which counts as an axiom of T (where the universal closure of a wff is the result of prefixing it with enough universal quantifiers to bind its free variables). So now (R) itself won't count as an axiom, but its universal closure does:

$$\text{(R')} \quad \forall v(\{R(v,0) \land \forall x(R(v,x) \to R(v,Sx))\} \to \forall x R(v,x)).$$

And then in one step we can recover (R) for use in a formalized version of our informal induction argument involving a relation.

This, then, will be our standard way of giving a theory some induction:

> Take any suitable $\varphi(x)$ which has 'x' free and perhaps other variables free too: then the universal closure of the corresponding instance of the Induction Schema is an axiom.

12.2 Introducing IΔ_0

But what makes a wff $\varphi(x)$ 'suitable' for appearing in an instance of the induction schema? We said: we want wffs which express genuine numerical properties (and relations), for these fall under the intuitive induction principle. So the question becomes: which wffs express genuine numerical properties and relations?

Let's start *very* cautiously. Suppose $\varphi(x)$ is a wff of L_A which has just 'x' free and which has at most bounded quantifiers; in other words, suppose $\varphi(x)$ is Δ_0. Then such a wff surely does express an entirely determinate monadic property. Indeed, by Theorem 11.2, for any n, we can effectively determine whether n has that property, i.e. we can simply calculate whether $\varphi(\bar{n})$ is true or not.

Likewise, for analogous reasons, a Δ_0 wff with two or more free variables surely expresses an entirely determinate relation.

So in this chapter, we'll consider what should be the quite uncontroversial use of induction for Δ_0 wffs. Here then is a standard definition:

> IΔ_0 is the first-order theory whose language is L_A, whose deductive logic is classical, and whose axioms are those of Q plus the following extra axioms: the (universal closures of) all instances of the Induction Schema where φ is Δ_0.[1]

---

[1] Predictably, given that Δ_0 and Σ_0 are alternative notations (as noted in Section 11.5, fn. 8), this theory is also commonly known as IΣ_0. Detailed investigations of its scope and limits are surprisingly late: one of the first is by Parikh (1971).

12.3 What IΔ_0 can prove

(a)   We begin with a portmanteau theorem.

Theorem 12.1 *The following general claims about the successor function, addition and ordering are all provable in* IΔ_0:

    1. $\forall x(x \neq Sx)$

    2. $\forall x(0 + x = x)$

    3. $\forall x \forall y(Sx + y = S(x + y))$

    4. $\forall x \forall y(x + y = y + x)$

    5. $\forall x \forall y \forall z(x + (y + z) = (x + y) + z)$

    6. $\forall x \forall y \forall z(x + y = x + z \rightarrow y = z)$

    7. $\forall x \forall y(x \leq y \lor y \leq x)$

    8. $\forall x \forall y((x \leq y \land y \leq x) \rightarrow x = y)$

    9. $\forall x \forall y \forall z((x \leq y \land y \leq z) \rightarrow x \leq z)$.

There are similar elementary results involving multiplication too, but that's enough to be going on with.

We'll leave giving a full derivation of these results as a rather unexciting exercise for those who have a taste for such things. Here are just a few hints and outlines, and even these can be skipped: don't get bogged down in detail. (Remember, although our official internal logic for first-order arithmetics will be a Hilbert-style system, for ease of understanding we will allow ourselves to give proof sketches in an equivalent natural deduction system.)

Proof sketch for (1) Take $\varphi(x)$ to be $x \neq Sx$. Then Q proves $\varphi(0)$ because that's Axiom 1, and proves $\forall x(\varphi(x) \rightarrow \varphi(Sx))$ by contraposing Axiom 2. We then can invoke the corresponding induction axiom $\{\varphi(0) \land \forall x(\varphi(x) \rightarrow \varphi(Sx))\} \rightarrow \forall x \varphi(x)$, and deduce $\forall x \varphi(x)$, i.e. no number is a self-successor. ☒

Recall that our deviant interpretation which makes the axioms of Q true while making $\forall x(0 + x = x)$ false had Kurt Gödel himself as a self-successor (see Section 10.5). A smidgen of induction, we have now seen, rules out self-successors.[2]

Proof sketch for (2) We showed in Section 10.5 that the wff $\forall x(0 + x = x)$ is *not* provable in Q. But we can derive it once we have induction on the scene.

For take $\varphi(x)$ to be $(0 + x = x)$, which is of course Δ_0. Q implies $\varphi(0)$, since that's just an instance of Axiom 4.

Next we show Q entails $\forall x(\varphi(x) \rightarrow \varphi(Sx))$. Arguing inside Q, it is enough to suppose $\varphi(a)$ and derive $\varphi(Sa)$. So suppose $\varphi(a)$, i.e. $0 + a = a$. But Axiom 5 entails $0 + Sa = S(0 + a)$. So by Leibniz's Law, $0 + Sa = Sa$, i.e. $\varphi(Sa)$.

---

[2]Don't be lulled into a false sense of security, though! While induction axioms may rule out deviant interpretations based on self-successors, they don't rule out some other deviant interpretations. See Kaye (1991) for a very rich exploration of 'non-standard models'.

And now an instance of the Induction Schema tells us that if we have both $\varphi(0)$ and $\forall x(\varphi(x) \to \varphi(Sx))$ we can deduce $\forall x \varphi(x)$, i.e. $\forall x(0 + x = x)$.   ⊠

Proof sketch for (4)   Take the universal closure of the instance of the Induction Schema for $\varphi(x) = x + y = y + x$. Instantiate with an arbitrary parameter to get:

$$\{0 + a = a + 0 \land \forall x(x + a = a + x \to Sx + a = a + Sx)\}$$
$$\to \; \forall x(x + a = a + x)$$

To proceed, we can obviously derive $0 + a = a + 0$, the first conjunct in the curly brackets, by using (2) plus Q's Axiom 4.

Now suppose $b + a = a + b$. Then $Sb + a = S(b + a) = S(a + b) = a + Sb$, by appeal to the unproved Result (3), our supposition, and Axiom 5 in turn. So by Conditional Proof $b + a = a + b \to Sb + a = a + Sb$. Generalizing gets us the second conjunct in curly brackets.

Detach the consequent and then generalize again to get what we want.   ⊠

Proof sketch for (7)   Swapping variables, let's show $\forall y \forall x(y \leq x \lor x \leq y)$. Take the universal closure of the instance of induction for $\varphi(x) = y \leq x \lor x \leq y$, and instantiate to get:

$$\{(a \leq 0 \lor 0 \leq a) \land \forall x((a \leq x \lor x \leq a) \to (a \leq Sx \lor Sx \leq a))\}$$
$$\to \; \forall x(a \leq x \lor x \leq a)$$

So again we need to prove each conjunct in the curly brackets; then we can detach the consequent of the conditional, and generalize to get the desired result.

It's trivial to prove the first conjunct, $(a \leq 0 \lor 0 \leq a)$, since its second disjunct always obtains, by (O1) of Section 11.8.

Next we show that if we suppose $a \leq b \lor b \leq a$, we can derive $a \leq Sb \lor Sb \leq a$ for arbitrary b, which is enough to prove the second conjunct. Argue by cases.

Suppose first $a \leq b$, i.e. $\exists v(v + a = b)$. But if for some c, $c + a = b$, then $S(c + a) = Sb$, so by Result (3), $Sc + a = Sb$, whence $\exists v(v + a = Sb)$, i.e. $a \leq Sb$, so $(a \leq Sb \lor Sb \leq a)$.

Suppose secondly $b \leq a$, i.e. $\exists v(v + b = a)$. Then for some c, $c + b = a$. By Q's Axiom 3, either $c = 0$, or $c = Sc'$ for some c'. In the first case $0 + b = a$, so by Result (2) above $a = b$, from which it is trivial that $a \leq Sb$. In the second case, $Sc' + b = a$, and using Result (3) and Axiom 5 we get $c' + Sb = a$, and hence $Sb \leq a$. So $(a \leq Sb \lor Sb \leq a)$ again.

So we can infer $\forall x(a \leq x \lor x \leq a)$. Generalizing gives us the desired result. ⊠

Enough already! This all goes to show that fully-written-out formal proofs by induction will soon get *very* tiresome (even in a natural deduction logic). Carrying on in the same way, however, we can in fact prove many of the most obvious general properties of addition and of multiplication too, using only (the universal closures of) instances of the Induction Schema where φ is a Δ_0 wff.

(b)   Here, though, is one implication of Result (7) which is worth pointing up for later use:

Theorem 12.2 *In any theory which is at least as strong as* IΔ_0, *a wff starting with $n > 1$ unbounded existential quantifiers is provably equivalent to a wff starting with just a* single *unbounded quantifier.*

Proof   First we show that the sentences

   i. $\exists x \exists y \varphi(x, y)$

   ii. $\exists w (\exists x \leq w)(\exists y \leq w) \varphi(x, y)$

(which intuitively are true in the same cases) are provably equivalent in IΔ_0.

Assume (i) and reason inside IΔ_0. So we suppose that for some a, b, $\varphi(a, b)$. But we know from Result (7) above that $a \leq b \vee b \leq a$. Assume $a \leq b$: then we can derive $a \leq b \wedge b \leq b \wedge \varphi(a, b)$, which in turn implies $(\exists x \leq b)(\exists y \leq b)\varphi(x, y)$, and hence (ii). Assume $b \leq a$: we can similarly derive (ii). So, arguing by cases and then $\exists$-elimination, (i) implies (ii).

For the converse, unpacking the abbreviations involved in the bounded quantifiers is enough to show that (ii) logically entails (i). So (i) and (ii) are indeed provably equivalent in IΔ_0.

Using that equivalence,

   iii. $\exists x \exists y \exists z \psi(x, y, z)$

is provably equivalent in IΔ_0 to $\exists w (\exists x \leq w)(\exists y \leq w) \exists z \psi(x, y, z)$ which is logically equivalent to $\exists w \exists z (\exists x \leq w)(\exists y \leq w) \psi(x, y, z)$; and now we do the same trick as before to end up with

   iv. $\exists u (\exists w \leq u)(\exists z \leq u)(\exists x \leq w)(\exists y \leq w) \psi(x, y, z)$

which is the provably equivalent to (iii).

Repeated use of the same trick enables us to reduce any number of initial unbounded existential quantifiers to one. And the argument still applies even if the core wff has additional variables left unbound.   ☒

12.4   IΔ_0 is not complete

In Section 10.5, we showed that Q is (very unsurprisingly) a negation-incomplete theory, and we did not need to appeal to Gödelian considerations to prove this. We can use another non-Gödelian argument to show that IΔ_0 is also (again pretty unsurprisingly) incomplete, assuming it is sound.

Enthusiasts might be interested in an outline proof: non-mathematicians or impatient readers willing to take the result on trust can very cheerfully skip again. The headline news is as follows:

1. Although exponentiation isn't built into the language L_A, we can – with some rather unobvious trickery – find a Δ_0 wff $\varepsilon(x, y)$ which expresses the exponential function 2^m, so $\varepsilon(\overline{m}, \overline{n})$ is true iff $2^m = n$.

2. The sentence $\forall x \exists y \varepsilon(x, y)$ therefore expresses the arithmetical truth that exponentiation is a total function, i.e. for every m, 2^m has a value. This is a Π_2 sentence (what does that mean? – see Section 11.5(d)).

3. Using some (relatively) straightforward model theory, Rohit Parikh (1971) shows that, for any Δ_0 wff $\varphi(x, y)$, if IΔ_0 proves $\forall x \exists y \varphi(x, y)$, then there is an L_A-term $\tau(x)$ such that IΔ_0 also proves $\forall x (\exists y \leq \tau(x)) \varphi(x, y)$.

4. Reflect that an L_A-term $\tau(x)$ has to be built up using only successor, addition and multiplication, and therefore expresses a polynomial function (why?). So Parikh's theorem tells us that if IΔ_0 proves $\forall x \exists y \varphi(x, y)$ (where φ is Δ_0, and expresses the function f), then the rate of growth of $f(m)$ is bounded by a polynomial.

5. But the exponential 2^m grows faster than any polynomial. So it follows that IΔ_0 cannot prove $\forall x \exists y \varepsilon(x, y)$.

6. But since $\forall x \exists y \varepsilon(x, y)$ is indeed true on the standard interpretation, IΔ_0 can't prove its negation either (assuming the theory is sound).[3]

12.5   On to IΣ_1

(a)   The most natural way of plugging the identified gap in IΔ_0 is, of course, to allow *more* induction. For note that the unprovable-in-IΔ_0 truth that we have found is the universal quantification of the Σ_1 expression $\exists y \varepsilon(x, y)$: in order to prove it, it is enough to now allow Σ_1 predicates to feature in instances of the Induction Schema. If we *only* stretch induction as far as allowing such predicates, this gives us the arithmetic known as IΣ_1.

(b)   We won't pause to show that IΣ_1 can indeed prove $\exists y \varepsilon(x, y)$. Instead we note something simpler:

Theorem 12.3 *If φ is Δ_0, then IΣ_1 proves that (α) $(\forall x \leq u) \exists y \varphi(x, y)$ is equivalent to (β) $\exists v (\forall x \leq u)(\exists y \leq v) \varphi(x, y)$.*

We of course can't swap the order of *unbounded* quantifiers of different flavours; but we can drag an unbounded existential forward past a *bounded* universal (leaving behind a bounded existential). The argument is elementary but messy – so again feel free to skim past it to get to the explanation of why it matters.

Proof sketch   To show β logically implies α is easy. For the difficult direction, the plan is to assume α and use induction on the Σ_1 wff

$$\psi(w) =_{\text{def}} (u \leq w \wedge u \neq w) \vee \exists v (\forall x \leq w)(\exists y \leq v) \varphi(x, y),$$

---

[3]Neat! Filling in the details, however, requires ideas which go far beyond those needed for this book. For the record, the materials can be found in Buss (1998) and the encyclopaedic Hájek and Pudlák (1993), ch. 5.

which is equivalent to $w \leq u \to \exists v (\forall x \leq w)(\exists y \leq v)\varphi(x, y)$, to prove $\forall w \varphi(w)$, which then entails β.

Assume $(\forall x \leq u)\exists y \varphi(x, y)$. Given α, the base case (i) $\psi(0)$ quickly follows. So now we just need to establish the induction step (ii) $\forall w(\psi(w) \to \psi(Sw))$. So suppose $\psi(w)$ and $Sw \leq u$ and aim to derive $\exists v (\forall x \leq Sw)(\exists y \leq v)\varphi(x, y)$.

Since $Sw \leq u$, we a fortiori have $w \leq u$, therefore given $\psi(w)$ it follows that (1) $\exists v (\forall x \leq w)(\exists y \leq v)\varphi(x, y)$.

And from α and $Sw \leq u$ we get (2) $\exists y \varphi(Sx, y)$.

We then proceed by assuming $(\forall x \leq w)(\exists y \leq a)\varphi(x, y)$ and $\varphi(Sx, b)$. We know that we have $a \leq b \lor b \leq a$ and it is easy to then get $\exists v (\forall x \leq Sw)(\exists y \leq v)\varphi(x, y)$ either way.

Finally, $\exists$-elimination using (1) and (2) to discharge our last two assumptions delivers the desired conclusion, and our proof is done!　　　　$\boxtimes$

But why on earth go through the palaver of that messy proof? Because the resulting theorem now easily delivers the following useful result:

> **Theorem 12.4** *In any theory at least as strong as* IΣ_1, *any wff* φ *which is* Σ_1 *by our definition is provably equivalent to a wff of the form* $\exists \xi \psi(\xi)$ *with the same free variables as* φ *and where* ψ *is* Δ_0.

Proof sketch Take a Σ_1 wff. Start pulling the *unbounded* existential quantifiers out towards the front of the wff. Always work on the currently innermost one.

When we pull it outside a conjunction or disjunction or a bounded existential, this unproblematically takes us to a logically equivalent wff (after relabelling bound variables if necessary to avoid clashes).

When it meets another unbounded existential, collapse the two into one using Theorem 12.2, and we are again left with an equivalent wff.

These moves will always leave our innermost existential quantifier governing a Δ_0 wff (think about it!). So if we now need to pull it past a bounded universal quantifier, we can invoke the last theorem to get an equivalent wff with an existential quantifier still applying to a Δ_0 wff.

Keep on going, until we are left with a wff with just a single initial existential quantifier applied to a Δ_0 wff.[4]　　　　$\boxtimes$

This theorem explains, then, a common simpler definition of Σ_1 wffs as those wffs having the form $\exists \xi \psi(\xi)$ where ψ is Δ_0. In the presence of a bit of induction, the definitions characterize equivalent wffs.

(c)　　There's more to be said, for IΣ_1 indeed has notable technical interest (we will encounter it again when we come to discuss Gödel's Second Incompleteness Theorem). But – to press the question that you have no doubt been impatient to ask – why stop with induction for Σ_1 wffs? Why hobble ourselves like this? Why not be much more generous with induction?

Well, that's our topic in the next chapter.

---

[4]Exercise: make this into a nice proof by induction.

13 First-order Peano Arithmetic

In the last chapter, we considered the theory $I\Delta_0$ built in the language L_A, whose axioms are those of Q, plus (the universal closures of) all instances of the Induction Schema for Δ_0 predicates. Now we lift that restriction on induction, and allow *any* L_A predicate to appear in instances of the Schema. The result is (first-order) Peano Arithmetic.

13.1 Being generous with induction

(a)   Given what we said in Section 9.1(a) about the motivation for the induction principle, any instance of the Induction Schema will be intuitively acceptable as an axiom, so long as we replace φ in the Schema by a suitable open wff which expresses a genuine property/relation.

We argued at the beginning of the last chapter that Δ_0 wffs are eminently suitable, and we considered the theory you get by adding to Q the instances of the Induction Schema involving such wffs. But why should we be so very restrictive?

Take any open wff φ of L_A at all. This will be built from no more than the constant term '0', the familiar successor, addition and multiplication functions, plus identity and other logical apparatus. Therefore – you might very well suppose – it ought also to express a perfectly determinate arithmetical property or relation. So why not be generous and allow *any* open L_A wff to be substituted for φ in the Induction Schema? The result of adding to Q (the universal closures of) every instance of the Schema is PA – *First-order Peano Arithmetic*.[1]

(b)   Now, there is considerable *technical* interest in seeing just how many basic truths of arithmetic can in fact be proved by adding some limited amount of induction to Q (e.g. induction for Δ_0 or Σ_1 wffs), and also in seeing what *can't* be proved in this or that restricted theory. But could there be any *philosophically principled* reason for stopping at some point between the theory $I\Delta_0$ with minimal induction and the inductively generous PA, and for saying that some intermediate theory with a restricted induction principle is all that we should really accept?

If our explanatory remarks about induction were right, then there is no mileage in the idea some complex wffs $\varphi(\mathsf{x})$ of L_A might express genuine numerical

---

[1] The name is conventional. Giuseppe Peano did publish a list of axioms for arithmetic in Peano (1889). But they weren't first-order, only explicitly governed the successor relation, and – as he acknowledged – had already been stated by Richard Dedekind (1888).

properties yet somehow induction doesn't apply to them. Hence, if we are to philosophically motivate a restriction on induction, it must be grounded in the thought that some predicates that we can syntactically construct (and which we thought our classical semantics $\mathcal{I}_A$ gave sense to) in fact don't express kosher properties at all. But why should we suppose *that*?

One route to such a thought would be to endorse a very radical 'constructivism' that says that a predicate only expresses a genuine property if we can effectively decide whether it applies to a given number or not. Now, in general, if $\varphi(x)$ embeds unrestricted quantifiers, we can't mechanically decide whether $\varphi(\bar{n})$ holds or not: so this radical constructivism would have to say that such wffs typically don't express properties. We shouldn't therefore use these predicates-that-express-no-property in a contentful arithmetic, and hence we shouldn't deploy induction axioms involving such predicates. This would leave us with a very weak arithmetic which doesn't even have full Σ_1 induction. A close cousin of this line of thought has been famously defended by Edward Nelson (1986), though few have been persuaded.

We can't digress, however, to discuss versions of constructivism, intriguing though they are: that would take a book in itself. Here, we can only note that few philosophers and even fewer working mathematicians endorse such a highly revisionary position. The standard view is that L_A-predicates indeed make good sense and express determinate properties (even if, in the general case, we can't always decide whether a given number has the property or not). But then generosity with induction seems entirely compelling.

13.2   Summary overview of PA

The inductively generous theory PA is very rich and powerful. Given its very natural motivation, it is the benchmark axiomatized first-order theory built in L_A. Just for neatness, then, let's bring together all the elements of its specification in one place.

But first, a quick observation. PA allows, in particular, induction for the Σ_1 formula

$$\varphi(x) =_{\text{def}} (x = 0 \lor \exists y(x = Sy)).$$

Now note that the corresponding $\varphi(0)$ is a trivial logical theorem. $\forall x \varphi(Sx)$ is an equally trivial theorem, and that boringly entails $\forall x(\varphi(x) \to \varphi(Sx))$. So we can use an instance of the Induction Schema inside PA to derive $\forall x \varphi(x)$. However, that's equivalent to Axiom 3 of Q.[2] So our initial presentation of PA – as consisting in Q plus all the instances of the Induction Schema – involves a little redundancy.

---

[2] As we saw in Section 11.8, Axiom 3 enables us to prove some important general claims in Q, despite the absence of the full range of induction axioms. It, so to speak, functions as a very restricted surrogate for induction in certain proofs.

Bearing that in mind, here's our overview of PA.

1. PA's *language* is L_A, a first-order language whose non-logical vocabulary comprises just the constant '0', the one-place function symbol 'S', and the two-place function symbols '+', '×', and whose intended interpretation is the obvious one.

2. PA's official deductive *proof system* is a Hilbert-style axiomatic version of classical first-order logic with identity (the differences between various presentations of first-order logic of course don't make any difference to what sentences can be proved in PA, and our official choice is just for later metalogical convenience).

3. Its non-logical *axioms* – eliminating the redundancy from our original specification – are the following sentences:

Axiom 1 $\forall x(0 \neq Sx)$

Axiom 2 $\forall x \forall y(Sx = Sy \to x = y)$

Axiom 3 $\forall x(x + 0 = x)$

Axiom 4 $\forall x \forall y(x + Sy = S(x + y))$

Axiom 5 $\forall x(x \times 0 = 0)$

Axiom 6 $\forall x \forall y(x \times Sy = (x \times y) + x)$

plus every sentence that is the universal closure of an instance of the following

Induction Schema $\{\varphi(0) \land \forall x(\varphi(x) \to \varphi(Sx))\} \to \forall x \varphi(x)$

where $\varphi(x)$ is an open L_A wff that has 'x', and perhaps other variables, free.

Plainly, it is still decidable whether any given wff has the right shape to be one of the induction axioms, so PA is indeed an effectively formalized theory.

13.3   Hoping for completeness

Our examples in the last chapter of proofs by induction in $I\Delta_0$ are, trivially, also examples of proofs in PA, and there is little to be learnt from giving more particular examples of PA-proofs. So we turn to generalities.

(a)   In Section 10.5, we used an elementary model-theoretic argument to show that Q cannot prove the wff $\forall x(0 + x = x)$ and lots of familiar Π_1 truths like it. But we can repair *that* shortcoming in Q by allowing induction over Δ_0 wffs and adopting the theory $I\Delta_0$. In Section 12.4 we outlined a fancier model-theoretic argument which shows that $I\Delta_0$ is not able to prove a certain Π_2 truth of the form $\forall x \exists y \varepsilon(x, y)$. But *this* shortcoming can be repaired by going on to allow induction over Σ_1 wffs. The pattern repeats: we repair more and more gaps in weaker arithmetics by allowing induction using more and more complex formulae. And if we had never heard of Gödel's Theorems we might reasonably have hoped that, when we go to the limit as in PA and allow instances of the Induction Schema for *any* wff $\varphi(x)$, however complex, then we close all the gaps and get a negation-complete theory that proves any L_A truth.

(b)   Here's another observation that might reasonably encourage the hope that PA is complete.

Suppose we define the language L_P to be L_A without the multiplication sign. Take P to be the theory couched in the language L_P, whose axioms are Q's now familiar axioms for successor and addition, plus the universal closures of all instances of the Induction Schema that can be formed in L_P. In short, P is PA minus multiplication. Now,

> **Theorem 13.1** P *is a negation-complete theory of successor and addition.*

We are not going to prove this result here. The argument uses a standard model-theoretic method called 'elimination of quantifiers' which isn't hard and was known pre-Gödel, but it would just take too long to explain. Note, though, that the availability of a complete formalized theory for successor and addition for integers positive and negative was proved by this method as early as 1929 by Mojżesz Presburger.[3]

So the situation is as follows. (i) There is a complete formalized theory BA whose theorems are exactly the quantifier-free truths expressible using successor, addition and multiplication (and the connectives). (ii) There is another complete formalized theory (equivalent to PA minus multiplication) whose theorems are exactly the first-order truths expressible using just successor and addition. Against this background, the Gödelian result that adding multiplication in order to get full PA gives us a theory which is incomplete and incompletable (if consistent) comes as a rather nasty surprise. It certainly wasn't obviously predictable that multiplication would make all the difference. Yet it does.[4]

---

[3]Enthusiasts will find an accessible outline proof of our theorem in Fisher (1982, ch. 7), which can usefully be read in conjunction with Boolos et al. (2002, ch. 24).

For Presburger's own completeness proof – produced when a young graduate student – see his (1930). A few years later, Hilbert and Bernays (1934) showed that his methods could be applied to the natural number theory P, and it is the latter which is these days typically referred to as 'Presburger Arithmetic'.

[4]Note, it isn't that multiplication is in itself somehow intrinsically intractable. In 1929 (the proof was published in his 1930), Thoralf Skolem showed that there is a complete theory for the truths expressible in a suitable first-order language with multiplication but lacking addition (or the successor function). Why then does putting multiplication together with addition and successor produce incompleteness? The answer will emerge over the coming chapters.

(c)   Now, PA certainly isn't 'straightforwardly' incomplete in the way that Q and IΔ_0 are: there is no obvious way of coming up with an L_A-sentence which can be shown to be undecidable in PA by a reasonably elementary model-theoretic argument. But PA extends Q, and Q is sufficiently strong (as we claimed in Section 10.6, but can't yet prove), so PA is sufficiently strong too. Therefore Theorem 7.2 applies: hence – assuming it is consistent – PA must indeed *be* incomplete. Starting in the next chapter we will begin to put together the ingredients for a version of the original Gödelian argument for this conclusion (an argument that doesn't require the idea of 'sufficient strength').

Now, if an effectively axiomatized theory extends PA it will still be sufficiently strong, so if it remains consistent it will be again incomplete by Theorem 7.2. Given that our main topic is incompleteness, there is therefore a sense in which we don't really need to give extensive coverage of effectively axiomatized theories stronger than PA; what goes for PA as far as the incompleteness phenomenon is concerned must go for them too.

However, we *will* pause later to say just a little about some of those stronger theories. After all, we saw in Section 9.4(b) that there is an attractive way of framing the intuitive principle of arithmetical induction which makes it a *second-order* claim, quantifying over all numerical properties. For the moment, we have been keeping everything within a first-order framework and so we introduced the Induction Schema to handle induction. But perhaps we shouldn't hobble ourselves like this. Why not instead add to Q a formal *second-order* Induction Axiom, so we get some version of second-order arithmetic (closer indeed to Peano's original intentions)?

That is a good question. But it would be distracting to discuss it now: we will leave it for Chapter 29. For the moment, first-order PA can serve as our canonical example of a very powerful arithmetic which isn't obviously incomplete but which Gödelian arguments show is in fact both incomplete and *incompletable*, assuming consistency. And since no effectively axiomatized consistent theory which contains PA can be complete, a (properly axiomatized, consistent) second-order extension of it will be incomplete too.

13.4   Is PA consistent?

(a)   PA is incomplete, *if* it is consistent. But *is* it? Let's pause to briefly consider the issue – *not* because there's really a sporting chance that PA might actually be in trouble, but because it gives us an opportunity to mention again the consistency theme which we touched on briefly in Section 1.6 and which will occupy us later, from Chapter 31 on.

In fact, our motivation for endorsing PA grounds a very natural informal argument that the theory *is* consistent (on a par with our informal argument that Q is sound and so consistent at the beginning of Section 10.5).

In plodding detail:

Take the given interpretation $\mathcal{I}_A$ which we built into PA's language $L_A = \langle \mathcal{L}_A, \mathcal{I}_A \rangle$: on this interpretation, '0' has the value zero; 'S' represents the successor function, etc. Then, on $\mathcal{I}_A$, (1) the first two axioms of PA are core truths about the operation that takes one number to its successor. And the next four axioms are equally fundamental truths about addition and multiplication. (2) The informal induction principle is warranted by our understanding of the structure of the natural number sequence (and in particular, by the lack of 'stray' numbers outside the sequence of successors of zero). And that informal induction principle warrants, in particular, the instances of PA's Induction Schema; so they too will also be true on $\mathcal{I}_A$. But (3) the classical first-order deductive logic of PA is truth-preserving so – given that the axioms are true and PA's logical apparatus is in good order – *all its theorems are true on* $\mathcal{I}_A$. Hence (4), there cannot be pairs of PA-theorems of the form φ and $\neg\varphi$ (for these of course couldn't both be true together on $\mathcal{I}_A$). Therefore PA is consistent.

This intuitive argument may look compelling. However, it appeals to our supposed grasp of the structure of the natural numbers and also to the idea that all open wffs of L_A express genuine numerical properties and relations (which we can therefore apply induction to). And the latter idea could perhaps be challenged (see Section 13.1(b)).[5]

(b)   But even if we don't positively endorse some radically revisionary challenge, a hyper-cautious philosopher might still remark that an argument for a theory's consistency which appeals to our alleged grasp of some supposedly intuitive truths (however appealing) *can* lead us badly astray. And she might refer to one of the most famous episodes in the history of logic, which concerns the fate of the German logician Gottlob Frege's *Grundgesetze der Arithmetik*.[6]

Frege aimed to construct a formal system in which first arithmetic and then the theory of the real numbers can be rigorously developed by deducing them from logic-plus-definitions (cf. Section 1.1). He has a wide conception of what counts as logic, which embraces axioms for what is in effect a theory of classes, so that the number sequence can be identified as a certain sequence of classes, and then rational and real numbers can be defined via appropriate classes of these classes.[7] Frege takes as his fifth Basic Law the assumption, in effect, that

---

[5]True, we could – if we liked – rigorize our intuitive reasoning inside a rich enough framework such as a set theory that formalizes the interpretation $\mathcal{I}_A$. But would that actually be any more compelling? After all, are our intuitions about the basic laws governing a richer universe of sets *more* secure than those about the basic laws governing the much sparser universe of natural numbers?

[6]The first volume of *Grundgesetze* was published in 1893, the second in 1903. For a partial translation, as *The Basic Laws of Arithmetic*, see Frege (1964).

[7]When talking about the views of Frege and Russell, it seems more appropriate to use Russell's favoured term 'class' rather than 'set', if only because the latter has become so very

for every well-constructed open wff $\varphi(\mathsf{x})$ of his language, there is a class (possibly empty) of exactly those things that satisfy this wff. And what could be more plausible? If we can coherently express some condition, then we should surely be able to talk about the (possibly empty) collection of just those things that satisfy that condition. Can't we just 'see' that that's true?

But, famously, the assumption is disastrous (at least when combined with the assumption that classes are themselves things that can belong to classes). As Bertrand Russell pointed out in a letter which Frege received as the second volume of *Grundgesetze* was going through the press, the plausible assumption leads to contradiction.[8] Take for example the condition R expressed by '...is a class which isn't a member of itself'. This is, on the face of it, a perfectly coherent condition (the *class of people*, for example, satisfies the condition: the class of people contains only people, so it doesn't contain any classes, so doesn't contain itself in particular). And condition R is expressible in the language of Frege's system. So on Frege's assumptions, there will be a class of things that satisfy R. In other words, there is a class Σ_R of all the classes which aren't members of themselves. But now ask: is Σ_R a member of itself? A moment's reflection shows that it is if it isn't, and isn't if it is: contradiction! So there can be no such class as Σ_R; hence Russell's paradox shows that Frege's assumptions cannot be right, despite their intuitive appeal, and his formal system which embodies them is inconsistent.

Intuitions of mathematical truth can indeed be sadly mistaken.

(c)   But let's not rush to make too much of this. In the end, *any* argument has to take *something* as given – in the way we took it that Q's axioms are true in arguing for the theories soundness. And the fact that we *can* make mistakes in arguing for the cogency of a formal system on the basis of our supposed grasp of an intended interpretation isn't any evidence at all that we *have* made a mistake in our argument for the consistency of PA. Moreover, Peano Arithmetic and many stronger theories that embed it have been intensively explored for a century and no contradiction has been exposed. Is there any good reason to suspect it to be guilty of inconsistency?

'Still, the assumption of PA's consistency is notably stronger than the assumption that Q is consistent. So can't we do better,' you might still ask, 'than make the negative point that no contradiction has been found (yet): can't we perhaps *prove* that PA is consistent in some other way than by appealing to our supposed grasp of an interpretation (or by appealing to a richer theory like set theory)?'

Yes, there *are* other proofs. However, for now we'll have to put further discussion of this intriguing issue on hold until after we have said more about Gödel's *Second* Incompleteness Theorem. For that Theorem is all about consistency proofs and it will put some interesting limits on the possibilities here.

---

closely linked to a specific post-Russellian idea, namely the iterative conception of sets as explained e.g. in Potter (2004, §3.2).

[8] See Russell (1902).

14 Primitive recursive functions

The formal theories of arithmetic that we've looked at so far have (at most) the successor function, addition and multiplication built in. But why stop there? Even high school arithmetic acknowledges many more numerical functions. This chapter describes a very wide class of such functions, the so-called primitive recursive ones. But in Chapter 15 we will be able to show that in fact L_A can already express all of them. Then in Chapter 17 we prove that Q, and hence PA, already has the resources to capture all these functions too.

14.1 Introducing the primitive recursive functions

We start by considering two more functions that are very familiar from elementary arithmetic. First, take the factorial function $y!$, where e.g. $4! = 1 \times 2 \times 3 \times 4$. This can be defined by the following two equations:

$$0! = S0 = 1$$
$$(Sy)! = y! \times Sy.$$

The first clause tells us the value of the function for the argument $y = 0$; the second clause tells us how to work out the value of the function for Sy once we know its value for y (assuming we already know about multiplication). So by applying and reapplying the second clause, we can successively calculate $1!$, $2!$, $3!$, $4!$, ... Our two-clause definition therefore fixes the value of $y!$ for all numbers y.

For our second example – this time a two-place function – consider the exponential, standardly written in the form 'x^y'. This can be defined by a similar pair of equations:

$$x^0 = S0$$
$$x^{Sy} = (x^y \times x).$$

Again, the first clause gives the function's value for any given value of x and for $y = 0$; and – keeping x fixed – the second clause gives the function's value for the argument Sy in terms of its value for y.

We've seen this two-clause pattern before, of course, in our formal Axioms for the addition and multiplication functions. Informally, and now presented in the style of everyday mathematics (leaving quantifiers to be understood), we have:

$$x + 0 = x$$
$$x + Sy = S(x + y)$$

$$x \times 0 = 0$$
$$x \times Sy = (x \times y) + x.$$

Three comments about our examples so far:

i. In each definition, the second clause fixes the value of a function for argument Sn by invoking the value of the *same* function for argument n. This kind of procedure is standardly termed 'recursive' – or more precisely, 'primitive recursive'. And our two-clause definitions are examples of *definition by primitive recursion*.[1]

ii. Note, for example, that $(Sn)!$ is defined as $n! \times Sn$, so it is evaluated by evaluating $n!$ and Sn and then feeding the results of these computations into the multiplication function. This involves, in a word, the *composition* of functions, where evaluating a composite function involves taking the output(s) from one or more functions, and treating these as inputs to another function.

iii. Our series of examples, then, illustrates two short *chains* of definitions by recursion and functional composition. Working from the bottom up, addition is defined in terms of the successor function; multiplication is then defined in terms of successor and addition; then the factorial (or, on the other chain, exponentiation) is defined in terms of multiplication and successor.

Here's another little definitional chain:

$$P(0) = 0$$
$$P(Sx) = x$$

$$x \mathbin{\dot-} 0 = x$$
$$x \mathbin{\dot-} Sy = P(x \mathbin{\dot-} y)$$

$$|x - y| = (x \mathbin{\dot-} y) + (y \mathbin{\dot-} x).$$

'P' signifies the predecessor function (with zero being treated as its own predecessor); '$\dot-$' signifies 'subtraction with cut-off', i.e. subtraction restricted to the non-negative integers (so $m \mathbin{\dot-} n$ is zero if $m < n$). And $|m - n|$ is of course the absolute difference between m and n. This time, our third definition doesn't involve recursion, only a simple composition of functions.

---

[1] Strictly speaking, we need a proof of the claim that primitive recursive definitions really *do* well-define functions: such a proof was first given by Richard Dedekind (1888, §126) – for a modern version see, e.g., Moschovakis (2006, pp. 53–56).

There are also other, more complex, kinds of recursive definition – i.e. other ways of defining a function's value for a given argument in terms of its values for smaller arguments. Some of these kinds of definition turn out in fact to be equivalent to definitions by a simple, primitive, recursion; but others, such as the double recursion we meet in defining the Ackermann–Péter function in Section 38.3, are not. For a classic treatment see Péter (1951).

These simple examples motivate the following initial gesture towards a definition:

> A *primitive recursive function* is one that can be similarly characterized using a chain of definitions by recursion and composition.[2]

That is a quick-and-dirty characterization, though it should be enough to get across the basic idea. Still, we really need to pause to do better. In particular, we need to nail down more carefully the 'starter pack' of functions that we are allowed to take for granted in building a definitional chain.

14.2 Defining the p.r. functions more carefully

(a) Consider the recursive definition of the factorial again:

$$0! = 1$$
$$(Sy)! = y! \times Sy.$$

This is an example of the following general scheme for defining a one-place function f:

$$f(0) = g$$
$$f(Sy) = h(y, f(y)).$$

Here, g is just a number, while h is – crucially – a function we are assumed already to know about (i.e. know about prior to the definition of f). Maybe that's because h is an 'initial' function like the successor function that we are allowed just to take for granted. Or perhaps it is because we've already given recursion clauses to define h. Or perhaps h is a composite function constructed by plugging one known function into another – as in the case of the factorial, where $h(y, u) = u \times Sy$.

Likewise, with a bit of massaging, the recursive definitions of addition, multiplication and the exponential can all be treated as examples of the following general scheme for defining two-place functions:

$$f(x, 0) = g(x)$$
$$f(x, Sy) = h(x, y, f(x, y))$$

where now g and h are both functions that we already know about. Three points about this:

i. To get the definition of addition to fit this pattern, we have to take $g(x)$ to be the trivial identity function $I(x) = x$.

---

[2]The basic idea is there in Dedekind and highlighted by Skolem (1923). But the modern terminology 'primitive recursion' seems to be due to Rósza Péter (1934); and 'primitive recursive function' was first used in Stephen Kleene's classic (1936a).

ii. To get the definition of multiplication to fit the pattern, $g(x)$ has to be treated as the even more trivial zero function $Z(x) = 0$.

iii. Again, to get the definition of addition to fit the pattern, we have to take $h(x, y, u)$ to be the function Su. As this illustrates, we must allow h not to care what happens to some of its arguments. One neat way of doing this is to help ourselves to some further trivial identity functions that serve to select out particular arguments. Suppose, for example, we have the three-place function $I_3^3(x, y, u) = u$ to hand. Then, in the definition of addition, we can put $h(x, y, u) = SI_3^3(x, y, u)$, so h is defined by composition from previously available functions.

So with these points in mind, we now give the official 'starter pack' of functions we are allowed to take for granted:

The *initial functions* are

i. the successor function S,

ii. the zero function $Z(x) = 0$,

iii. all the k-place identity functions, $I_i^k(x_1, x_2, \ldots, x_k) = x_i$ for each k, and for each i, $1 \leq i \leq k$.[3]

(b)   We next want to generalize the idea of recursion from the case of one-place and two-place functions. There's a standard notational device that helps to put things snappily: we write $\vec{x}$ as short for the array of k variables $x_1, x_2, \ldots, x_k$. Then we can generalize as follows:

Suppose that

$$f(\vec{x}, 0) = g(\vec{x})$$
$$f(\vec{x}, Sy) = h(\vec{x}, y, f(\vec{x}, y));$$

then f *is defined from g and h by primitive recursion.*

This covers the case of one-place functions $f(y)$ like the factorial if when $\vec{x}$ is empty we take $g(\vec{x})$ to be some constant g.

(c)   Finally, we need to tidy up the idea of definition by composition. The basic idea, to repeat, is that we form a composite function f by treating the output value(s) of one or more given functions g, g', g'', ... as the input argument(s) to another function h. For example, we set $f(x) = h(g(x))$. Or, to take a slightly more complex case, we could set $f(x, y, z) = h(g'(x, y), g''(y, z))$.

There's a number of equivalent ways of covering the manifold possibilities of compounding multi-place functions. But one standard way is to define what we might call one-at-a-time composition (where we just plug *one* function g into another function h), thus:

---

[3]The identity functions are also often called *projection* functions. They 'project' the k-vector $\vec{x}$ with components $x_1, x_2, \ldots, x_k$ onto the i-th axis.

If $g(\vec{y})$ and $h(\vec{x}, u, \vec{z})$ are functions – with $\vec{x}$ and $\vec{z}$ possibly empty – then f *is defined by composition by substituting g into h* just if $f(\vec{x}, \vec{y}, \vec{z}) = h(\vec{x}, g(\vec{y}), \vec{z})$.

We can then think of generalized composition – where we plug more than one function into another function – as just iterated one-at-a-time composition. For example, we can substitute the function $g'(x, y)$ into $h(u, v)$ to define the function $h(g'(x, y), v)$ by composition. Then we can substitute $g''(y, z)$ into the defined function $h(g'(x, y), v)$ to get the composite function $h(g'(x, y), g''(y, z))$.

(d)   To summarize. We informally defined the primitive recursive functions as those that can be defined by a chain of definitions by recursion and composition. Working backwards down a definitional chain, it must bottom out with members of an initial 'starter pack' of trivially simple functions. At the outset, we highlighted the successor function among the given simple functions. But we have since noted that, to get our examples to fit our official account of definition by primitive recursion, we need to acknowledge some other, even more trivial, initial functions.

So putting everything together, let's now offer this more formal characterization of the primitive recursive functions – or the *p.r. functions* (as we'll henceforth call them for short):[4]

1. The initial functions S, Z, and I_i^k are p.r.
2. If f can be defined from the p.r. functions g and h by composition, substituting g into h, then f is p.r.
3. If f can be defined from the p.r. functions g and h by primitive recursion, then f is p.r.
4. Nothing else is a p.r. function.

Note again, we allow g in clauses (2) and (3) to be 'zero-place': when $\vec{x}$ is empty $g(\vec{x})$, i.e. g, just denotes some constant number.

Evidently, the initial functions are total functions of numbers, defined for every numerical argument; also, primitive recursion and composition both build total functions out of total functions. Which means that *all p.r. functions are total functions*, defined for all natural number arguments.

14.3   An aside about extensionality

(a)   We should pause here for a clarificatory remark about the identity conditions for functions, which we will then apply to p.r. functions in particular. The general remark is this (taking the case of one-place numerical functions):

---

[4]Careful! Many books use 'p.r.' to abbreviate 'partial recursive', which is a quite different idea. Our abbreviatory usage is, however, also a common one.

If f and g are one-place total numerical functions, we count them as being the *same* function iff, for each n, $f(n) = g(n)$.

Recall that in Section 5.4(d) we defined the extension of a function f to be the set of pairs $\langle m, n \rangle$ such that $f(m) = n$. So we can also put it this way: we count f and g as the same function if they have the same extension. The point generalizes to many-place numerical functions: we count them as the same if they match up however-many arguments to values in the same way.

In a word, then, we are construing talk of functions *extensionally*. Of course, one and the same function can be presented in different ways, e.g. in ways that reflect different rules for calculating it. For a trivial example, the function $f(n) = 2n + 1$ is the same function as $g(n) = (n+1)^2 - n^2$; but the two different 'modes of presentation' indicate different routines for evaluating the function.

(b)   Now, a p.r. function is by definition one that *can* be specified by a certain sort of chain of definitions. And an ideal way of presenting such a function will be by indicating a definitional chain for it (in a way which makes it transparent that the function *is* p.r.). But the same function can be presented in other ways; and some modes of presentation can completely disguise the fact that the given function is recursive. For a dramatic example, consider the function

$fermat(n) = n$ if there are solutions to $x^{n+3} + y^{n+3} = z^{n+3}$ (with
   x, y, z positive integers);
$fermat(n) = 0$ otherwise.

This definition certainly doesn't reveal on its face whether the function is primitive recursive. But we know now – thanks to Andrew Wiles's proof of Fermat's Last Theorem – that *fermat* is in fact p.r., for it is none other than the trivially p.r. function $Z(n) = 0$.

Note too that other modes of presentation may make it clear that a function is p.r., but still not tell us *which* p.r. function is in question. Consider, for example, the function defined by

$julius(n) = n$ if Julius Caesar ate grapes on his third birthday;
$julius(n) = 0$ otherwise.

There is no way (algorithmic or otherwise) of settling what Caesar ate on his third birthday! But despite that, the function $julius(n)$ is plainly primitive recursive. Why so? Well, either it is the trivial identity function $I(n) = n$, or it is the zero function $Z(n) = 0$. So we know that $julius(n)$ must be a p.r. function, though we can't determine *which* function it is from our style of definition.

The key observation is this: *primitive recursiveness is a feature of a function itself, irrespective of how it happens to be presented to us.*

(c)   We count numerical *functions* as being the same or different, depending on whether their extensions are the same. In the same vein, it is rather natural to treat numerical *properties* as the same or different, depending on whether *their* extensions are the same. And this will be our official line too.

Indeed, if you accept the thesis of Frege (1891), then we indeed have to treat properties and functions in the same way here. For Frege urges us to regard properties as just a special kind of function – so a numerical property, in particular, is a function that maps a number to the truth-value *true* (if the number has the property) or *false* (otherwise). Which comes very close to identifying a property with its characteristic function – see Section 2.2.

14.4   The p.r. functions are computable

To repeat, a p.r. function f is one that *can* be specified by a chain of definitions by recursion and composition, leading back ultimately to initial functions. But (a) it is trivial that the initial functions S, Z, and I_i^k are effectively computable. (b) The composition of two effectively computable functions g and h is also computable (you just feed the output from whatever algorithmic routine evaluates g as input into the routine that evaluates h). And (c) – the key observation – if g and h are effectively computable, and f is defined by primitive recursion from g and h, then f is effectively computable too. So as we build up longer and longer chains of definitions for p.r. functions, we always stay within the class of effectively computable functions.

To illustrate (c), return once more to our example of the factorial. Here is its p.r. definition again:

$$0! = 1$$
$$(Sy)! = y! \times Sy$$

The first clause gives the value of the function for the argument 0; then – as we said – you can repeatedly use the second recursion clause to calculate the function's value for $S0$, then for $SS0$, $SSS0$, etc. So the definition encapsulates an algorithm for calculating the function's value for any number, and corresponds exactly to a certain simple kind of computer routine.

Thus compare our definition with the following schematic program:

1. $fact := 1$
2.   for $y = 0$ to $n - 1$ do
3. $fact := (fact \times Sy)$
4.   end do

Here *fact* indicates a memory register that we initially prime with the value of $0!$. Then the program enters a loop: and the crucial thing about executing a 'for' loop is that the total number of iterations to be run through is fixed in advance. The program numbers the loops from 0, and on loop number k the program replaces the value in the register *fact* with Sk times the previous value (we'll assume the computer already knows how to find the successor of k and do the multiplication). When the program exits the loop after a total of n iterations, the value in the register *fact* will be $n!$.

More generally, for any one-place function f defined by primitive recursion in terms of g and the computable function h, the same program structure always does the trick for calculating $f(n)$. Thus compare

$$f(0) = g$$
$$f(Sy) = h(y, f(y))$$

with the corresponding program telling us what to put in a register *func*:

1. *func* $:= g$
2. for $y = 0$ to $n - 1$ do
3.   *func* $:= h(y, func)$
4. end do

So long as h is already computable, the value of $f(n)$ will be computable using this 'for' loop, which terminates with the required value now in the register *func*.

Exactly similarly, of course, for many-place functions. For example, the value of the two-place function defined by

$$f(x, 0) = g(x)$$
$$f(x, Sy) = h(x, y, f(x, y))$$

is calculated, for given first argument m, by the procedure

1. *func* $:= g(m)$
2. for $y = 0$ to $n - 1$ do
3.   *func* $:= h(m, y, func)$
4. end do

which is algorithmic so long as g and h are computable, and which terminates with the value for $f(m, n)$ in *func*.

Now, our mini-program for the factorial calls the multiplication function which can itself be computed by a similar 'for' loop (invoking addition). And addition can in turn be computed by another 'for' loop (invoking the successor). So reflecting the downward chain of recursive definitions

factorial $\Rightarrow$ multiplication $\Rightarrow$ addition $\Rightarrow$ successor

there is a program for the factorial containing *nested* 'for' loops, which ultimately calls the primitive operation of incrementing the contents of a register by one (or other operations like setting a register to zero, corresponding to the zero function, or copying the contents of a register, corresponding to an identity function).

The point obviously generalizes, establishing

> **Theorem 14.1** *Primitive recursive functions are effectively computable, and computable by a series of (possibly nested) 'for' loops.*

The converse is also true. Take a 'for' loop which computes the value of a one-place function f for argument n by going round a loop n times, on each circuit calling the function h and applying h again to the output of the previous loop. Then if h is p.r., f will be p.r. too. And generalizing, if a function can be computed by a program using just 'for' loops as its main programming structure – with the program's 'built in' functions all being p.r. – then the newly defined function will also be primitive recursive.

This gives us a quick-and-dirty (but reliable!) way of convincing ourselves that a new function is p.r.: *sketch out a routine for computing it and check that it can all be done with a succession of (possibly nested) 'for' loops which only invoke already known p.r. functions; then the new function will be primitive recursive.*[5]

14.5 Not all computable numerical functions are p.r.

(a) We have seen that any p.r. function is algorithmically computable. *But not all effectively computable numerical functions are primitive recursive.* In this section, we first make the claim that there are computable-but-not-p.r. numerical functions look plausible. Then we will cook up an example.[6]

We start, then, with some plausibility considerations. We have just seen that the values of a given primitive recursive function can be computed by a program involving 'for' loops as its main programming structure, where each such loop goes through a specified number of iterations. However, back in Section 3.1 we allowed procedures to count as computational even when they don't have nice upper bounds on the number of steps involved. In other words, we allowed computations to involve *open-ended searches*, with no prior bound on the length of search. We made essential use of this permission in Section 4.6, when we showed that negation-complete theories are decidable – for we allowed the process 'enumerate the theorems and wait to see which of φ or $\neg\varphi$ turns up' to count as a computational decision procedure.

---

[5]We can put all that just a bit more carefully by considering a simple programming language LOOP. A particular LOOP program operates on a set of registers. At the most basic level, the language has instructions for setting the contents of a register to zero, copying contents from one register to another, and incrementing the contents of a register by one. And the *only* important programming structure is the 'for' loop. Such a loop involves setting a register with some initial contents (at the zero-th stage of the loop) and then iterating a LOOP-defined process n times (where on each loop, the process is applied to the result of its own previous application), which has just the effect of a definition by recursion. Such loops can be nested. And sets of nested LOOP commands can be concatenated so that e.g. a loop for evaluating a function g is followed by a loop for evaluating h: concatenation evidently corresponds to composition of functions. Even without going into any more details, it is very easy to see that every LOOP program will define a p.r. function, and every p.r. function is defined by a LOOP program. For a proper specification of LOOP and proofs see Tourlakis (2002); the idea of such programs goes back to Meyer and Ritchie (1967).

[6]Our cooked-up example, however, isn't one that might be encountered in ordinary mathematical practice; it requires a bit of ingenuity to come up with a 'natural' example – see Section 38.3 where we introduce so-called Ackermann functions.

Standard computer languages of course have programming structures which implement just this kind of unbounded search. Because as well as 'for' loops, they allow 'do until' loops (or equivalently, 'do while' loops). In other words, they allow some process to be iterated until a given condition is satisfied – *where no prior limit is put on the number of iterations to be executed.*

Given that we count what are presented as unbounded searches as computations, then it looks very plausible that not everything computable will be primitive recursive.

(b)   But we can do better than a mere plausibility argument. We will now *prove*

Theorem 14.2 *There are effectively computable numerical functions which aren't primitive recursive.*

Proof   The set of p.r. functions is effectively enumerable. That is to say, there is an effective way of numbering off functions f_0, f_1, f_2, ..., such that each of the f_i is p.r., and each p.r. function appears somewhere on the list.

This holds because, by definition, every p.r. function has a full 'recipe' in which it is defined by recursion or composition from other functions which are defined by recursion or composition from other functions which are defined ... ultimately in terms of some primitive starter functions. So choose some standard formal specification language for representing these recipes. Then we can effectively generate 'in alphabetical order' all possible strings of symbols from this language; and as we go along, we select the strings that obey the rules for being a full recipe for a p.r. function (that's a mechanical procedure). That generates a list of recipes which effectively enumerates the p.r. functions f_i, repetitions allowed.

	0	1	2	3	...
f_0	$f_0(0)$	$f_0(1)$	$f_0(2)$	$f_0(3)$	...
f_1	$f_1(0)$	$f_1(1)$	$f_1(2)$	$f_1(3)$	...
f_2	$f_2(0)$	$f_2(1)$	$f_2(2)$	$f_2(3)$	...
f_3	$f_3(0)$	$f_3(1)$	$f_3(2)$	$f_3(3)$	...
...	...	...	...	...	↘

Now consider our table. Down the table we list off the p.r. functions f_0, f_1, f_2, ... An individual row then gives the values of f_n for each argument. Let's define the corresponding *diagonal* function, by putting 'going down the diagonal and adding one', i.e. we put $\delta(n) = f_n(n) + 1$ (cf. Section 2.5). To compute $\delta(n)$, we just run our effective enumeration of the recipes for p.r. functions until we get to the recipe for f_n. We follow the instructions in that recipe to evaluate that function for the argument n. We then add one. Each step is entirely mechanical. So our diagonal function is effectively computable, using a step-by-step algorithmic procedure.

By construction, however, the function δ can't be primitive recursive. For suppose otherwise. Then δ must appear somewhere in the enumeration of p.r. functions, i.e. be the function f_d for some index number d. But now ask what the value of $\delta(d)$ is. By hypothesis, the function δ is none other than the function f_d, so $\delta(d) = f_d(d)$. But by the initial definition of the diagonal function, $\delta(d) = f_d(d) + 1$. Contradiction.

So we have 'diagonalized out' of the class of p.r. functions to get a new function δ which is effectively computable but not primitive recursive. $\boxtimes$

'But hold on! *Why* is the diagonal function not a p.r. function?' Well, consider evaluating $d(n)$ for increasing values of n. For each new argument, we will have to evaluate a *different* function f_n for that argument (and then add 1). We have no reason to expect there will be a nice pattern in the successive computations of all the different functions f_n which enables them to be wrapped up into a single p.r. definition. And our diagonal argument in effect shows that this can't be done.

14.6   Defining p.r. properties and relations

We have defined the class of p.r. *functions*. Next, without further ado, we extend the scope of the idea of primitive recursiveness and introduce the ideas of *p.r. (numerical) properties* and *relations*.

> A *p.r. property* is a property with a p.r. characteristic function, and likewise a *p.r. relation* is a relation with a p.r. characteristic function.

Given that any p.r. function is effectively computable, p.r. properties and relations are among the effectively decidable ones.

14.7   Building more p.r. functions and relations

(a)   The last two sections of this chapter give some general principles for building new p.r. functions and relations out of old ones, and then give examples of some of these principles at work.

These are more 'trivial but tiresome' details which you could in fact cheerfully skip, since we only pick up their details in other sections that you can also skip. True, in proving Gödel's Theorems, we will need to claim that a variety of key functions and relations are p.r.; and our claims will seem more evidently plausible if you have already worked through some simpler cases. It is therefore probably worth skimming through these sections: but if you have no taste for this sort of detail, don't worry. *Don't get bogged down!*

(b)   A couple more definitions before the real business gets under way. First, we introduce the *minimization* operator 'μx', to be read: 'the least x such that . . . '.

Much later, in Section 38.1, we'll be considering the general use of this operator; but here we will be concerned with *bounded minimization*. So we write

$$f(n) = (\mu x \leq n)P(x)$$

when f takes the number n as argument and returns as value the least number $x \leq n$ such that $P(x)$ if such an x exists, or returns n otherwise. Generalizing,

$$f(n) = (\mu x \leq g(n))P(x)$$

returns as value the least number $x \leq g(n)$ such that $P(x)$ if such an x exists, or returns $g(n)$ otherwise.

Second, suppose that the function f is defined in terms of $k + 1$ other p.r. functions f_i as follows

$$f(n) = f_0(n) \text{ if } C_0(n)$$
$$f(n) = f_1(n) \text{ if } C_1(n)$$
$$\vdots$$
$$f(n) = f_k(n) \text{ if } C_k(n)$$
$$f(n) = a \text{ otherwise}$$

where the conditions C_i are mutually exclusive and express p.r. properties (i.e. have p.r. characteristic functions c_i), and a is a constant. Then f is said to be *defined by cases* from other p.r. functions.

(c)   Consider the following claims:

A. If $f(\vec{x})$ is an n-place p.r. function, then the corresponding relation expressed by $f(\vec{x}) = y$ is an $n + 1$-place p.r. relation.

B. Any truth-functional combination of p.r. properties and relations is p.r.

C. Any property or relation defined from a p.r. property or relation by bounded quantifications is also p.r.

D. If P is a p.r. property, then the function $f(n) = (\mu x \leq n)P(x)$ is p.r. And generalizing, suppose that $g(n)$ is a p.r. function, and P is a p.r. property; then $f'(n) = (\mu x \leq g(n))P(x)$ is also p.r.

E. Any function defined by cases from other p.r. functions is also p.r.

In each case, we have a pretty obvious claim about what can be done using 'for' loops. For example, claim (A) comes to this (applied to one-place functions): if you can evaluate $f(m)$ by an algorithmic routine using 'for' loops, then you can check whether $f(m) = n$ by an algorithmic routine using 'for' loops. The other claims should all look similarly evident. But let's make this an official theorem:

Theorem 14.3 *The claims (A) to (E) are all true.*

Proof for (A) Start with a preliminary result. Put $sg(n) = 0$ for $n = 0$, and $sg(n) = 1$ otherwise. Then sg is primitive recursive. For we just note that

$$sg(0) = 0$$
$$sg(Sy) = SZ(sg(y))$$

where $SZ(u)$ is p.r. by composition, and $SZ(sg(y)) = S0 = 1$. Also, let $\overline{sg}(n) = 1$ for $n = 0$, and $\overline{sg}(n) = 0$ otherwise. Then $\overline{sg}$ is similarly shown to be p.r.

We now argue for (A) in the case where f is a one-place function (generalizing can be left as an exercise). The characteristic function of the relation expressed by $f(x) = y$ – i.e. the function $c(x, y)$ whose value is 0 when $f(x) = y$ and is 1 otherwise[7] – is given by

$$c(x, y) = sg(|f(x) - y|).$$

The right-hand side is a composition of p.r. functions, so c is p.r.   ☒

Proof for (B) Suppose $p(x)$ is the characteristic function of the property P. It follows that $\overline{sg}(p(x))$ is the characteristic function of the property *not-P*, since $\overline{sg}$ simply flips the two values 0 and 1. But by simple composition of functions, $\overline{sg}(p(x))$ is p.r. if $p(x)$ is. Hence if P is a p.r. property, so is *not-P*.

Similarly, suppose that $p(x)$ and $q(x)$ are the characteristic functions of the properties P and Q respectively. $p(n) \times q(n)$ takes the value 0 so long as either n is P or n is Q, and takes the value 1 otherwise. So $p(x) \times q(x)$ is the characteristic function of the disjunctive property of being either P or Q; and by composition, $p(x) \times q(x)$ is p.r. if both $p(x)$ and $q(x)$ are. Hence the disjunction of p.r. properties is another p.r. property.

But any truth-functional combination of properties is definable in terms of negation and disjunction. Which completes the proof.   ☒

Proof for (C) Just reflect that checking to see whether e.g. $(\exists x \leq n)Px$ involves using a 'for' loop to check through the cases from 0 to n to see whether any satisfy Px. Likewise, if f is p.r., checking to see whether $(\exists x \leq f(n))Px$ involves calculating $f(n)$ and then using a 'for' loop to check through the cases from 0 to $f(n)$ to see whether Px holds. It follows that, if f is p.r., then so are both of

$$K(n) =_{\text{def}} (\exists x \leq n)Px$$
$$K'(n) =_{\text{def}} (\exists x \leq f(n))Px.$$

More carefully, suppose that $p(x)$ is P's p.r. characteristic function. And by composition define the p.r. function $h(u, v) = (p(Su) \times v)$. Put

$$k(0) = p(0)$$
$$k(Sy) = h(y, k(y))$$

---

[7]For those who are forgetful or have been skipping, this isn't a misprint. We are following the minority line of taking '0' to be the positive value for a characteristic function, indicating that the relation in question holds. See Section 2.2.

so we have

$$k(n) = p(n) \times p(n-1) \times \ldots \times p(1) \times p(0).$$

Then k is K's characteristic function – i.e. the function such that $k(n) = 1$ until we get to an n such that n is P, when $k(n)$ goes to zero, and thereafter stays zero. Since k is p.r., K is p.r. by definition.

And to get the generalized result, we just note that $K'(n)$ iff $K(f(n))$, so K' has the p.r. characteristic function $k(f(n))$.

We also have further similar results for bounded universal quantifiers (exercise!). Note too that we can apply the bounded quantifiers to relations as well as monadic properties; and in the bounded quantifiers we could equally use '<' rather than '≤'. ⊠

Proof for (D) Again suppose p is the characteristic function of P, and define k as in the last proof. Then consider the function defined by

$$f(0) = 0$$
$$f(n) = k(n-1) + k(n-2) + \ldots + k(1) + k(0), \text{ for } n > 0.$$

Now, $k(j) = 1$ for each j that isn't P, and $k(j)$ goes to zero and stays zero as soon as we hit a j that is P. So $f(n) = (\mu x \leq n)P(x)$, i.e. $f(n)$ returns either the least number that is P, or n, whichever is smaller. So we just need to show that f so defined is primitive recursive. Well, use composition to define the p.r. function $h'(u, v) = (k(u) + v)$, and then put

$$f(0) = 0$$
$$f(Sy) = h'(y, f(y)).$$

Which proves the first, simpler, part of Fact D. For the generalization, just note that by the same argument we have $f(g(n)) = (\mu x \leq g(n))P(x)$ is p.r. if g is, so we can put $f'(n) = f(g(n))$ and we are done. ⊠

Proof for (E) Just note that

$$f(n) = \overline{sg}(c_0(n))f_0(n) + \overline{sg}(c_1(n))f_1(n) + \ldots + \overline{sg}(c_k(n))f_k(n) + c_0(n)c_1(n)\ldots c_k(n)a$$

since $\overline{sg}(c_i(n)) = 1$ when $C_i(n)$ and is otherwise zero, and the product of all the $c_i(n)$ is 1 just in case none of $C_i(n)$ are true, and is zero otherwise. ⊠

14.8 Further examples

With our shiny new tools to hand, we can finish the chapter by giving a few more examples of p.r. functions, properties and relations:

1. The relations $m = n$, $m < n$ and $m \leq n$ are primitive recursive.

2. The relation $m|n$ that holds when m is a factor of n is primitive recursive.

3. Let $Prime(n)$ be true just when n is a prime number. Then $Prime$ is a p.r. property.[8]

4. List the primes as $\pi_0, \pi_1, \pi_2, \ldots$. Then the function $\pi(n)$ whose value is π_n is p.r.

5. Let $exf(n, i)$ be the possibly zero *ex*ponent of π_i – the $i + 1$-th prime – in the *f*actorization of n. Then exf is a p.r. function.[9]

6. Let $len(0) = len(1) = 0$; and when $n > 1$, let $len(n)$ be the 'length' of n's factorization, i.e. the number of distinct prime factors of n. Then len is again a p.r. function.

You should pause here to convince yourself that all these claims are true by the quick-and-dirty method of sketching out how you can compute the relevant (characteristic) functions without doing any unbounded searches, just by using 'for' loops.

But – if you insist – we can also do this the hard way:

> **Theorem 14.4** *The properties, relations and functions listed in (1) to (6) are indeed all primitive recursive.*

Proof for (1) The characteristic function of $m = n$ is $sg(|m - n|)$, where $|m - n|$ is the absolute difference function we showed to be p.r. in Section 14.1. The characteristic functions of $m < n$ and $m \leq n$ are $sg(Sm \mathbin{\dot-} n)$ and $sg(m \mathbin{\dot-} n)$ respectively. These are all compositions of p.r. functions, and hence themselves primitive recursive. $\boxtimes$

Proof for (2) We have

$$m|n \leftrightarrow (\exists y \leq n)(0 < y \wedge 0 < m \wedge m \times y = n).$$

The relation expressed by the subformula after the quantifier is a truth-functional combination of p.r. relations (multiplication is p.r., so the last conjunct is p.r. by Fact A of the last section). So that relation is p.r. by Fact B. Hence $m|n$ is a p.r. relation by Fact C. $\boxtimes$

Proof for (3) The property of being *Prime* is p.r. because

$$Prime(n) \leftrightarrow n \neq 1 \wedge (\forall u \leq n)(\forall v \leq n)(u \times v = n$$
$$\rightarrow (u = 1 \vee v = 1))$$

---

[8] Remember the useful convention: capital letters for the names of predicates and relations, small letters for the names of functions.

[9] We should perhaps note that a more conventional notation for this function is, simply, $(n)_i$, a usage which goes back to e.g. Kleene (1952, p. 230). But this isn't pretty – we end up with too many nested brackets – nor is it very memorable.

and the right-hand side is built up from p.r. components by truth-functional combination and restricted quantifiers. (Here we rely on the trivial fact that the factors of n cannot be greater than n.) ⊠

Proof for (4) Recall the familiar fact that the next prime after k is no greater than $k! + 1$ (for either the latter is prime, or it has a prime factor which must be greater than k). Put $h(k) = (\mu x \le k! + 1)(k < x \land Prime(x))$. Then h is p.r., by the generalized version of Fact D, and returns the next prime after k.

But then the function $\pi(n)$, whose value is the $n + 1$-th prime π_n, is defined by

$$\pi(0) = 2$$
$$\pi(Sn) = h(\pi(n)),$$

therefore π is p.r. ⊠

Proof for (5) By the Fundamental Theorem of Arithmetic, which says that numbers have a unique factorization into primes, this function is well-defined. And no exponent in the prime factorization of n is larger than n itself, so

$$exf(n, i) = (\mu x \le n)\{(\pi_i^x | n) \land \lnot(\pi_i^{x+1} | n)\}.$$

That is to say, the desired exponent of π_i is the number x such that π_i^x divides n but π_i^{x+1} doesn't: note that $exf(n, k) = 0$ when π_k isn't a factor of n. Again, our definition of exf is built out of p.r. components by operations that yield another p.r. function. ⊠

Proof for (6) $(Prime(m) \land m|n)$ holds when m is a prime factor of n. This is a p.r. relation (being a conjunction of p.r. properties/relations). So it has a p.r. characteristic function which we'll abbreviate $pf(m, n)$. Now consider the function

$$p(m, n) = \overline{sg}(pf(m, n)).$$

Then $p(m, n) = 1$ just when m is a prime factor of n and is zero otherwise. So

$$len(n) = p(0, n) + p(1, n) + \ldots + p(n - 1, n) + p(n, n).$$

So to give a p.r. definition of *len*, we can first put

$$l(x, 0) = p(0, x)$$
$$l(x, Sy) = (p(Sy, x) + l(x, y))$$

and then finally put $len(n) = l(n, n)$. ⊠

And that's *quite* enough to be going on with. All good clean fun if you like that kind of thing. But as I said before, don't worry if you don't. For having shown that these kinds of results *can* be proved, you can now cheerfully forget the tricksy little details of how to do it.

15 L_A can express every p.r. function

The built-in resources of L_A – the first-order language of basic arithmetic which we first introduced in Section 5.2 – are minimal: there are no non-logical predicates, and just three primitive functions, successor, addition and multiplication.

We have previously noted, though, a few examples of what else L_A can express ('express' in the sense of Section 5.4). We now radically extend our list of examples by proving that L_A can in fact express *any* primitive recursive function.

15.1 Starting the proof

Our target, then, is to prove

Theorem 15.1 *Every p.r. function can be expressed in L_A.*

The proof strategy   Suppose that the following three propositions are all true:

1. L_A can express the initial functions.
2. If L_A can express the functions g and h, then it can also express a function f defined by composition from g and h.
3. If L_A can express the functions g and h, then it can also express a function f defined by primitive recursion from g and h.

Now, any p.r. function f must be specifiable by a chain of definitions by composition and/or primitive recursion, building up from initial functions. So as we follow through the full chain of definitions which specifies f, we start with initial functions which are expressible in L_A, by (1). By (2) and (3), each successive definitional move takes us from expressible functions to expressible functions. So, given (1) to (3) *are* true, f must therefore also be expressible in L_A. Hence in order to prove Theorem 15.1, it is enough to prove (1) to (3).

Proof for (1)   There are three easy cases to consider:

i. The successor function $Sx = y$ is expressed by the wff $\mathsf{S}\mathsf{x} = \mathsf{y}$.
ii. The zero function $Z(x) = 0$ is expressed by $\mathsf{Z}(\mathsf{x}, \mathsf{y}) =_{\text{def}} (\mathsf{x} = \mathsf{x} \wedge \mathsf{y} = 0)$.
iii. Finally, the three-place identity function $I_2^3(x, y, z) = y$, to take just one example, is expressed by the wff $\mathsf{I}_2^3(\mathsf{x}, \mathsf{y}, \mathsf{z}, \mathsf{u}) =_{\text{def}} (\mathsf{x} = \mathsf{x} \wedge \mathsf{y} = \mathsf{u} \wedge \mathsf{z} = \mathsf{z})$. Likewise for all the other identity functions.   ⊠

Note that this shows that all the initial functions are expressible by Δ_0 wffs.

Proof for (2)   Suppose g and h are one-place functions, expressed by the wffs $G(x,y)$ and $H(x,y)$ respectively. Then, the function $f(x) = h(g(x))$ is evidently expressed by the wff $\exists z(G(x,z) \wedge H(z,y))$. Other cases where g and/or h are multi-place functions can be handled similarly.   ⊠

Starting the proof for (3)   Now for the fun part. Consider the primitive recursive definition of the factorial function again:

$$0! = 1$$
$$(Sx)! = x! \times Sx.$$

The multiplication and successor functions here are of course expressible in L_A: but how can we express our defined function in L_A?

Think about the p.r. definition for the factorial in the following way. It tells us how to construct a sequence of numbers $0!, 1!, 2!, \dots, x!$, where we move from the u-th member of the sequence (counting from zero) to the next by multiplying by Su. Putting $x! = y$, the p.r. definition thus says

A.   There is a sequence of numbers $k_0, k_1, \dots, k_x$ such that: $k_0 = 1$, and if $u < x$ then $k_{Su} = k_u \times Su$, and $k_x = y$.

So the question of how to reflect the p.r. definition of the factorial inside L_A comes to this: how can we express facts about finite sequences of numbers using the limited resources of L_A?

15.2   The idea of a β-function

Let's pause the proof at this point, and think first about the *kind* of trick we could use here.

Suppose $\pi_0, \pi_1, \pi_2, \pi_3, \dots$ is the series of prime numbers $2, 3, 5, 7, \dots$. Now consider the number

$$c = \pi_0^{k_0} \cdot \pi_1^{k_1} \cdot \pi_2^{k_2} \cdot \dots \cdot \pi_n^{k_n}.$$

Here, c can be thought of as encoding the whole sequence $k_0, k_1, k_2, \dots, k_n$. For we can extract the coded sequence from c by using the function $exf(c,i)$ which returns the exponent of the prime number π_i in the unique factorization of c (we met this function in Section 14.8, and showed it to be primitive recursive). By the construction of c, $exf(c,i) = k_i$ for $i \leq n$.

So this gives us an example of how we might use a number as a code for a sequence of numbers, and then have a decoding function that enables us to extract members of the sequence from the code.

Let's generalize. We will say that

A two-place β-*function* is a numerical function $\beta(c,i)$ such that, for *any* finite sequence of natural numbers $k_0, k_1, k_2, \dots, k_n$ there is a code number c such that for every $i \leq n$, $\beta(c,i) = k_i$.

In other words, for any finite sequence of numbers you choose, you can select a corresponding code number c. Then, if you feed c as first argument to β, this function will decode it and spit out the members of the required sequence in order as its second argument is increased.

We've just seen that there is nothing in the least bit magical or mysterious about the idea of a β-function: exf is a simple example. However, this first illustration of a β-function is defined in terms of the exponential function, which *isn't* built into L_A.[1] So the obvious next question is: can we construct a β-function just out of the successor, addition and multiplication functions which *are* built into L_A? Then we can use this β-function to deal with facts about sequences of numbers while still staying within L_A.

It turns out to simplify things if we liberalize our notion of a β-function just a little. So we'll now also consider *three*-place β-functions, which take *two* code numbers c and d, as follows:

> A three-place β-function is a function of the form $\beta(c, d, i)$ such that, for *any* finite sequence of natural numbers $k_0, k_1, k_2, \ldots, k_n$ there is a pair of code numbers c, d such that for every $i \leq n$, $\beta(c, d, i) = k_i$.

A three-place β-function will do just as well as a two-place function to help us deal with facts about finite sequences.

Even with this liberalization, though, it still isn't obvious how to define a β-function in terms of the functions built into basic arithmetic. But Gödel neatly solved our problem as follows. Put

$$\beta(c, d, i) =_{\text{def}} \text{the remainder left when } c \text{ is divided by } d(i+1)+1.^2$$

Then, given any sequence $k_0, k_1, \ldots, k_n$, we can find a suitable pair of numbers c, d such that for $i \leq n$, $\beta(c, d, i) = k_i$.[3]

This claim should look intrinsically plausible. As we divide c by $d(i + 1) + 1$ for different values of i ($0 \leq i \leq n$), we'll get a sequence of $n + 1$ remainders. Vary c and d, and the sequence of $n + 1$ remainders will vary. The permutations

---

[1]Note that we could have started by taking our fundamental language of arithmetic at the outset to be not L_A but L_A^+, i.e. the language you get by adding the exponential function to L_A. Correspondingly, we could have taken as our basic theories Q^+ and PA^+, which you get from Q and PA by adding the obvious recursion axioms for the exponential.

Then we'd have a very easily constructed β-function available in L_A^+ and we could have avoided all the fuss in the rest of this section (and in particular the need for the non-obvious argument of fn. 4 below).

As so very often in logic, we have a trade-off. We are making life quite a bit harder for ourselves at this point by working with L_A rather than L_A^+ (and with Q/PA rather than $\mathsf{Q}^+/\mathsf{PA}^+$). The pay-off is that our eventual incompleteness theorems show that there is no nice complete theory even for the basic arithmetic of L_A truths.

[2]So $\beta(c, d, i)$ is the unique r, $0 \leq r < k$, such that for some n, $c = nk + r$, where $k = d(i + 1) + 1$.

[3]This is shown in Gödel's 1931 paper: but the function is first labelled as 'β' in his Princeton Lectures (1934, p. 365).

as we vary c and d without limit *appear* to be simply endless. We just need to check, then, that appearances don't deceive, and we *can* always find a (big enough) c and a (smaller) d which makes the sequence of remainders match a given $n + 1$-term sequence of numbers.[4]

But now reflect that the concept of a remainder on division can be elementarily defined in terms of multiplication and addition. Thus consider the following open wff (using memorable variables):

$$\mathsf{B(c,d,i,y)} =_{def} (\exists u \leq c)[c = \{S(d \times Si) \times u\} + y \wedge y \leq (d \times Si)].$$

This, as we want, expresses our Gödelian β-function in L_A (and shows that it can be expressed by a Δ_0 wff).

15.3 Finishing the proof

Continuing the proof for (3) Suppose we have some three-place β-function to hand. So, given any sequence of numbers $k_0, k_1, \ldots, k_x$, there are code numbers c, d such that for $i \leq x$, $\beta(c, d, i) = k_i$. Then we can reformulate

---

[4] Here is how to check that claim (this is, however, an exercise in elementary number theory, not logic, which is why we relegate it to the small print, a footnote for enthusiasts).

Let $rm(c, d)$ be the function which gives the value r when r is the remainder when c is divided by d. Then we have the following preliminary

> **Lemma** Suppose x, y are *relatively prime*, i.e. have no common factor other than 1. Then there is a number $b < y$ such that $rm(bx, y) = 1$.

Proof Suppose that there are $b_1 < b_2 < y$ such that $rm(b_1 x, y) = rm(b_2 x, y)$. Then $(b_2 - b_1)x$ must be an exact multiple of y. Which is impossible, since y doesn't divide x, and $(b_2 - b_1) < y$. So that means that for each different value of $b < y$, $rm(bx, y)$ gives a different result less than y. Therefore one of those remainders must be 1. ⊠

That little lemma now allows us to prove a version of the ancient

> **Chinese Remainder Theorem** Suppose $m_0, m_1, m_2, \ldots, m_n$ are pairwise relatively prime. Then for any sequence of numbers $k_0, k_1, k_2, \ldots, k_n$ (where for each $i \leq n$, $k_i \leq m_i$) there is a number c such $k_i = rm(c, m_i)$.

Proof Define M_i as the product of all the m_j except for m_i. Note that M_i and m_i are also relatively prime. So for each i there is a number b_i such that $rm(b_i M_i, m_i) = 1$.

So now put $c = \sum_{j=0}^{n} k_j b_j M_j$. If we divide c by m_i, what's the remainder? Well, for $j \neq i$, each term $k_j b_j M_j$ in the sum divides exactly by m_i and there is no remainder. So that just leaves the remainder of $k_i b_i M_i$ when divided by m_i, which is exactly k_i (since $k_i \leq m_i$). Which shows that $k_i = rm(c, m_i)$. ⊠

We can now prove

> **Gödel's β-function Lemma** For any sequence of numbers $k_0, k_1, k_2, \ldots, k_n$, there are numbers c, d, such that $rm(c, d(i + 1) + 1) = k_i$.

Proof Let u be the maximum of $n + 1, k_0, k_1, k_2, \ldots, k_n$, and put $d = u!$. For $i \leq n$ put $m_i = d(i + 1) + 1$. Now it is easily seen that the m_i are pairwise prime.

For suppose otherwise. Then for some prime p, and some a, b such that $1 \leq a < b \leq n + 1$, p divides both $da + 1$ and $db + 1$, so that p divides $d(b - a)$. But $(b - a)$ is a factor of d, so this means that p divides d without remainder. However that means p can't exactly divide $da + 1$ or $db + 1$ after all, contradicting our supposition.

So, since the m_i, i.e. $d(i + 1) + 1$ (for $0 \leq i \leq n$), are pairwise prime, and $k_i < m_i$, we can use the Chinese Remainder Theorem to find a c such that $rm(c, d(i + 1) + 1) = k_i$. ⊠

A. There is a sequence of numbers $k_0, k_1, \ldots, k_x$ such that: $k_0 = 1$, and if $u < x$ then $k_{Su} = k_u \times Su$, and $k_x = y$,

as follows:

B. There is some pair c, d such that: $\beta(c, d, 0) = 1$, and if $u < x$ then $\beta(c, d, Su) = \beta(c, d, u) \times Su$, and $\beta(c, d, x) = y$.

But we've seen that there's a three-place β-function which can be expressed in L_A by the open wff we abbreviated B. So fixing on this β-function, we can render (B) into L_A as follows:

C. $\exists c \exists d \{ B(c, d, 0, 1) \land$
$(\forall u \leq x)[u = x \lor \exists v \exists w \{ (B(c, d, u, v) \land B(c, d, Su, w)) \land w = v \times Su \}] \land$
$B(c, d, x, y) \}$.

Abbreviate all that by '$F(x, y)$', and we've arrived. For this evidently expresses the factorial function.

Concluding the proof for (3) We need to show that we can use the same β-function trick and prove more generally that, if the function f is defined by recursion from functions g and h which are already expressible in L_A, then f is also expressible in L_A. So here, just for the record, is the entirely routine generalization we need.

We are assuming that

$$f(\vec{x}, 0) = g(\vec{x})$$
$$f(\vec{x}, Sy) = h(\vec{x}, y, f(\vec{x}, y)).$$

This definition amounts to fixing the value of $f(\vec{x}, y) = z$ thus:

A* There is a sequence of numbers $k_0, k_1, \ldots, k_y$ such that: $k_0 = g(\vec{x})$, and if $u < y$ then $k_{u+1} = h(\vec{x}, u, k_u)$, and $k_y = z$.

So using a three-place β-function again, that comes to

B* There is some c, d such that: $\beta(c, d, 0) = g(\vec{x})$, and if $u < y$ then $\beta(c, d, Su) = h(\vec{x}, u, \beta(c, d, u))$, and $\beta(c, d, y) = z$.

Suppose we can already express the n-place function g by a $(n + 1)$-variable expression G, and the $(n+2)$-variable function h by the $(n+3)$-variable expression H. Then – using '$\vec{x}$' to indicate a suitable sequence of n variables – (B*) can be rendered into L_A by

C* $\exists c \exists d \{ \exists k [B(c, d, 0, k) \land G(\vec{x}, k)] \land$
$(\forall u \leq y)[u = y \lor \exists v \exists w \{ (B(c, d, u, v) \land B(c, d, Su, w)) \land H(\vec{x}, u, v, w) \}] \land$
$B(c, d, y, z) \}$.

Abbreviate this wff $\varphi(\vec{x}, y, z)$; it is then evident that φ will serve to express the p.r. defined function f. Which gives us the desired general claim (3). ⊠

So, we have shown how to establish each of the claims (1), (2) and (3) made at the start of Section 15.1. Hence every p.r. function can be expressed in L_A. Theorem 15.1 is in the bag!

15.4   The p.r. functions and relations are Σ_1-expressible

(a)   Reviewing the proof we have just given, it is easy to see that it in fact establishes something stronger:

Theorem 15.2 *Every p.r. function can be expressed by a Σ_1 wff.*

Proof   We have just given a method for building up a wff which expresses a given p.r. function (by the dodge of recapitulating the stages of its full definition as a p.r. function in terms of simpler p.r. functions, which are in turn defined in terms of yet simpler ones, etc., with the chain of definitions ultimately ending with initial functions). But now note:

1. The initial functions are expressed by Δ_0 (hence Σ_1) wffs.

2. Compositions are expressed using existential quantifiers applied to previous wffs expressing the functions we want to compose: hence if each of those earlier functions is expressed by a Σ_1 wff so too is their composition.

3. A function f defined by recursion from g and h is expressed by a wff like (C*) built up from wffs G and H expressing g and h together with the Δ_0 wff B expressing the β-function. Inspection reveals that if the wffs G and H are both Σ_1, so is the wff (C*) which expresses f.

So, as we track through the full p.r. definition of a given function f, writing down wffs to express the functions defined at each interim stage, we start from Σ_1 wffs and then reflect definitions by composition and recursion by writing down more Σ_1 wffs. So we end up with a Σ_1 wff expressing f. ⊠

(b)   We should note a very simple corollary of all this. In Section 14.6 we defined a property to be primitive recursive iff its characteristic function is primitive recursive. But we have just proved that L_A can express all primitive recursive functions, and hence all primitive recursive characteristic functions, using Σ_1 wffs. And in Section 5.4(e) we noted that if $\psi(x, y)$ expresses the function c_P, then $\psi(x, 0)$ expresses P (likewise of course for relations). Putting all that together, it follows that

Theorem 15.3 *Every p.r. property and relation can be expressed by a Σ_1 wff.*

16 Capturing functions

We have seen that the language L_A, despite its meagre built-in expressive resources, can in fact *express* every p.r. function. In the next chapter, we will show that even the simple theory Q, despite its meagre proof-resources, can in fact *capture* every p.r. function and thereby correctly evaluate it for arbitrary inputs.

This short intervening chapter does some necessary groundwork, pinning down the requisite notion of 'capturing a function'. It is unavoidably a bit tedious: bear with the fiddly details!

16.1 Capturing defined

(a)   To get straight to the business, here is our official definition:

> A one-place numerical function f is *captured* by $\varphi(\mathsf{x}, \mathsf{y})$ in theory T iff, for any m, n,
>   i. if $f(m) = n$, then $T \vdash \varphi(\overline{\mathsf{m}}, \overline{\mathsf{n}})$,
>   ii. $T \vdash \exists! \mathsf{y}\, \varphi(\overline{\mathsf{m}}, \mathsf{y})$.[1]

So this says that T captures f by the open wff φ just if, (i) when $f(m) = n$, T can prove the appropriate instance of φ, and also (ii) T 'knows' that the value for a given argument m is unique.

This definition now generalizes in the obvious way to cover many-place functions (and we won't keep repeating this observation for future definitions in this chapter).

(b)   Assume that T contains Baby Arithmetic and is consistent. Then the definition of capturing for one-place functions is easily seen to be equivalent to this, another standard formulation in the literature:

> The one-place numerical function f is captured by $\varphi(\mathsf{x}, \mathsf{y})$ in theory T iff, for any m, n, if $f(m) = n$, then $T \vdash \forall \mathsf{y}(\varphi(\overline{\mathsf{m}}, \mathsf{y}) \leftrightarrow \mathsf{y} = \overline{\mathsf{n}})$.

(c)   Let's show that our definition dovetails with the notion of capturing properties by proving

> **Theorem 16.1** *On minimal assumptions, theory T captures a property P iff T captures its characteristic function c_P.*[2]

---

[1] Recall the standard uniqueness quantifier '$\exists!\mathsf{y}$', to be read 'there is exactly one y such that …', so '$\exists!\mathsf{y}\, \varphi(\mathsf{y})$' can be treated as short for '$\exists \mathsf{y}(\varphi(\mathsf{y}) \land \forall \mathsf{v}(\varphi(\mathsf{v}) \to \mathsf{v} = \mathsf{y}))$', where '$\mathsf{v}$' is a variable new to φ.

[2] See again Section 2.2, and recall that for us the value 0 indicates truth.

Proof of the 'only if' direction   Suppose $\varphi(\mathsf{x})$ captures P, and consider the wff $\varphi'(\mathsf{x}, \mathsf{y}) =_{\text{def}} (\varphi(\mathsf{x}) \wedge \mathsf{y} = 0) \vee (\neg\varphi(\mathsf{x}) \wedge \mathsf{y} = 1)$.

If $c_P(m) = 0$, i.e. m has property P, then $T \vdash \varphi(\overline{m})$ and $T \vdash 0 = 0$, so $T \vdash \varphi'(\overline{m}, 0)$. If $c_P(m) = 1$, i.e. m lacks property P, then $T \vdash \neg\varphi(\overline{m})$ and $T \vdash 1 = 1$, so $T \vdash \varphi'(\overline{m}, 1)$. Since 0 and 1 are the only values for c_P, that establishes (i), for any m, n, if $c_P(m) = n$, $T \vdash \varphi'(\overline{m}, \overline{n})$.

The proof of $T \vdash \exists! \mathsf{y}\, \varphi'(\overline{m}, \mathsf{y})$ for arbitrary m can be left as an exercise.   ☒

Proof of the 'if' direction   Suppose $\psi(\mathsf{x}, \mathsf{y})$ captures c_P. If m has P, i.e. $c_P(m) = 0$, then $T \vdash \psi(\overline{m}, 0)$. If m doesn't have P, so $c_P(m) = 1$, then both $T \vdash \psi(\overline{m}, 1)$ and $T \vdash \exists! \mathsf{y}\, \varphi(\overline{m}, \mathsf{y})$, which entail $T \vdash \neg\psi(\overline{m}, 0)$. So $\psi(\mathsf{x}, 0)$ captures P.   ☒

Exercise: what minimal assumptions about T are made in this proof?

(d)   That is the main news about the idea of capturing a function. However, we do need to say more, for two reasons.

First, this is one of those points at which different presentations offer different definitions of different strengths, not always with much comment. So, to aid comparisons, we had better explain what is going on.

Second, and more importantly, in a pivotal proof in Section 17.3 we will need the little construction which we are about to introduce in the next section that takes us from φ to $\widetilde{\varphi}$.

16.2   'Weak' capturing

In Section 5.5(c), we noted the natural first-shot proposal that T captures the (one-place) function f just if it captures the corresponding functional relation which has the same extension. Spelling that out gives us

> A one-place numerical function f is *weakly captured* by $\varphi(\mathsf{x}, \mathsf{y})$ in theory T just if, for any m, n,
> i.   if $f(m) = n$, then $T \vdash \varphi(\overline{m}, \overline{n})$,
> ii. if $f(m) \neq n$, then $T \vdash \neg\varphi(\overline{m}, \overline{n})$.

What our preferred new definition in the previous section adds is that T 'knows' that the captured relation is indeed functional.

What is to choose between our two notions of capturing? First, note:

> **Theorem 16.2** *Suppose the relevant theory T is at least as strong as* Q: *then,*
>
> 1. *if φ captures f in T, it weakly captures f too.*
> 2. *if φ weakly captures f in T, then there is a closely related wff $\widetilde{\varphi}$ which captures f.*

Proof for (1)   Assume (i) and the original unique existence clause (ii). Suppose $f(m) \neq n$ because $f(m) = k$, where $n \neq k$. By (i), $T \vdash \varphi(\overline{m}, \overline{k})$. Hence by (ii),

$T \vdash \overline{n} \neq \overline{k} \rightarrow \neg\varphi(\overline{m}, \overline{n})$. But if $n \neq k$, then $T \vdash \overline{n} \neq \overline{k}$. Hence $T \vdash \neg\varphi(\overline{m}, \overline{n})$. Which establishes the new (ii): if $f(m) \neq n$ then $T \vdash \neg\varphi(\overline{m}, \overline{n})$. ◻

Now, isn't the converse equally straightforward? If φ weakly captures f, and $f(m) = n$, then (i) $T \vdash \varphi(\overline{m}, \overline{n})$, and (ii) for all k, $T \vdash \overline{k} \neq \overline{n} \rightarrow \neg\varphi(\overline{m}, \overline{k})$. But doesn't (ii) give us (iii) $T \vdash \forall y(y \neq \overline{n} \rightarrow \neg\varphi(\overline{m}, y))$? And (i) and (iii) imply $T \vdash \exists! y\, \varphi(\overline{m}, y)$, showing φ captures f.

This argument limps badly, however. The move from (ii) to (iii) is illegitimate. In fact, T may not even have quantifiers! But even if it does (e.g. because it contains Q), T's inference rules might allow us to prove all the particular numerical instances of a certain generalization without enabling us to prove the generalization itself. In fact, we will later find that this situation is endemic in formal theories of arithmetic (see Section 21.6 on the idea of ω-incompleteness).

So, we can't just infer capturing from weak capturing. However, we do have the next best thing:

Proof sketch for (2) Suppose that φ weakly captures the one-place function f in T. And now consider

$$\widetilde{\varphi}(x, y) =_{\mathrm{def}} \varphi(x, y) \land (\forall z \leq y)(\varphi(x, z) \rightarrow z = y).$$

Then, for a given m, $\widetilde{\varphi}(\overline{m}, y)$ is satisfied by a *unique* n, i.e. the smallest n such that $\varphi(\overline{m}, \overline{n})$ is true. And we can show that $\widetilde{\varphi}$ captures f, so long as T can prove facts about bounded quantifiers. But we know from Section 11.3 that even Q is good at proving results involving bounded quantifiers. ◻

Tedious proof details for (2) – just for completists! Still on the assumption that φ weakly captures f in T, we need to prove

i. if $f(m) = n$ then $T \vdash \widetilde{\varphi}(\overline{m}, \overline{n})$,
ii. for every m, $T \vdash \exists! y\, \widetilde{\varphi}(\overline{m}, y)$.

So suppose $f(m) = n$. Since the value of $f(m)$ is unique, that means $f(m) \neq k$ for all $k < n$. Because φ weakly captures f in T, that means (a) $T \vdash \varphi(\overline{m}, \overline{n})$, and (b) for $k < n$, $T \vdash \neg\varphi(\overline{m}, \overline{k})$. But (a) and (b) together imply (c): for $k \leq n$, $T \vdash \varphi(\overline{m}, \overline{k}) \rightarrow \overline{k} = \overline{n}$. And (c) and (O4) of Section 11.3 in turn entail (d): $T \vdash (\forall x \leq \overline{n})(\varphi(\overline{m}, x) \rightarrow x = \overline{n})$. Putting (a) and (d) together, that means $T \vdash \widetilde{\varphi}(\overline{m}, \overline{n})$, which establishes (i).

Since $T \vdash \widetilde{\varphi}(\overline{m}, \overline{n})$, to establish (ii) it is now enough to show that, for arbitrary a, $T \vdash \widetilde{\varphi}(\overline{m}, a) \rightarrow a = \overline{n}$. So, arguing in T, suppose $\widetilde{\varphi}(\overline{m}, a)$, i.e. (e) $\varphi(\overline{m}, a) \land (\forall z \leq a)(\varphi(\overline{m}, z) \rightarrow z = a)$. By (O8) of Section 11.3, $a \leq \overline{n} \lor \overline{n} \leq a$. If the first, (d) yields $\varphi(\overline{m}, a) \rightarrow a = \overline{n}$, and so $a = \overline{n}$. If the second, (e) gives $\varphi(\overline{m}, \overline{n}) \rightarrow \overline{n} = a$, so $\overline{n} = a$. So either way $a = \overline{n}$. Discharge the supposition, and we're done. ◻

Given this theorem, then, for some purposes there is little to choose between our two notions of capturing. But our preferred somewhat stronger notion is the one that we will need to work with in smoothly proving the key Theorem 17.1. So that's why we concentrate on it.

Finally, we have just proved one uniqueness result, that with $\widetilde{\varphi}$ as defined, then for any m, $T \vdash \exists!y\, \widetilde{\varphi}(\overline{m}, x)$. Let's now add a companion uniqueness result which we will need later:

Theorem 16.3 *Suppose T includes Q, and $\widetilde{\varphi}$ is as defined above. Then, for any n, $T \vdash \forall x \forall y(\widetilde{\varphi}(x, \overline{n}) \wedge \widetilde{\varphi}(x, y) \to y = \overline{n})$.*

Proof   Suppose, for arbitrary a, b, $\widetilde{\varphi}(a, \overline{n}) \wedge \widetilde{\varphi}(a, b)$.

Since T includes Q, it can prove $b \leq \overline{n} \vee \overline{n} \leq b$ (by (O8) of Section 11.3 again). So now argue by cases.

Suppose $b \leq \overline{n}$. From $\widetilde{\varphi}(a, \overline{n})$ we have, by definition, $(\forall z \leq \overline{n})(\varphi(a, z) \to z = \overline{n})$ and hence $\varphi(a, b) \to b = \overline{n}$. And from $\widetilde{\varphi}(a, b)$ we have $\varphi(a, b)$. Whence $b = \overline{n}$.

Suppose $\overline{n} \leq b$. Then $b = \overline{n}$ follows similarly.

So either way, $b = \overline{n}$. Hence $\widetilde{\varphi}(a, \overline{n}) \wedge \widetilde{\varphi}(a, b) \to b = \overline{n}$. Generalizing on the arbitrary parameters a, b, the theorem follows.   ⊠

16.3   'Strong' capturing

(a)   We are not quite done yet (though you can certainly skip this last section on a first reading, though do glance at the final footnote).

Suppose that $\varphi(x, y)$ captures the (one-place) function f in theory T. It would be convenient, if only for ease of notation, to expand T's language L to L' by adding a corresponding function symbol 'f', and to expand T to T' by adding a new definitional axiom

$$\forall x \forall y\{f(x) = y \leftrightarrow \varphi(x, y)\}.$$

However, augmenting T by what is intended to be a mere notational convenience shouldn't make any difference to which wffs of the original, unextended, language L are provable. In a probably familiar jargon, this extension of T to T' needs to be a *conservative* one. And the condition for it to be harmlessly conservative to add the function symbol 'f' with its definitional axiom is that T can prove $\forall x \exists! y \varphi(x, y)$. That's an entirely standard result about first-order theories and we won't pause to prove it here.[3]

But note that T's proving $\forall x \exists! y \varphi(x, y)$ is a stronger condition than the our original unique existence condition (ii), which just requires T to prove each instance of $\exists! y \varphi(\overline{m}, y)$ case by case. So let's now say that

A one-place numerical function f is *strongly captured* by $\varphi(x, y)$ in theory T just if,
  i. for any m, n, if $f(m) = n$, then $T \vdash \varphi(\overline{m}, \overline{n})$,
  ii. $T \vdash \forall x \exists! y\, \varphi(x, y)$.

---

[3] See e.g. Enderton (2002, Theorem 27A, p. 165) or Mendelson (1997, pp. 103–104).

Then, if f is strongly captured in T, either T already has a function symbol f for f, (so that if $f(m) = n$, then $T \vdash \mathsf{f}(\overline{m}) = \overline{n}$ and the uniqueness condition is trivially satisfied), or we can conservatively extend T so that it has one.[4]

(b)   Trivially, if φ strongly captures f, then φ captures f. And again, we have an almost-converse to that:

> **Theorem 16.4** *If φ captures f in T, then there is another closely related wff $\vec{\varphi}$ which* strongly *captures f in T.*

In fact

$$\vec{\varphi}(\mathsf{x}, \mathsf{y}) =_{\text{def}} \{\varphi(\mathsf{x}, \mathsf{y}) \land \exists! \mathsf{u} \varphi(\mathsf{x}, \mathsf{u})\} \lor \{\mathsf{y} = 0 \land \neg \exists! \mathsf{u} \varphi(\mathsf{x}, \mathsf{u})\}$$

does the trick.

Proof for enthusiasts   Since φ captures f, it is immediate that if $f(m) = n$ then $T \vdash \varphi(\overline{m}, \overline{n})$ and $T \vdash \exists! \mathsf{u} \varphi(\overline{m}, \mathsf{u})$, and hence (i) $T \vdash \vec{\varphi}(\overline{m}, \overline{n})$.

So we need to show (ii) $T \vdash \forall \mathsf{x} \exists! \mathsf{y} \vec{\varphi}(\mathsf{x}, \mathsf{y})$. Logic tells us that for arbitrary a, we have $\exists! \mathsf{u} \varphi(\mathsf{a}, \mathsf{u}) \lor \neg \exists! \mathsf{u} \varphi(\mathsf{a}, \mathsf{u})$.

Suppose (1) $\exists! \mathsf{u} \varphi(\mathsf{a}, \mathsf{u})$. So for some b, $\varphi(\mathsf{a}, \mathsf{b})$. Then, trivially, $\vec{\varphi}(\mathsf{a}, \mathsf{b})$. And if $\vec{\varphi}(\mathsf{a}, \mathsf{b}')$ then (1) rules out the second disjunct of its definition, so $\varphi(\mathsf{a}, \mathsf{b}')$, so using (1) again $\mathsf{b} = \mathsf{b}'$. Giving us $\exists! \mathsf{y} \vec{\varphi}(\mathsf{a}, \mathsf{y})$.

Suppose (2) $\neg \exists! \mathsf{y} \varphi(\mathsf{a}, \mathsf{y})$. That plus the theorem $0 = 0$ yields $\vec{\varphi}(\mathsf{a}, 0)$ and hence $\exists \mathsf{y} \vec{\varphi}(\mathsf{a}, \mathsf{y})$. Then we again need to show that $\vec{\varphi}(\mathsf{a}, \mathsf{b})$ and $\vec{\varphi}(\mathsf{a}, \mathsf{b}')$ imply $\mathsf{b} = \mathsf{b}'$. This time, (2) rules out the first disjuncts in their definitions, so we have $\mathsf{b} = 0$ and $\mathsf{b}' = 0$. So we can again conclude $\exists! \mathsf{y} \vec{\varphi}(\mathsf{a}, \mathsf{y})$.

Arguing by cases, we have $\exists! \mathsf{y} \vec{\varphi}(\mathsf{a}, \mathsf{y})$. Finally, since a was arbitrary, we can infer $\forall \mathsf{x} \exists! \mathsf{y} \vec{\varphi}(\mathsf{x}, \mathsf{y})$.   ⊠

In sum, once we are dealing with arithmetics as strong as Q, if a function is weakly capturable at all it is also capturable in both stronger senses. Which is why different treatments can offer different definitions for capturing (or 'representing') a function, depending on the technical context, and often without very much comment.

But enough already! Let's move on to putting our preferred official notion of capturing to real work.

---

[4]As we said, we are labouring over these variant definitions of capturing in part to expedite comparisons with discussions elsewhere. But terminology in the literature varies widely. For the pair of ideas 'weakly capturing' and 'capturing' we find e.g. 'weakly defines'/'strongly defines' (Smullyan, 1992, p. 99), 'defines'/'represents' (Boolos et al., 2002, p. 207), 'represents'/'functionally represents' (Cooper, 2004, pp. 56, 59). While those who highlight our idea of capturing sometimes use e.g. 'defines' for *that* notion (Lindström, 2003, p. 9), plain 'represents' seems most common (Mendelson, 1997, p. 171; Epstein and Carnielli, 2000, p. 192). Finally, when e.g. Hájek and Pudlák (1993, p. 47) or Buss (1998, p. 87) talk of a formula defining a function *they* mean what we are calling strongly capturing a function.

So again the moral is very plain: when reading other discussions, do carefully check the local definitions of the jargon!

17   Q is p.r. adequate

In Chapter 7, we defined a theory to be 'sufficiently strong' iff it captures all effectively decidable numerical properties. We later remarked that even Q turns out to be sufficiently strong. We can't show that yet, however, because we do not have to hand a *general* account of effective computability, and hence of effective decidability.

However, we *do* now know about a large class of effectively computable numerical functions, namely the primitive recursive ones; and we know about a correspondingly large class of the effectively decidable numerical properties, again the primitive recursive ones. And in this chapter, we take a big step towards showing that Q is indeed sufficiently strong, by showing that it can in fact capture all p.r. functions and properties. In a phrase, Q is *p.r. adequate*.

17.1   The idea of p.r. adequacy

(a)   As we have said before, the whole aim of formalization is to systematize and regiment what we can already do. Since we can informally calculate the value of a p.r. function for a given input in an entirely mechanical way – ultimately by just repeating lots of school-arithmetic operations – then we will surely want to aim for a formal arithmetic which is able to track these informal calculations.

So it seems that we will want a formal arithmetic T worth its keep to be able to calculate the values of p.r. functions for specific arguments. And it seems a very modest additional requirement that it can also recognize that those values are unique. Which motivates the following definition:

> A theory T is *p.r. adequate* iff, for every p.r. function f, there is a corresponding φ that captures f in T.

It immediately follows (from the definition of a p.r. property and Theorem 16.1) that if T is p.r. adequate, then it also captures every p.r. property and relation.

(b)   There is a brute-force way of constructing a p.r. adequate theory. Just build into the language a primitive function expression for each p.r. function alongside those for successor, addition, and multiplication; and then add to the theory axioms defining each primitive recursive function in terms of simpler ones.[1]

---

[1]An example of such a theory is Primitive Recursive Arithmetic, which has every p.r. function built in and also induction for quantifier-free wffs. This theory has been credited with particular philosophical significance; but discussion of this would take us too far afield. See the opening essays in Tait (2005).

However, we can box more cleverly. We already know that unaugmented L_A has – perhaps surprisingly – the resources to express every p.r. function. In this chapter we now establish the pivotal result that even Q has the resources to capture every p.r. function using such suitably expressive wffs.

17.2 Starting the proof

(a) One strategy – indeed the obvious and natural one – is to try to echo the way we proved that L_A could express every p.r. function (and do so using a Σ_1 wff). Suppose, then, that we can show:

1. Q can capture the initial functions.

2. If Q can capture the functions g and h, then it can also capture a function f defined by composition from g and h.

3. If Q can capture the functions g and h, then it can also capture a function f defined by primitive recursion from g and h.

If we can establish these claims, then – by exactly parallel reasoning to that at the beginning of Section 15.1 – we evidently have a proof that every p.r. function can be captured. Which establishes

Theorem 17.1 Q *is p.r. adequate.*

(b) The first two stages in the outlined proof strategy are pleasingly simple to execute. So let's get them out of the way.

Proof for (1) This is straightforward:

i. The successor function $Sx = y$ is captured by the open wff $\mathsf{S x = y}$.

ii. The zero function $Z(x) = 0$ is captured by $\mathsf{Z(x, y)} =_{\mathrm{def}} (\mathsf{x = x} \land \mathsf{y = 0})$.

iii. The three-place identity function $I_2^3(x, y, z) = y$, to take just one example, is captured by the wff $\mathsf{I_2^3(x, y, z, u)} =_{\mathrm{def}} (\mathsf{x = x} \land \mathsf{y = u} \land \mathsf{z = z})$. Likewise for all the other identity functions.

Note that the formal proofs of the uniqueness clauses for capturing are in these cases more or less trivial (we just invoke the standard laws for identity built into Q). ⊠

Proof for (2) Suppose g and h are one-place numerical functions, captured by the wffs $\mathsf{G(x, y)}$ and $\mathsf{H(x, y)}$ respectively. We want to prove that the composite function $h(g(\cdot))$ is captured by $\exists \mathsf{z(G(x, z)} \land \mathsf{H(z, y))}$.

(i) Suppose $h(g(m)) = n$. For some k, therefore, $g(m) = k$, and $h(k) = n$. By hypothesis then, $\mathsf{Q} \vdash \mathsf{G(\overline{m}, \overline{k})}$ and $\mathsf{Q} \vdash \mathsf{H(\overline{k}, \overline{n})}$. Simple logic then gives $\mathsf{Q} \vdash \exists \mathsf{z(G(\overline{m}, z)} \land \mathsf{H(z, \overline{n}))}$, as we want.

(ii) We now need to show $\mathsf{Q} \vdash \exists! \mathsf{y} \exists \mathsf{z(G(\overline{m}, z)} \land \mathsf{H(z, y))}$. Given what we have just proved it is enough to show that Q proves $\exists \mathsf{z(G(\overline{m}, z)} \land \mathsf{H(z, a))} \to \mathsf{a = \overline{n}}$

for arbitrary a. And to show *that*, it is enough to show, arguing in Q, that the supposition (α) $G(\overline{m}, b) \land H(b, a)$ implies (β) $a = \overline{n}$ for arbitrary b.

Note that the first half of (α), the already derived $G(\overline{m}, \overline{k})$, and the uniqueness condition on G together yield $b = \overline{k}$. So using the second half of (α) we can infer $H(\overline{k}, a)$. And this together with the already derived $H(\overline{k}, \overline{n})$ and the uniqueness condition on H yields (β) $a = \overline{n}$, as required.

More complex cases where g and/or h are multi-place functions can be handled similarly.   ⊠

(c)   So far, so good. Now for step (3). In Section 15.3, we used the β-function trick to build a wff which expresses f out of wffs which express g and h (where f is defined by primitive recursion from g and h). We will show that the same construction – with a minor tweak – gives us a wff that captures f, built out of wffs which capture g and h.

This is inevitably going to be just a bit messy, however, given that the construction in question is a little involved. So you have two options. Take the claim that the tweaked construction works on trust, and skip to the last section below (perhaps the sensible option!). Or tackle the details in the next section. The choice is yours.

17.3   Completing the proof

(a)   We are going to be making use again of Gödel's β-function. By way of reminder,

$$\beta(c, d, i) =_{\mathrm{def}} \text{the remainder left when } c \text{ is divided by } d(i+1)+1.$$

As we noted in Section 15.2, this relation can be expressed in L_A by the Δ_0 wff

$$B(c, d, i, y) =_{\mathrm{def}} (\exists u \le c)[c = \{S(d \times Si) \times u\} + y \land y \le (d \times Si)].$$

We first note that B *weakly* captures β in Q. For Theorem 11.3 tells us that Q can correctly decide all Δ_0 sentences. So, as we want (i) if $\beta(c, d, i) = n$, then $Q \vdash B(\overline{c}, \overline{d}, \overline{i}, \overline{n})$. Likewise (ii) if $\beta(c, d, i) \ne n$, then $B(\overline{c}, \overline{d}, \overline{i}, \overline{n})$ is a Δ_0 falsehood, hence $Q \vdash \neg B(\overline{c}, \overline{d}, \overline{i}, \overline{n})$.[2]

In the argument that follows, however, we are going to need to use a wff that *captures* β (not just weakly captures). But that is easily found. We just appeal to the construction in Theorem 16.2. We put

---

[2] You won't need reminding by now. But overlined letters like '$\overline{c}$' stand in for numerals denoting particular numbers; plain letters like 'c' are variables (perhaps used as parameters). Keep this clearly in mind and you won't get lost in passages that use e.g. both '$B(c, d, i, y)$' and '$B(\overline{c}, \overline{d}, \overline{i}, \overline{n})$'. We could have avoided the small possibility of confusion here by choosing another notation, e.g. by keeping firmly to late in the alphabet for our choice of variables – but then anonymous 'x's and 'y's can make it difficult to keep tabs on which variable serves which purpose in a formula. It's swings and roundabouts: every notational choice has its costs and benefits.

$$\widetilde{\mathsf{B}}(\mathsf{c},\mathsf{d},\mathsf{i},\mathsf{y}) =_{\mathrm{def}} \mathsf{B}(\mathsf{c},\mathsf{d},\mathsf{i},\mathsf{y}) \wedge (\forall \mathsf{z} \leq \mathsf{y})(\mathsf{B}(\mathsf{c},\mathsf{d},\mathsf{i},\mathsf{z}) \to \mathsf{z} = \mathsf{y})$$

and we are done. And note that since B is Δ_0, $\widetilde{\mathsf{B}}$ is also Δ_0 (pause to check that claim!).

(b)   To avoid distracting clutter, let's start by considering the very simplest non-trivial case, where the one-place function f is defined by primitive recursion using another one-place-function h as follows:

$$f(0) = g$$
$$f(Sx) = h(f(x)).$$

Suppose that the function h is captured in Q by $\mathsf{H}(\mathsf{x},\mathsf{y})$. We are going to prove that the wff constructed out of H according to the recipe of Section 15.3 for *expressing* f, tweaked by using $\widetilde{\mathsf{B}}$ rather than plain B, in fact *captures* the function too.

Recall the construction. We start by noting that $f(m) = n$ if and only if

A. there is a sequence of numbers $k_0, k_1, \ldots, k_m$ such that: $k_0 = g$, and if $u < m$ then $k_{Su} = h(k_u)$, and $k_m = n$.

Using the β-function trick, that is equivalent to

B. There is some pair c, d such that: $\beta(c, d, 0) = g$, and if $u < m$ then $\beta(c, d, Su) = h(\beta(c, d, u))$, and $\beta(c, d, m) = n$.

This can now be rendered into L_A by

C. $\exists \mathsf{c} \exists \mathsf{d} \{ \widetilde{\mathsf{B}}(\mathsf{c},\mathsf{d},0,\overline{\mathsf{g}}) \wedge$
$\quad (\forall \mathsf{u} \leq \overline{\mathsf{m}})[\mathsf{u} = \overline{\mathsf{m}} \vee \exists \mathsf{v} \exists \mathsf{w} \{ (\widetilde{\mathsf{B}}(\mathsf{c},\mathsf{d},\mathsf{u},\mathsf{v}) \wedge \widetilde{\mathsf{B}}(\mathsf{c},\mathsf{d},S\mathsf{u},\mathsf{w})) \wedge \mathsf{H}(\mathsf{v},\mathsf{w}) \}] \wedge$
$\quad \widetilde{\mathsf{B}}(\mathsf{c},\mathsf{d},\overline{\mathsf{m}},\overline{\mathsf{n}}) \}.$

Abbreviate this by $\mathsf{F}(\overline{\mathsf{m}},\overline{\mathsf{n}})$ – and note that *this* is a Σ_1 wff (check that claim too).

(c)   To show that F captures f, we need to show

(i) if $f(m) = n$, then $\mathsf{Q} \vdash \mathsf{F}(\overline{\mathsf{m}},\overline{\mathsf{n}})$,
(ii) for every m, $\mathsf{Q} \vdash \exists! \mathsf{y}\, \mathsf{F}(\overline{\mathsf{m}},\mathsf{y})$.

The first of these is easy. If $f(m) = n$, then $\mathsf{F}(\overline{\mathsf{m}},\overline{\mathsf{n}})$ by construction is a true Σ_1 wff. But Q is Σ_1 complete (by Theorem 11.5). So as required $\mathsf{Q} \vdash \mathsf{F}(\overline{\mathsf{m}},\overline{\mathsf{n}})$.

So we turn to (ii). The existence part immediately follows from what we've just shown, so we just need to prove uniqueness, i.e. that $\mathsf{Q} \vdash \forall \mathsf{y}(\mathsf{F}(\overline{\mathsf{m}},\mathsf{y}) \to \mathsf{y} = \overline{\mathsf{n}})$.

Proof   Arguing inside Q we aim to show $\mathsf{F}(\overline{\mathsf{m}},\mathsf{e}) \to \mathsf{e} = \overline{\mathsf{n}}$, for arbitrary parameter e. So let's assume for arbitrary c, d that we have

D. $\widetilde{\mathsf{B}}(\mathsf{c},\mathsf{d},0,\overline{\mathsf{g}}) \wedge$
$\quad (\forall \mathsf{u} \leq \overline{\mathsf{m}})[\mathsf{u} = \overline{\mathsf{m}} \vee \exists \mathsf{v} \exists \mathsf{w} \{ (\widetilde{\mathsf{B}}(\mathsf{c},\mathsf{d},\mathsf{u},\mathsf{v}) \wedge \widetilde{\mathsf{B}}(\mathsf{c},\mathsf{d},S\mathsf{u},\mathsf{w})) \wedge \mathsf{H}(\mathsf{v},\mathsf{w}) \}] \wedge$
$\quad \widetilde{\mathsf{B}}(\mathsf{c},\mathsf{d},\overline{\mathsf{m}},\mathsf{e}).$

127

If $m = 0$ and so $n = g$, then our assumption gives us $\widetilde{B}(c, d, 0, \overline{n}) \wedge \widetilde{B}(c, d, 0, e)$. Theorem 16.3 tells us that $e = \overline{n}$.

So now assume $m > 0$. The middle term of (D) gives us (α):

$$\exists v \exists w \{(\widetilde{B}(c, d, 0, v) \wedge \widetilde{B}(c, d, 1, w)) \wedge H(v, w)\}.$$

Temporarily suppose we have (β) $(\widetilde{B}(c, d, 0, a) \wedge \widetilde{B}(c, d, 1, b)) \wedge H(a, b)$. But by Theorem 16.3, given $\widetilde{B}(c, d, 0, \overline{g})$ and $\widetilde{B}(c, d, 0, a)$ it follows that $a = \overline{g}$, so $H(\overline{g}, b)$. Since $h(g) = k_1$, and H captures h, Q proves $H(g, \overline{k_1})$. And by the uniqueness of capturing, we now have $b = \overline{k_1}$. Hence, still on supposition (β), we can infer $\widetilde{B}(c, d, 1, \overline{k_1})$. So, given ($\alpha$) and that proof from the supposition (β), we can discharge the supposition (β) by a double appeal to $\exists$-elimination to infer

$$\widetilde{B}(c, d, 1, \overline{k_1}).$$

Now off we go again. This time note that the middle term in assumption (D) yields (α'):

$$\exists v \exists w \{(\widetilde{B}(c, d, 1, v) \wedge \widetilde{B}(c, d, 2, w)) \wedge H(v, w)\}.$$

So this time we suppose (β') $(\widetilde{B}(c, d, 1, a) \wedge \widetilde{B}(c, d, 2, b)) \wedge H(a, b)$. And by exactly similar reasoning we can eventually discharge the supposition to infer by $\exists$-elimination

$$\widetilde{B}(c, d, 2, \overline{k_2}).$$

Keep going in the same way until we establish

$$\widetilde{B}(c, d, \overline{m}, \overline{k_m}), \text{ i.e. } \widetilde{B}(c, d, \overline{m}, \overline{n}).$$

Now make one last appeal to Theorem 16.3. Given $\widetilde{B}(c, d, \overline{m}, \overline{n})$ and the last conjunct of (D), i.e. $\widetilde{B}(c, d, \overline{m}, e)$, it follows that $e = \overline{n}$.

So: for any m, the supposition (D) implies $e = \overline{n}$. Therefore, since c and d were arbitrary, the existential quantification of (D), i.e. the original $F(\overline{m}, e)$ implies $e = \overline{n}$. Which is what needed to be proved. $\boxtimes$

(d)   We have shown, then, that in the simple case where the one-place function f is defined from the one-place function h by primitive recursion, then if Q captures h (by some $H(x, y)$) then it can capture f (by that complex wff $F(x, y)$).

However, to establish (3), the last stage of our three-step adequacy proof, we do need to generalize the argument to cases where a many-place function f is defined by primitive recursion from the many-place functions g and h. But now you know how the proof works in the simple case, it is easy to see that it is just a routine matter to generalize. It can therefore be left as a tiresome book-keeping exercise to fill in the details.

So that establishes Theorem 17.1: Q is indeed p.r. adequate. Phew.[3]

---

[3]Can the fiddly detail in this section be avoided? For an indication of another way of showing that Q is 'recursively adequate' and hence p.r. adequate, see Chapter 39, fn. 2. And for a third, more devious, route to the same conclusion, see Boolos et al. (2002, p. 206) – which was indeed the route taken in the first edition of this book.

17.4 All p.r. functions can be captured in Q by Σ_1 wffs

(a)   In Section 15.4 we noted that the proof we had given that L_A can express all p.r. functions delivers more – it in fact also proves that Σ_1 wffs are enough to do the job.

In an exactly parallel way, we now remark that our proof that Q can capture all p.r. functions again delivers more:

Theorem 17.2 Q *can capture any p.r. function using a* Σ_1 *wff.*

That's because (1) Q captures the initial functions by Σ_1 wffs, (2) as we define new functions from old by composition, Q capture these too by Σ_1 wffs, and similarly (3) if Q captures the functions g and h using Σ_1 wffs, then it can also capture a function f defined from them by primitive recursion using a Σ_1 wff basically along the lines of the wff (C) in the previous section.

(b)   We finally note another feature of our adequacy proof. We have not just established the bald existence claim that, for any p.r. function f, there is a Σ_1 wff that captures it in Q: we have actually shown how to construct a particular wff that does the capturing job when we are presented with a full p.r. definition for f. In a word, we have proved our adequacy theorem *constructively*. The idea is essentially to recapitulate f's p.r. definition by building up a wff which reflects the stages in that definition (with the help of the β-function trick).

Let's highlight the point in a definition:

> A wff that captures a p.r. function f by being constructed so as to systematically reflect a full p.r. definition of f (as in our proof over the last two sections) will be said to *canonically capture* the function.

Three remarks. (i) A wff that canonically captures f will also express it (since we are just constructing again the wff which we used before, when we showed how to express any p.r. function, with the little tweak of using $\widetilde{\mathsf{B}}$ rather than B, which makes no expressive difference). (ii) There are actually going to be many ways of canonically capturing a given p.r. function f in Q. Boringly, we can re-label variables, and shift around the order of conjuncts; more importantly, there will always be alternative ways of giving p.r. definitions for f (if only by including redundant detours). (iii) There will of course be innumerable non-canonical ways of capturing any p.r. function f: just recall the remark in Section 5.5(b) which points out in effect that if $\varphi(\vec{\mathsf{x}}, \mathsf{y})$ captures f in some theory, so does $\varphi(\vec{\mathsf{x}}, \mathsf{y}) \wedge \theta$, for *any* theorem θ as redundant baggage.

18   Interlude: A very little about *Principia*

In the last Interlude, we gave a five-stage map of our route to Gödel's First Incompleteness Theorem. The first two stages we mentioned are now behind us. They involved (1) introducing the standard theories Q and PA, then (2) defining the p.r. functions and – the hard bit! – proving Q's p.r. adequacy. In order to do the hard bit, we have already used one elegant idea from Gödel's epoch-making 1931 paper, namely the β-function trick. But most of his proof is still ahead of us: at the end of this Interlude, we will review the stages that remain.

But first, let's relax for a moment after all our labours, and pause to take a very short look at some of the scene-setting background. We will say more about the historical context in a later Interlude (Chapter 37). But for now, we'll say enough to explain the *title* of Gödel's great paper: 'On formally undecidable propositions of *Principia Mathematica* and related systems I'.[1]

18.1   *Principia*'s logicism

Frege aimed in his *Grundgesetze der Arithmetik* to reconstruct arithmetic (and some analysis too) on a secure footing by deducing it from logic plus definitions. But as we noted in Section 13.4, Frege's overall logicist project – in its original form – founders on his disastrous fifth Basic Law. And the fatal contradiction that Russell exposed in Frege's system was not the only paradox to bedevil early treatments of the theory of classes.

Various responses to these contradictions were proposed at the beginning of the twentieth century. One was to try to avoid paradox and salvage logicism by, in effect, keeping much of Frege's logic while avoiding the special assumption about classes that gets him into disastrous trouble.

To explain: Frege's general logical system involves a kind of *type hierarchy*. It very carefully distinguishes 'objects' (things, in a broad sense) from properties from properties-of-properties from properties-of-properties-of-properties, etc, and insists that every item belongs to a determinate level of the hierarchy. Then the claim is – plausibly enough – that it only makes sense to attribute properties which belong at level l ($l > 0$) to items at level $l - 1$. For example, the property of *being wise* is a level-1 property, while Socrates is an item at level 0; and it makes sense to attribute the level-1 property to Socrates, i.e. to claim that Socrates is wise. Likewise, the property of *being instantiated by some people*

---

[1]That's a roman numeral one at the end of the title! Gödel originally planned a Part II, fearing that readers would not, in particular, accept the very briskly sketched Second Theorem without further elaboration. But Gödel's worries proved groundless and Part II never appeared.

is a level-2 property, and it makes sense to attribute that property to the level-1 property of being wise, i.e. to claim that the property of being wise is instantiated by some people. But you get nonsense if, for example, you try to attribute that level-2 property to Socrates and claim that Socrates is instantiated.

Note that this strict stratification of items into types blocks the derivation of the property analogue of Russell's paradox about classes. The original paradox concerned the class of all classes that are not members of themselves. So now consider the putative property of *being a property that doesn't apply to itself*. Does this apply to itself? It might seem that the answer is that it does if it doesn't, and it doesn't if it does – contradiction! But on Frege's hierarchical theory of properties, there is no real contradiction to be found here: (i) Every genuine property belongs to some particular level of the hierarchy, and only applies to items at the next level down. A level-l property therefore can't sensibly be attributed to any level-l property, including itself. (ii) However, there is no generic property of 'being a property that doesn't apply to itself' shared by every property at any level. No genuine property can be type-promiscuous in that way.

One way to avoid class-theoretic paradox is to stratify the universe of classes into a type-hierarchy in the way that Frege stratifies the universe of properties. So suppose we now distinguish classes from classes-of-classes from classes-of-classes-of-classes, and so forth; and on one version of this approach we then insist that classes at level l $(l > 0)$ can only have as members items at level $l - 1$.[2] Frege himself doesn't take this line: his disastrous Basic Law V in effect flattens the hierarchy for classes and puts them all on the same level. However, Bertrand Russell and Alfred North Whitehead do adopt the hierarchical view of classes in their monumental *Principia Mathematica* (1910–13). They take over and develop ('ramify') Frege's stratification of properties and then link this to the stratification of classes in a very direct way, by treating talk about classes as in effect just lightly disguised talk about their corresponding defining properties. The resulting system is – as far as we know – consistent.

Having proposed a paradox-blocking theory of types as their logical framework, Russell and Whitehead set out in *Principia* – like Frege in his *Grundgesetze*, and following a similar strategy – to derive all of arithmetic from definitional axioms (compare Section 1.1). Indeed, the project is even more ambitious: the ultimate aim, as Russell described it earlier, is to prove that

> all pure mathematics deals exclusively with concepts definable in terms of a very small number of fundamental logical concepts, and ... all its propositions are deducible from a very small number of fundamental logical principles. (Russell, 1903, p. xv)

Yes, 'all'! Let's concentrate, however, on the more limited but still ambitious project of deriving just arithmetic from logic plus definitions.

---

[2]Exercise: why does this block Russell's paradox from arising? An alternative approach – the now dominant Zermelo-Fraenkel set theory – is more liberal: it allows sets formed at level l to contain members from *any* lower level. In the jargon, we get a *cumulative* hierarchy.

But how could anyone think that project is even possible? Well, consider the following broadly Fregean chain of ideas (we will have to ride roughshod over subtleties; but it would be a pity to say nothing, as the ideas are so pretty!).

i. We'll say that the Fs and Gs are *equinumerous* just in case we can match up the Fs and the Gs one-to-one (i.e., in the jargon of Section 2.1, if there is a bijection between the Fs and the Gs). Thus: the knives and the forks are equinumerous if we can pair them up, one to one, with none left over.

   Now, this idea of there being a one-one correspondence between the Fs and the Gs can be defined using quantifiers and identity (so you can see why it might be thought to count as a purely logical notion). Here's one definition, in words: there's a one-one correspondence if there is a relation R such that every F has relation R to a unique G, and for every G there is a unique F which has relation R to it (the function f such that $f(x) = y$ iff Rxy will then be a bijection). In symbols, there's a one-one correspondence between the Fs and the Gs when

$$\exists R\{\forall x(Fx \to \exists!y(Gy \wedge Rxy)) \wedge \forall y(Gy \to \exists!x(Fx \wedge Rxy))\}$$

   Here, '$\exists!$' is the uniqueness quantifier again; and the initial quantifier is a second-order quantifier ranging over two-place relations.

ii. Intuitively, the number of Fs is identical with the number of Gs just in case the Fs and Gs are equinumerous in the logical sense just defined. This claim is nowadays – with only tenuous justification – called *Hume's Principle*. Any account of what numbers are should surely respect it.

iii. Here's another, equally intuitive, claim – call it the *Successor Principle*: the number of Fs is the successor of the number of Gs if there is an object o which is an F, and the remaining things which are F-but-not-identical-with-o are equinumerous with the Gs.

iv. What though *are* numbers? Here's a brute-force way of identifying them while respecting Hume's Principle. Take *the number of Fs* to be *the class of all classes C such that the Cs are equinumerous with the Fs*. Then, as we want, the number of Fs is identical with the number of Gs just if the class of all classes with as many members as there are Fs is identical with the class of all classes with as many members as there are Gs, which holds just if the Fs and Gs are equinumerous.

v. Taking this brute-force line on identifying numbers, we can define 0 to be the class of all classes equinumerous with the non-self-identical things. For assuredly, zero *is* the number of x such that $x \neq x$. And, on the most modest of assumptions, zero will then exist – it is the class of all empty classes (but there is only one empty class since classes with the same members are identical). Pressing on, we next put 1 to be the number of

x such that $x = 0$. It is then easy to see that, by our definitions, 1 is the class of all singletons and is the successor of 0 (and also $0 \neq 1$). Likewise, we identify 2 as the number of x such that $x = 0 \vee x = 1$: our definitions make two the class of all pairs and the successor of 1. Predictably, we now put 3 as the number of x such that $x = 0 \vee x = 1 \vee x = 2$. And so it goes.

vi. Finally, we need an account of what the finite *natural* numbers are (for note that our basic definition of *the number of Fs* applies equally when there is an infinite number of Fs). Well, let's say that a property F is *hereditary* if, whenever a number has it, so does its successor. Then *a number is a natural number if it has all the hereditary properties that zero has*. Which in effect defines the natural numbers as those for which the now familiar induction principle holds.

We have a story, then, about what numbers themselves are. We have a story about zero, one, two, three and so on. We have a story about what it is for one number to be the successor of another. We have a story about which numbers are natural numbers, and why induction holds. So suppose that you buy the (rather large!) assumption that the talk about 'classes' in the story so far still counts as *logical* talk, broadly construed. Then, quite wonderfully, we are launched on our way towards (re)constructing arithmetic in logical terms.

Note that, given Frege's cunning construction, we won't run out of numbers. The number n is the class of all n-membered classes. And there *must* be n-membered classes and so the number n must exist, because the class of *preceding* numbers $\{0, 1, 2, \ldots, n-1\}$ has exactly n members. So given that the numbers less than n exist, so does n. Since zero exists, then, the rest of the natural numbers must exist. Which is ingenious, though – you might well think – possibly a bit *too* ingenious. It smacks suspiciously of conjuring the numbers into existence, one after the other.[3]

In Frege's hands, however, this construction all proceeds inside his flawed logical system. What if we instead try to develop these ideas within *Principia*'s logic? Well now note that, intuitively, there can be n-membered classes at different levels, and so n (thought of as the class of all n-membered classes) would have members at different levels, which offends against *Principia*'s paradox-blocking stratification of classes into different levels. And Frege's explanation of why we don't run out of numbers also offends.[4] So Russell and Whitehead have to complicate their version of the logicist story. And to ensure that the numbers don't give out they make the bald assumption that there are indeed an unlimited number of things available to be collected into classes of every different finite size. But then this 'Axiom of Infinity' hardly looks like a *logically* given truth.

---

[3]But for a vigorous neo-Fregean defence, see Wright (1983), Hale and Wright (2001).

[4]If zero, the class containing just the empty class, is two levels up, and the class containing just zero is at level three, then one, conceived of as the class of all classes like $\{0\}$ would be at level four, and there then couldn't be a class containing both zero and one.

However, let's ignore philosophical worries about whether *Principia*'s system counts as pure logic. For now enter Gödel, with a devastating *formal* objection to the idea that we can continue the story in any way that would reveal *all* arithmetical truths to be derivable from *Principia*'s assumptions (whether those assumptions count as pure logic or not).

18.2   Gödel's impact

What the First Incompleteness Theorem shows is that, despite its great power, Russell and Whitehead's system still can't capture even all truths of basic arithmetic, at least assuming it *is* consistent. As Gödel puts it:

> The development of mathematics toward greater precision has led, as is well known, to the formalization of large tracts of it, so that one can prove any theorem using nothing but a few mechanical rules. The most comprehensive formal systems that have been set up hitherto are the system of *Principia Mathematica* ... on the one hand and the Zermelo-Fraenkel axiom system for set theory ... on the other. These two systems are so comprehensive that in them all methods of proof today used in mathematics are formalized, that is, reduced to a few axioms and rules of inference. One might therefore conjecture that these axioms and rules of inference are sufficient to decide *any* mathematical question that can at all be formally expressed in these systems. It will be shown below that this is not the case, that on the contrary there are in the two systems mentioned relatively simple problems in the theory of integers which cannot be decided on the basis of the axioms. This situation is not in any way due to the special nature of the systems that have been set up, but holds for a very wide class of formal systems; .... (Gödel, 1931, p. 145)

Now, to repeat, Russell and Whitehead's system is built on a logic that allows quantification over properties, properties-of-properties, properties-of-properties-of-properties, and so on up the hierarchy. Hence the language of *Principia* is *immensely* richer than the language L_A of first-order PA (where we can only quantify over individuals). It perhaps wouldn't be a surprise, then, to learn that Russell and Whitehead's axioms don't settle every question that can be posed in their immodest formal language. What *is* a great surprise is that there are 'relatively simple' propositions which are 'formally undecidable' in *Principia* – by which Gödel means that there are wffs φ in effect belonging to L_A, the language of basic arithmetic, such that we can't prove either φ or $\neg\varphi$ from the axioms. Even if we buy all the assumptions of *Principia* (including the Axiom of Infinity), we *still* don't get the very minimum the logicist hoped to get, i.e. a complete theory of basic arithmetic.

18.3   Another road-map

As Gödel himself notes, however, his incompleteness proof only needs to invoke some fairly elementary features of the full-blooded theories of *Principia* and of Zermelo-Fraenkel set theory ZF, and these features are equally shared by PA. So we can now largely forget about *Principia* and pretend henceforth that Gödel was really talking about PA all along.

In what follows, there will be other deviations from the details of his original proof; but the basic lines of argument in the next four chapters are all in his great paper. Not surprisingly, other ways of establishing his results (and generalizations and extensions of them) have been discovered since 1931, and we will be mentioning some of these later. But there still remains much to be said for introducing the incompleteness theorems by something close to Gödel's own arguments.

Here, then, is an abbreviated guide to the three stages in our Gödelian proof which remain ahead of us:

1. First, we look at Gödel's great innovation – the idea of systematically associating expressions of a formal arithmetic with numerical codes. We'll stick closely to Gödel's original type of numbering scheme. With a coding scheme in place, we can reflect key properties and relations of expressions of PA (to concentrate on that theory) by properties and relations of their Gödel numbers. For a pivotal example, we can define the numerical relation $Prf(m, n)$ which holds when m codes for a sequence of wffs that is a PA proof, and n codes the closed wff that is thereby proved. Gödel proves that such arithmetical properties and relations are primitive recursive. (Chapters 19, 20)

2. Since $Prf(m, n)$ is p.r., it can be expressed – indeed, can be captured – in PA. We can now use this fact to construct a sentence G that, given the coding scheme, is true if and only if it is unprovable in PA. We can then show that G is unprovable, assuming no more than that PA is consistent. So we've found an arithmetical sentence which is true but unprovable in PA. (And given a slightly stronger assumption than PA's consistency, ¬G must also be unprovable in PA.) Moreover, it turns out that this unprovable sentence is in one respect a pretty simple one: it is in fact trivially equivalent to a Π_1 wff. (Chapter 21)

3. As Gödel notes, the true-but-unprovable sentence G for PA is generated by a method that can be applied to any other arithmetic that satisfies some modest conditions. Which means that PA is not only incomplete but incompletable. Indeed, *any* properly axiomatized theory containing Q and satisfying certain consistency conditions is incompletable. (Chapter 22)

So, map in hand, on we go ...!

19 The arithmetization of syntax

In the main part of this chapter, we introduce Gödel's simple but wonderfully powerful idea of associating the expressions of a formal theory with code numbers. In particular, we will fix on a scheme for assigning code numbers first to expressions of L_A and then to proof-like sequences of such expressions. This coding scheme will correlate various *syntactic* properties with purely *numerical* properties – in a phrase, the scheme *arithmetizes syntax*.

For example, take the syntactic property of being a term of L_A. We can define a corresponding numerical property *Term*, where *Term*(n) holds just when n codes for a term. Likewise, we can define *Atom*(n), *Wff*(n), and *Sent*(n) which hold just when n codes for an atomic wff, a wff, or a closed wff (sentence) respectively. It will be easy to see – at least informally – that these numerical properties are primitive recursive ones.

More excitingly, we can define the numerical relation *Prf*(m, n) which holds just when m is the code number in our scheme of a PA-derivation of the sentence with number n. It will also be easy to see – still in an informal way – that this relation too is primitive recursive.

The short second part of the chapter introduces the idea of the *diagonalization* of a wff. This is basically the idea of taking a wff $\varphi(\mathsf{y})$, and substituting (the numeral for) its own code number in place of the free variable. We can think of a code number as a way of referring to a wff. So in Chapter 21 we are going to use a special case of 'diagonalization' to form a wff that – putting it roughly – refers to itself via the Gödel coding and says 'I am unprovable in PA'. Corresponding to the operation of diagonalization there will be a function *diag* which maps the code number of a wff to the code number of its diagonalization. This function too is primitive recursive.

Along the way, we introduce some cute standard notation for Gödel numbers.

19.1 Gödel numbering

(a)   Here is one classic, Gödelian, way of numbering expressions of L_A.

Suppose that our version of L_A has the usual logical symbols (connectives, quantifiers, identity, brackets), and symbols for zero and for the successor, addition and multiplication functions. We will associate all those with odd numbers (different symbol, different number, of course).

L_A will also have an inexhaustible but countable supply of variables. Order those in some way as the variables $v_0, v_1, v_2, \ldots$ and associate these with even numbers, so v_k is coded by the number $2(k + 1)$.

So, to fix details, here is our preliminary series of *basic codes*:

$$\neg \quad \wedge \quad \vee \quad \rightarrow \quad \leftrightarrow \quad \forall \quad \exists \quad = \quad (\quad) \quad 0 \quad S \quad + \quad \times \quad x \quad y \quad z \quad \ldots$$

$$1 \quad 3 \quad 5 \quad 7 \quad 9 \quad 11 \quad 13 \quad 15 \quad 17 \quad 19 \quad 21 \quad 23 \quad 25 \quad 27 \quad 2 \quad 4 \quad 6 \quad \ldots$$

Our Gödelian numbering scheme for expressions is now defined in terms of this table of basic codes as follows:

> Let expression e be the sequence of $k+1$ symbols and/or variables $s_0, s_1, s_2, \ldots, s_k$. Then e's *Gödel number* (g.n.) is calculated by taking the basic code number c_i for each s_i in turn, using c_i as an exponent for the $i+1$-th prime number π_i, and then multiplying the results, to get
>
> $$\pi_0^{c_0} \cdot \pi_1^{c_1} \cdot \pi_2^{c_2} \cdot \ldots \cdot \pi_k^{c_k}.$$

For example:

i. The single symbol 'S' has the g.n. 2^{23} (the first prime raised to the appropriate power as read off from our correlation table of basic codes).

ii. The standard numeral SS0 has the g.n. $2^{23} \cdot 3^{23} \cdot 5^{21}$ (the product of the first three primes raised to the appropriate powers).

iii. The wff

$$\exists y \, (S0 + y) = SS0$$

has the g.n.

$$2^{13} \cdot 3^4 \cdot 5^{17} \cdot 7^{23} \cdot 11^{21} \cdot 13^{25} \cdot 17^4 \cdot 19^{19} \cdot 23^{15} \cdot 29^{23} \cdot 31^{23} \cdot 37^{21}.$$

(b) That last number is, of course, *enormous*. So when we say that it is elementary to decode the resulting g.n. by taking the exponents of prime factors, we don't mean that the computation is quick and easy. We mean that the computational routine required for decoding – namely, repeatedly extracting prime factors – involves no more than the mechanical operations of school-room arithmetic.

In fact, we have already met the main engine for decoding the Gödel number for an expression. Recall, the p.r. function $exf(n, i)$ returns the exponent of the $i+1$-th prime in the factorization of n. So given a number n which numbers the sequence of symbols and/or variables $s_0, s_1, s_2, \ldots, s_k$, $exf(n, i)$ returns the basic code for the symbol/variable s_i. Look up the table of basic codes and you will have recovered s_i.

(c) Three quick remarks. First, we earlier allowed the introduction of abbreviatory symbols into L_A (for example, '$\leq$' and '3'); take the g.n. of an expression including such symbols to be the g.n. of the unabbreviated version.

Second, we will later be assuming there are similar numbering schemes for the expressions of other theories which use possibly different languages L. We can imagine each of these numbering schemes to be built up in the same way, but starting from a different table of preliminary basic codes to cope with the different basic symbols of L. We won't need to spell out the details.

Third, there are as you might expect other kinds of coding schemes which would serve our purposes just as well.[1] Some will in fact produce much smaller code numbers; but *that* is neither here nor there, for our concern is going to be with the principle of the thing, not with practicalities.

19.2 Acceptable coding schemes

Now for a fourth, much more important, remark. We are going to be introducing numerical properties like *Term* and proving them to be primitive recursive. But note, $Term(n)$ is to hold when n is the code number of a term *according to our Gödel-numbering scheme*. However, our numbering scheme was fairly arbitrarily chosen. We could, for example, shuffle around the preliminary assignment of basic codes to get a different numbering scheme; or (more radically) we could use a scheme that isn't based on powers of primes. So could it be that a property like *Term* is p.r. when defined in terms of our numbering scheme and not p.r. when defined in terms of some alternative but equally sensible scheme?

Well, what counts as 'sensible' here? The key feature of our Gödelian scheme is this: there is a pair of *algorithms*, one of which takes us from an L_A expression to its code number, the other of which takes us back again from the code number to the original expression – and moreover, in following through these algorithms, the length of the computation is a simple function of the length of the L_A expression to be encoded or the size of the number to be decoded. The algorithms don't involve open-ended computations using unbounded searches: in other words, the computations can be done just using 'for' loops.

So let S be any other comparable coding scheme, which similarly involves a pair of algorithmic methods for moving to and fro between L_A expressions and numerical codes (where the methods don't involve open-ended searches). And suppose S assigns code n_1 to a certain L_A expression. Consider the process of first decoding n_1 to find the original L_A expression and then re-encoding the expression using *our* Gödelian scheme to get the code number n_2.[2] By hypothesis, this process will combine two simple computations which just use 'for' loops. Hence, there will be a *primitive recursive* function which maps n_1 to n_2. Similarly, there will be another p.r. function which maps n_2 back to n_1.

Let's say, then, that a coding scheme S is *acceptable* iff there is a p.r. function tr which 'translates' code numbers according to S into code numbers under our

---

[1] See, for example, Smullyan (1992).

[2] Strictly, we need to build in a way of handling the 'waste' cases where n_1 isn't an S-code for any wff.

official Gödelian scheme, and another p.r. function tr^{-1} which converts code numbers under our scheme back into code numbers under scheme S. Then we've just argued that being acceptable in this sense is at least a *necessary* condition for being an intuitively 'sensible' numbering scheme.[3]

It is immediate that a property like *Term* defined using our scheme is p.r. if and only if the corresponding property $Term_S$ defined using scheme S is p.r., for any acceptable scheme S. Why? Well, let the characteristic functions of *Term* and $Term_S$ be $term$ and $term_S$ respectively. Then $term_S(n) = term(tr(n))$, hence $term_S$ will be p.r. by composition so long as $term$ is p.r.; and similarly $term(n) = term_S(tr^{-1}(n))$, hence $term$ is p.r. if $term_S$ is. So, in sum, *Term* is p.r. iff $Term_S$ is p.r.; the property's status as p.r. is *not* dependent on any particular choice of coding scheme (so long as it is acceptable).

19.3   Coding sequences

As we've already announced, the relation $Prf(m, n)$ will be crucial to what follows, where this is the relation which holds just when m codes for an array of wffs that is a PA proof, and n codes for the closed wff (sentence) that is thereby proved. But how *do* we code for proof-arrays?

The details will obviously depend on the kind of proof system we adopt for our preferred version of PA. But we said back in Sections 10.4 and 13.2 that our official line will be that PA has a Hilbert-style axiomatic system of logic. And in this rather old-fashioned framework, proof-arrays are simply *linear sequences* of wffs. A standard way of coding these is by what we will call *super Gödel numbers*. Given a sequence of wffs or other expressions

$$e_0, e_1, e_2, \ldots, e_n$$

we first code each e_i by a regular g.n. g_i, to yield a sequence of numbers

$$g_0, g_1, g_2, \ldots, g_n.$$

We then encode this sequence of regular Gödel numbers using a single *super g.n.* by repeating the trick of multiplying powers of primes to get

$$\pi_0^{g_0} \cdot \pi_1^{g_1} \cdot \pi_2^{g_2} \cdot \ldots \cdot \pi_n^{g_n}.$$

Decoding a super g.n. therefore involves two steps of taking prime factors: first find the exponents of its prime factors; then treat those exponents as themselves regular Gödel numbers, and take *their* prime factors to arrive back at the expressions in the sequence.

---

[3]Being acceptable in this sense does not rule out being ludicrously messy. But for our purposes, we needn't worry about what else might be needed by way of *sufficient* conditions for being intuitively 'sensible'.

19.4   *Term, Atom, Wff, Sent* and *Prf* are p.r.

Here, then, are some official definitions (highlighting for the moment just a few of the many numerical properties and relations induced by Gödel coding):

1. $Term(n)$ iff n is the g.n. (on our numbering scheme) for a term of L_A.

2. Likewise, $Atom(n)$, $Wff(n)$ or $Sent(n)$ iff n is the g.n., respectively, for an atomic wff, wff, or sentence of L_A.

3. $Prf(m, n)$ iff m is the super g.n. of a sequence of L_A expressions which constitutes a proof in (your favourite official version of) PA of the sentence with g.n. n.

We now need to convince ourselves of the following crucial result:

Theorem 19.1 Term, Atom, Wff, Sent *and* Prf *are all primitive recursive.*

And we can have a very much easier time of it than Gödel did. Writing at the very beginning of the period when concepts of computation were being forged, he couldn't expect his audience to take anything on trust about what was or wasn't '*rekursiv*' or – as we would now put it – primitive recursive. He therefore had to do all the hard work of explicitly showing how to define these properties by a long chain of definitions by composition and recursion.

However, assuming only a very modest familiarity with computer programs and the conception of p.r. functions as those computed without open-ended searches, we can perhaps short-cut all that effort and be entirely persuaded by the following:

Proof sketch   We can determine whether $Term(n)$ holds by proceeding as follows. Decode n: that's a mechanical exercise. Now ask: is the resulting expression – if there is one – a term? That is to say, is it '0', a variable, or built up from '0' and/or variables using just the successor, addition and multiplication functions? (see Section 5.2). That's algorithmically decidable too. And moreover the steps here involve simple bounded computations that don't involve any open-ended search. (The length of the required computation will be bounded by a p.r. function of the length of the expression we are testing.)

Similarly, we can mechanically decide whether $Atom(n)$, $Wff(n)$ or $Sent(n)$. Decode n again. Now ask: is the result an atomic wff, a wff, or a sentence of L_A? In each case, that's algorithmically decidable, without any open-ended searches.

To determine whether $Prf(m, n)$, proceed as follows. First doubly decode m: that's a mechanical exercise. Now ask: is the result a sequence of L_A wffs? That's algorithmically decidable (since it is decidable whether each separate string of symbols is a wff). If it does decode into a sequence of wffs, ask: is this sequence a properly constructed proof in the particular Hilbert-style system of logic that you have fixed on in your official version of PA? That's decidable too (check whether each wff in the sequence is either an axiom or is an immediate consequence of previous wffs by one of the rules of inference of PA's Hilbert-style logical system). If the sequence is a proof, ask: does its final wff have the g.n. n? That's again decidable. Finally ask whether $Sent(n)$ is true. Putting all that together, there is a computational procedure for telling whether $Prf(m, n)$ holds. Moreover, it is again clear that, at each and every stage, the computation involved is once more a bounded procedure. For example, at the stage where we are checking whether a sequence of expressions is a PA proof, the length of the check will be bounded by a simple p.r. function of the length of the sequence and the complexity of the expressions.

In sum, then: suppose that we set out to construct programs for determining whether $Term(n)$, $Atom(n)$, $Wff(n)$, $Sent(n)$ or $Prf(m, n)$. Then we will be able to do each of these using programming structures no more exotic than bounded 'for' loops (in particular, we don't need to use any of those open-ended 'do while'/'do until' structures that can take us outside the limits of the primitive recursive).

Now, the computations we've described informally involve shuffling strings of symbols, but – using ASCII codes or the like – these will be implemented on real computers by being coded up numerically. So the computations can be thought as in effect numerical ones. And that observation gives us our desired conclusion: if the relevant computations can therefore be done with 'for' loops ultimately operating on numbers, the numerical properties and relations which are decided by the procedures must be primitive recursive (see Section 14.4). $\boxtimes$

Well, that is indeed sketchy; but the argument ought to strike you as entirely convincing. And if you are now happy to take it on trust that – if we really want to – we can spell out official p.r. definitions of those properties and relations we mentioned, then that is fine. If you aren't, then the next chapter is for you.

19.5 Some cute notation for Gödel numbers

That's the main part of the chapter done. Before proceeding to talk about 'diagonalization', let's pause to introduce a really pretty bit of notation. Assume we have chosen some system for Gödel-numbering the expressions of some language L. Then

> If φ is an L-expression, then we will use '$\ulcorner\varphi\urcorner$' in our logicians' augmented English to denote φ's Gödel number.

Borrowing a species of quotation mark is appropriate because the number $\ulcorner\varphi\urcorner$ can be thought of as referring to the expression φ via our coding scheme.

We also want a notation for the standard numeral inside the formal language L for φ's Gödel number. There's a choice to be made. Overlining is a common way of signalling standard numerals: thus we are writing '$\overline{c}$' for the standard numeral for some particular number c (thereby making a visual distinction between the numeral and 'c' used as a variable, blocking possible misunderstandings). On the other hand we have written e.g. simply '2' to abbreviate 'SS0' rather than writing '$\overline{2}$', because here there is no possibility of confusion. So what shall we do here when we want to write the numeral for the g.n. of 'SS0'? Overline as in '$\overline{\ulcorner SS0 \urcorner}$' or avoid extra visual clutter and write simply '$\ulcorner SS0 \urcorner$'?

We take the cleaner second line: so

> In a definitional extension of L, allowing new abbreviations, '$\ulcorner \varphi \urcorner$' is shorthand for L's standard numeral for the g.n. of φ.

In other words, inside formal expressions of L_A-plus-abbreviations, '$\ulcorner \varphi \urcorner$' stands in, unambiguously, for the numeral for the number $\ulcorner \varphi \urcorner$, just as '2' unambiguously occurs in formal expressions as the numeral for the number 2.

So, to summarize:

1. 'SS0' is an L_A expression, the standard numeral for 2.

2. On our numbering scheme $\ulcorner SS0 \urcorner$, the g.n. of 'SS0', is $2^{23} \cdot 3^{23} \cdot 5^{21}$.

3. So, by our further convention, we can also use the expression '$\ulcorner SS0 \urcorner$' inside (a definitional extension of) L_A, as an abbreviation for the standard numeral for that g.n., i.e. as an abbreviation for 'SSS...S0' with $2^{23} \cdot 3^{23} \cdot 5^{21}$ occurrences of 'S'!

This double usage of 'corner quotes' – outside a formal language to denote a g.n. of a formal expression and inside a formal language to take the place of a standard numeral for that g.n. – should by this stage not give you any pause.

19.6   The idea of diagonalization

(a)   Gödel is going to tell us how to construct a wff G in PA that is true if and only if it is unprovable in PA. We now have an inkling of how he can do that: wffs can contain numerals that refer to numbers which – via Gödel coding – are correlated with wffs.

Gödel's construction involves taking an open wff that we'll abbreviate as U, or as U(y) when we want to emphasize that it contains just 'y' free. This wff has g.n. $\ulcorner U \urcorner$. And then – the crucial move – Gödel substitutes *the numeral for U's g.n.* for the free variable in U. So the key step involves forming the wff U($\ulcorner U \urcorner$).

This substitution operation is called *diagonalization*, which at first sight might seem an odd term for it. But in fact, Gödel's construction does involve something quite closely akin to the 'diagonal' constructions we encountered in e.g. Sections 7.2 and 14.5. In the first of those cases, we matched the index of a wff $\varphi_n(x)$ (in an enumeration of wffs with one free variable) with the numeral substituted for its free variable, to form $\varphi_n(\bar{n})$. In the second case, we matched the index of a function f_n (in an enumeration of p.r. functions) with the number the function takes as argument, to form $f_n(n)$. Here, in our Gödelian diagonal construction, we match U's Gödel number – and we can think of that as indexing the wff in a list of wffs – with the numeral substituted for its free variable, and this yields the Gödelian $U(\ulcorner U \urcorner)$.

(b)   Now note the following additional point. Given the wff $U(y)$, it can't matter much whether we do the Gödelian construction by forming (i) $U(\ulcorner U \urcorner)$ (as Gödel himself did in 1931) or alternatively by forming (ii) $\exists y(y = \ulcorner U \urcorner \wedge U(y))$. For (i) and (ii) are trivially equivalent. But it makes a couple of claims a bit simpler to prove if we do things the second way.

So that motivates our official definition:

The *diagonalization* of φ is $\exists y(y = \ulcorner \varphi \urcorner \wedge \varphi)$.

To make things run technically smoothly later, this generous definition applies even when φ doesn't contain the variable 'y' free. But of course we will be particularly interested in cases where we diagonalize wffs of the form $\varphi(y)$ which do have just that variable free: in such cases the diagonalization of $\varphi(y)$ is equivalent to $\varphi(\ulcorner \varphi \urcorner)$. In other cases, φ's diagonalization is trivially equivalent to φ itself.

It should go without saying that there is no special significance to using the variable 'y' for the quantified variable here!

(c)   Diagonalization is, evidently, a very simple mechanical operation on expressions. Arithmetizing it should therefore give us a simple computable function. In fact,

> **Theorem 19.2** *There is a p.r. function* diag *which, when applied to a number n which is the g.n. of some L_A wff, yields the g.n. of that wff's diagonalization.*

We can give an official proof of this in the next chapter. But the following informal argument should suffice.

Proof sketch   Consider this procedure. Decode the g.n. $n = \ulcorner \varphi \urcorner$ to get some expression φ (assume we have some sensible convention for dealing with 'waste' cases where we don't get an expression). Then form the standard numeral for that g.n., and go on to write down φ's diagonalization $\exists y(y = \ulcorner \varphi \urcorner \wedge \varphi)$. Then work out the g.n. of this expression to compute $diag(n)$. This procedure doesn't involve any unbounded searches. So we again will be able to program the procedure using just 'for' loops. Hence *diag* is p.r.   ⊠

20 Arithmetization in more detail

In the last chapter we gave informal but hopefully entirely persuasive arguments that key numerical properties and relations that arise from the arithmetization of the syntax of PA – such as *Term*, *Wff* and *Prf* – are primitive recursive.

Gödel, as we said, gives rigorous proofs of such results (or rather, he proves the analogues for his particular formal system). He shows how to define a sequence of more and more complex functions and relations by composition and recursion, eventually leading up to a p.r. definition of *Prf*. Inevitably, this is a laborious job: Gödel does it with masterly economy and compression but, even so, it takes him forty-five steps of function-building to show that *Prf* is p.r.

We have in fact already traced some of the first steps in Section 14.8. We showed, in particular, that extracting exponents of prime factors – the key operation used in decoding Gödel numbers – can be done by a p.r. function, *exf*. To follow Gödel further, we need to keep going in the same vein, defining ever more complex functions. What I propose to do in this chapter is to fill in the next few steps moderately carefully, and then indicate rather more briefly how the remainder go. This should be quite enough to give you a genuine feel for Gödel's demonstration and to indicate how it can be completed, without going into too much unnecessary detail.[1]

However, although we are only going to give a partial rehearsal of the full p.r. definition of *Prf*, by all means skip even this cut-down discussion, and jump to the next chapter where the real excitement starts. You will miss nothing of wider conceptual interest. Gödel's Big Idea is that *Prf* is primitive recursive. Working up a fully detailed demonstration of this Big Idea (to round out the informal argument of Section 19.4) just involves cracking some mildly interesting brain-teasers.

20.1 The concatenation function

(a)   Still aboard? Then quickly review Sections 14.7 and 14.8. In particular, recall two facts proved in Theorem 14.4:

> The function $exf(n, i)$ is p.r., where this returns the exponent of π_i in the prime factorization of n.

> The function $len(n)$ is p.r., where this returns the number of distinct prime factors of n.

---

[1]Masochists and completists are quite welcome to tackle the full general story in e.g. Mendelson (1997, pp. 193–198).

And now note that on our chosen scheme for Gödel numbering, we have:

1. If n is a g.n. (super or regular), then it consists of multiples of the first $len(n)$ primes, i.e. of the primes from π_0 to $\pi_{len(n)-1}$.

2. If n is the g.n. of an expression e which is a sequence of symbols/variables $s_0, s_1, s_2, \ldots, s_k$, then $exf(n, i)$ gives the basic code of s_i.

3. If n is the super g.n. of a sequence of expressions $e_0, e_1, e_2, \ldots, e_k$, then $exf(n, i)$ gives the g.n. of e_i.

(b)   We will now add another key p.r. function to our growing list:

> **Theorem 20.1** *There is a two-place concatenation function, stan-dardly represented by $m \star n$, such that (i) if m and n are the g.n. of two expressions then $m \star n$ is the g.n. of the result of writing the first expression followed by the second, and (ii) this function is primitive recursive.*

For example, suppose m is the g.n. of the expression '$\exists y$', i.e. (on our scheme, of course) $m = 2^{13} \cdot 3^4$, and n is the g.n. of '$y = 0$', i.e. $n = 2^4 \cdot 3^{15} \cdot 5^{21}$. Then $m \star n$ is to deliver the g.n. of the concatenation of those two expressions, i.e. the g.n. of '$\exists y\, y = 0$'; so we want $m \star n = 2^{13} \cdot 3^4 \cdot 5^4 \cdot 7^{15} \cdot 11^{21}$.

Proof   Look at what happens to the exponents in that example. Now generalize. So suppose m and n are Gödel numbers, $len(m) = j$ and $len(n) = k$. We want the function $m \star n$ to yield the value obtained by taking the first $j + k$ primes, raising the first j primes to powers that match the exponents of the j primes in the prime factorization of m and then raising the next k primes to powers that match the k exponents in the prime factorization of n. Then $m \star n$ will indeed yield the g.n. of the expression which results from stringing together the expression with g.n. m followed by the expression with g.n. n.

Recall that bounded minimization keeps us in the sphere of the primitive recursive (Section 14.7(b)). It is then readily seen we can define a p.r. function $m \star n$ which applies to Gödel numbers in just the right way by putting

$$m \star n = (\mu x \leq B(m,n))[(\forall i < len(m))\{exf(x, i) = exf(m, i)\} \wedge \\ (\forall i < len(n))\{exf(x, i + len(m)) = exf(n, i)\}]$$

so long as $B(m, n)$ is a suitable primitive recursive function whose values keep the minimization operator finitely bounded. $B(m, n) = \pi_{m+n}^{m+n}$ is certainly big enough to cover all eventualities.

Note that our definition does indeed make the star function $m \star n$ primitive recursive, since it is constructed using devices allowed by Theorem 14.3.[2]   ⊠

---

[2] A reality check: we are here doing *informal* mathematics. If we use the quantifier and conjunction symbols familiar from our formal languages, that is for brevity's sake. So we can e.g. freely use '$<$' as well as '$\leq$' in defining the star because this isn't a formal wffs! Cf. Section 5.1; also see the comment in the proof for (C) in Section 14.7.

(c)   Note first that

i. $((m \star n) \star o) = (m \star (n \star o))$,

which is why we can suppress bracketing with the star function. Now for some very simple examples:

ii. $\ulcorner S \urcorner \star \ulcorner 0 \urcorner$ yields $\ulcorner S0 \urcorner$, the g.n. of S0.

iii.   Suppose a is the g.n. of the wff $\exists x \, Sx = 0$; then $\ulcorner \neg \urcorner \star a$ is the g.n. of $\neg \exists x \, Sx = 0$.

iv.   Suppose b is the g.n. of 'S0 = 0'. Then $\ulcorner (\ulcorner \star a \star \ulcorner \rightarrow \urcorner \star b \star \ulcorner) \urcorner$ is the g.n. of $(\exists x \, Sx = 0 \rightarrow S0 = 0)$.

Note too that it falls out of our definition that

v. $0 \star n = n \star 0 = n$

so 0 can be thought of as the g.n. of the null expression.

(d)   Here are two more results that we can now easily prove:

Theorem 20.2 *(i) The function* num *which maps n to the g.n. of L_A's standard numeral for n is p.r. Hence (ii),* diag *is p.r.*

Proof for (i)   The standard numeral for Sn is of the form 'S' followed by the standard numeral for n. So we have

$$num(0) = \ulcorner 0 \urcorner = 2^{21}$$
$$num(Sx) = \ulcorner S \urcorner \star num(x) = 2^{23} \star num(x).$$

Hence *num* is primitive recursive.   ⊠

Proof for (ii)   The *diag* function from Section 19.6 maps n, the g.n. of φ, to the g.n. of φ's diagonalization $\exists y (y = \ulcorner \varphi \urcorner \wedge \varphi)$. So we can put

$$diag(n) = \ulcorner \exists y (y = \urcorner \star num(n) \star \ulcorner \wedge \urcorner \star n \star \ulcorner) \urcorner$$
$$= 2^{13} \cdot 3^4 \cdot 5^{17} \cdot 7^4 \cdot 11^{15} \star num(n) \star 2^3 \star n \star 2^{19}.$$

where *num* is as just defined, so *diag* is a composition of p.r. functions.   ⊠

Note that with *diag* defined like this, it deals smoothly with all the 'waste cases' where n isn't the g.n. of a wff, and still delivers an output for every input.

20.2   Proving that *Term* is p.r.

Prf will evidently be defined in part in terms of *Wff* and *Sent* (since the relation holds when a sequence of wffs with the right structure ends in the right sentence); *Wff* will be defined in terms of *Atom* (since a wff is built up out of atomic wffs);

and *Atom* will be defined in terms of *Term* (since an atomic wff is built up out of terms). So we will begin by showing that *Term* is primitive recursive.

A term, recall, is either '0', or a variable, or is built up from those using the function-symbols 'S', '+', '×'.

Let's say, then, that a 'constructional history' for a term, or a *term-sequence*, is a sequence of expressions $\langle \tau_0, \tau_1, \ldots, \tau_n \rangle$ such that each expression τ_k in the sequence either (i) is '0'; or else (ii) is a variable; or else (iii) has the form $S\tau_i$, where τ_i is an earlier expression in the sequence; or else (iv) has the form $(\tau_i + \tau_j)$, where τ_i and τ_j are earlier expressions in the sequence; or else (v) has the form $(\tau_i \times \tau_j)$, where τ_i and τ_j again are earlier expressions. Since any well-formed term must have the right kind of constructional history, we can say: *a term is an expression which is the last expression in some term-sequence.*

Now let's implement this idea and so prove the following three-part theorem:

> **Theorem 20.3** *(i) The property* Var, *which holds of n iff n is the g.n. of a variable, is p.r. (ii) The property* Termseq, *which holds of n iff n is the super g.n. for a term-sequence, is p.r. Hence (iii)* Term *is p.r.*

Proof for (i) We just note that we can put

$$Var(n) =_{\text{def}} (\exists x \leq n)\, n = 2^{2(x+1)}.$$

For the basic code for a variable always has the form $2n$, for $n \geq 1$, and the g.n. for a single expression is 2 to the power of that basic code. ⊠

Proof for (ii) *Termseq(n)* will hold if n is a super g.n. of 'length' $l = len(n)$ and when we decode using *exf*, each value of $exf(n, k)$ as k runs from 0 to $k - 1$ is the code either for 0 or for a variable or for the successor, sum, or product of earlier items. In symbols, then:[3]

$$
\begin{aligned}
Termseq(n) =_{\text{def}} (\forall k < len(n))\{ &exf(n, k) = \ulcorner 0 \urcorner \vee Var(exf(n, k)) \vee \\
&(\exists j < k)(exf(n, k) = \ulcorner S \urcorner \star exf(n, j)) \vee \\
&(\exists i, j < k)(exf(n, k) = \ulcorner(\urcorner \star exf(n, i) \star \ulcorner + \urcorner \star exf(n, j) \star \ulcorner) \urcorner) \vee \\
&(\exists i, j < k)(exf(n, k) = \ulcorner(\urcorner \star exf(n, i) \star \ulcorner \times \urcorner \star exf(n, j) \star \ulcorner) \urcorner) \}.
\end{aligned}
$$

This right-hand side is built up from p.r. components using disjunction and bounded quantifiers as allowed by Theorem 14.3, so *Termseq* is p.r. ⊠

Proof for (iii) Since a term has to be the final member of some term-sequence, we can say that n numbers a term if there is a super g.n. number x which numbers a term-sequence, and the last item of that sequence (as extracted by the *exf* function) has number n. So

$$Term(n) =_{\text{def}} (\exists x \leq B(n))(Termseq(x) \wedge n = exf(x, len(x) - 1)),$$

---

[3]To labour the obvious, '$(\exists i, j < n)$' is just convenient shorthand for '$(\exists i < n)(\exists j < n)$'.

where $B(n)$ is a suitably large p.r. bounding function. Given a term with g.n. n, and hence with $l = len(n)$ symbols and variables, there will be a term-sequence for it at most l long (a minimal sequence without any redundant junk thrown in). So the super g.n. of a term-sequence constructing it can be less than $B(n) = (\pi_l^n)^l$.   ⊠

20.3 Proving that *Atom*, *Wff* and *Sent* are p.r.

(a)   We have now proved that *Term* is p.r.: it follows quickly that

> **Theorem 20.4** *The property* Atom, *which holds of n when n is the g.n. of an atomic wff of L_A, is primitive recursive.*

Proof   The only atomic wffs of L_A are expressions of the kind $\sigma = \tau$, for terms σ and τ.[4] So we can put

$$Atom(n) =_{\text{def}} (\exists x, y < n)[Term(x) \wedge Term(y) \wedge n = (x \star \ulcorner = \urcorner \star y)].$$

Which confirms that *Atom* is primitive recursive.   ⊠

(b)   We can now proceed to repeat essentially the same trick we used in defining *Term* to define *Wff*. We first introduce the idea of a *formula-sequence* which is exactly analogous to the idea of a term-sequence. So: a formula-sequence is a sequence of expressions $\langle \varphi_0, \varphi_1, \ldots, \varphi_n \rangle$ such that each φ_k in the sequence is (i) is an atomic wff; or else (ii) is built from one or two previous wffs in the sequence by using a connective; or else (iii) is built from a previous wff in the sequence by prefixing $\forall \xi$ or $\exists \xi$ for some variable ξ.[5] We then set $Formseq(n)$ to be true when n is the super g.n. of a formula-sequence, and we can give an explicit definition modelled on our definition of $Termseq(n)$, which shows that as before what we end up with is primitive recursive. (Exercise: fill in the details.)

Then $Wff(n)$ iff there is a number x (under a suitable p.r. bound) such that $Formseq(x)$ and the last expression of the sequence coded by x has number n. (Exercise: again fill in the details.) This too is primitive recursive, so we now give ourselves

> **Theorem 20.5** *The property* Wff, *which holds of n when n is the g.n. of an L_A wff, is primitive recursive.*

(c)   A wff is a sentence iff it has no free variables. So, if we are to define the p.r. property *Sent*, we need to be able to handle facts about bound and free variables.

We start by defining $Bound(v, i, x)$, which is to be true when v numbers a variable which occurs *bound* in the wff numbered x, at least when it occurs at

---

[4] Recall, $\sigma \leq \tau$ is *not* an atomic formula, but the abbreviation for a quantified wff!

[5] To make things easy at this point, we can give L_A a liberal syntax which allows wffs with redundant quantifiers which don't bind any variables in their scope.

the $i + 1$-th 'position' in the wff. (Here, number positions by tallying symbols and variables from the left, so that $exf(x, i)$ gives the code for what's at position $i + 1$.) This does the trick:

$$Bound(v, i, x) =_{\text{def}} Var(v) \land Wff(x) \land$$
$$(\exists a, b, c < x)\{[x = a \star \ulcorner \exists \urcorner \star v \star b \star c \ \lor \ x = a \star \ulcorner \forall \urcorner \star v \star b \star c] \land$$
$$Wff(b) \land [len(a) \le i \ \land \ i \le (len(a) + len(b) + 1)]\}.$$

The middle clause says that the wff numbered x has the form '$\dots \exists \xi \varphi \dots$' or '$\dots \forall \xi \varphi \dots$' (where φ is the wff numbered by b, and ξ is some variable), and within that displayed component, ξ is evidently bound. The last clause ensures that the position i occurs within that component '$\exists \xi \varphi$'/'$\forall \xi \varphi$'.

Using the relation *Bound*, we can define a companion relation *Free* which holds when a variable occurs at a certain position in a wff but isn't bound there. But let's leave that as another exercise, and move on to prove

Theorem 20.6 *The property* Sent, *which holds of n when n is the g.n. of a closed L_A wff, is primitive recursive.*

Proof Evidently, n numbers a sentence just in case (i) n numbers a wff, and (ii) for any position along the wff, if there is a variable at that position it is bound there. Hence we can put

$$Sent(n) =_{\text{def}} Wff(n) \land$$
$$(\forall i < len(n))(Var(exf(n, i)) \to Bound(exf(n, i), i, n)),$$

which confirms *Sent* is p.r. ⊠

20.4 Towards proving *Prf* is p.r.

(a) So far so good. But how can we now prove that *Prf* is primitive recursive as claimed, and so complete our demonstration of Theorem 19.1 by the hard route?

The details will evidently depend on the details of the proof system for PA. But remember, in order to keep things simple, we have officially adopted an old-fashioned Hilbertian proof system (which doesn't allow temporary suppositions and sub-proofs, so proofs are therefore simple linear sequences of wffs).

In this kind of system, there are propositional axioms such as instances of the schema

$$(\varphi \to (\psi \to \varphi))$$

and quantificational axioms such as instances of the schema

$$(\forall \xi \varphi(\xi) \to \varphi(\tau)), \text{ where } \tau \text{ is a term 'free for' the variable } \xi \text{ in } \varphi(\xi).$$

And in standard versions there are just two rules of inference, modus ponens and the universal quantification rule that allows us to infer $\forall \xi \varphi$ from φ.[6]

Now, the relation $MP(m, n, o)$ – which holds when the wff with g.n. o follows from the conditional wff with g.n. m and its antecedent numbered n by modus ponens – is obviously p.r., for it is immediate that we can put

$$MP(m, n, o) =_{\text{def}} \{m = \ulcorner(\ulcorner \star n \star \ulcorner \to \urcorner \star o \star \ulcorner)\urcorner\} \land \mathit{Wff}(n) \land \mathit{Wff}(o).$$

Likewise the relation $Univ(m, n)$ – which holds when the wff with g.n. n is a universal quantification of the wff with g.n. m – is also obviously p.r., for we can put

$$Univ(m, n) =_{\text{def}} (\exists x \leq n)(var(x) \land \mathit{Wff}(m) \land \{n = \ulcorner \forall \urcorner \star x \star m\}).$$

Let us suppose, just for the moment, that we have also shown that the property *Axiom* is p.r., where $Axiom(n)$ holds just when n is the g.n. of a wff which is an axiom – whether a PA axiom or a logical axiom. Then we can echo again the pattern of our definition of *Termseq* in Section 20.2 and define $Proofseq(n)$ where this holds iff n is the super g.n. of a sequence of wffs forming a PA proof. For a PA proof is, of course, a sequence $\langle \varphi_0, \varphi_1, \ldots, \varphi_n \rangle$, each element of which is either an axiom or follows from previous wffs in the sequence by one of the two rules of inference. Hence we can put

$$\begin{aligned} Proofseq(n) =_{\text{def}} (\forall k &< len(n))\{Axiom(exf(n, k)) \lor \\ &(\exists i, j \leq k)MP(exf(n, i), exf(n, j), exf(n, k)) \lor \\ &(\exists j \leq k)Univ(exf(n, j), exf(n, k))\}. \end{aligned}$$

By inspection, we again see that *Proofseq* is p.r, given our supposition about *Axiom*.

(b)   So now let's think a bit about the final missing step here, the p.r. definition of *Axiom*. This will be messily disjunctive, in the form

$$Axiom(n) =_{\text{def}} Axiom_1(n) \lor Axiom_2(n) \lor \ldots \lor Axiom_{18}(n)$$

or thereabouts, with one clause for each different kind of axiom.

There will be simple clauses for the first six axioms of PA: for example

$$Axiom_1(n) =_{\text{def}} \{n = \ulcorner \forall x \neg 0 = Sx \urcorner\}$$

(using the official syntax of PA without the added brackets we allowed for readability, and without the contraction '$\neq$').

For $Axiom_7$, we need to specify the condition for being the g.n. of (the closure of) an instance of the Induction Schema for PA. This is going to involve, for a start, coding up the idea that φ has 'x' as a free variable, and the idea of substituting '0' or 'Sx' for 'x'. But we have seen in defining *Bound* how to handle

---

[6]See e.g. Mendelson (1997, p. 69).

the idea of a variable being bound at a position, and it is free there if it isn't bound. So working with that sort of construction will get us on the road to what we want. (Exercise for enthusiasts: think through the details of this!)

Now for logical axioms. Suppose the logic treats '$\rightarrow$', '$\neg$' and '$\forall$' as basic (a very traditional Hilbertian choice), with other logical operators being introduced by definitional axioms. To cover all instances of the basic logical axiom schema $(\varphi \rightarrow (\psi \rightarrow \varphi))$ is easy: we just put

$$Axiom_8(n) =_{\text{def}} (\exists j, k < n)\{n = \ulcorner(\neg \star j \star \ulcorner(\neg \star k \star \ulcorner\rightarrow\urcorner \star j \star \ulcorner))\urcorner\}.$$

Similarly for the other propositional axioms.

We then want $Axiom_{11}(n)$ to be true whenever n numbers a wff of the form $(\forall\xi\varphi(\xi) \rightarrow \varphi(\tau))$ but with the restriction that τ is free for the relevant variable. Again handling that restriction requires us to arithmetize claims about which variables are free where: but now we have a glimmering of how that can be done.

It will be similar when defining $Axiom_{12}(n)$, which we want to be true when n numbers an instance of the other standard Hilbertian quantification axiom schema, i.e. $(\forall\xi(\varphi \rightarrow \psi) \rightarrow (\varphi \rightarrow \forall\xi\psi))$, where the relevant ξ doesn't occur free in φ.

That leaves us to deal with the logical axioms for identity, plus the axioms for passing from the other connectives and the quantifier '$\exists$' to the basic '$\rightarrow$', '$\neg$' and '$\forall$'. (More exercises for enthusiasts: think through the details of these other cases.)

Let us suppose all this has been done. Then we are indeed entitled to claim that *Axiom* is p.r., and hence that *Proofseq* is p.r. too.

(c) With all that in place, we finally arrive at our intended destination:

> **Theorem 20.7** *The relation $Prf(m, n)$, which holds when m is the super g.n. of a* PA *proof of the sentence with g.n. n, is primitive recursive.*

Proof We just set

$$Prf(m, n) =_{\text{def}} Proofseq(m) \wedge \{exf(m, len(m) - 1) = n\} \wedge Sent(n)$$

and we are done! ⊠

(d) One *very* important final comment. Given the definition of *Prf*, we know that n numbers a provable sentence of PA just so long as, for some number m, $Prf(m, n)$. So let's put $Prov(n) =_{\text{def}} \exists v\, Prf(v, n)$: this holds when the sentence with g.n. n is a theorem (hence $Prov(\ulcorner\varphi\urcorner)$ iff PA $\vdash \varphi$).

We *can't*, however, read off from n some upper bound on the length of possible proofs for the sentence with g.n. n. So we *can't* define the provability property by some *bounded* quantification of the kind $(\exists v \leq B(n))Prf(v, n)$ with p.r. B. If we could, then the provability property would be p.r.: but it isn't – as we will show in Section 24.7.

21 PA is incomplete

The pieces we need are finally all in place. So in this chapter we at long last learn how to construct 'Gödel sentences' and use them to prove that PA is *incomplete*. Then in the next chapter, we show how our arguments can be generalized to prove that PA – and any other formal arithmetic satisfying very modest constraints – is not only incomplete but *incompletable*.

The beautiful – though quite easy! – proofs (pretty much Gödel's own) now come thick and fast: do savour them slowly.

21.1 Reminders

We start with some quick reminders – or bits of headline news, if you have impatiently been skipping forward in order to get to the exciting stuff.

Fix on some sensible scheme for mechanically associating wffs of PA with code numbers for them – i.e. with Gödel numbers ('g.n.') as they are standardly called. And extend it to similarly associate sequences of wffs with numbers – i.e. with what we have called 'super' Gödel numbers (see Sections 19.1, 19.3). Then,

   i. The diagonalization of the wff φ is $\exists y(y = \ulcorner\varphi\urcorner \land \varphi)$, where '$\ulcorner\varphi\urcorner$' here stands in for the standard numeral for φ's g.n. – the diagonalization of $\varphi(y)$ is thus equivalent to $\varphi(\ulcorner\varphi\urcorner)$. (Section 19.6)

   ii. *diag*(n) is a p.r. function which, when applied to a number n which is the g.n. of some wff φ, yields the g.n. of φ's diagonalization. (Sections 19.6, 20.1)

   iii. *Prf*(m, n) is the p.r. relation which holds iff m is the super g.n. of a sequence of wffs that is a PA proof of a sentence with g.n. n. (Sections 19.4, 20.4)

   iv. Any p.r. function or relation can be *expressed* by a Σ_1 wff of PA's language L_A. (See Section 11.5 for the definition of a Σ_1 wff, and Section 15.4 for the main result.)

   v. Any p.r. function or relation can be *captured* in Q, and hence in PA, by a Σ_1 wff. (Section 17.4)

   vi. In particular, we can choose a Σ_1 wff which '*canonically*' captures a given p.r. relation by recapitulating its full p.r. definition ultimately in terms of initial functions (or more strictly, by recapitulating the definition of the relation's characteristic function). And this wff which captures the given relation will express it too. (Section 17.4 (b))

For what follows, it isn't necessary that you know the *proofs* of the claims we've just summarized: but do ensure that you at least understand what they say.

21.2 'G is true if and only if it is unprovable'

We need just one new observation:

The relation $Gdl(m, n)$ – which holds just when m is the super g.n. for a PA proof of a sentence which is the diagonalization of the wff with g.n. n – is also p.r.

Proof $Gdl(m, n)$ holds, by definition, when $Prf(m, diag(n))$. The characteristic function of Gdl is therefore definable by composition from the characteristic function of Prf and the function $diag$, and hence is p.r., given facts (ii) and (iii) from our list of reminders. ☒

We now adopt the following notational convention:

$\mathsf{Gdl(x, y)}$ stands in for a Σ_1 wff with two free variables which canonically captures Gdl.

We know that there must be such a wff by fact (vi), and that $\mathsf{Gdl(x, y)}$ will also express Gdl. And we next follow Gödel in constructing the corresponding wff

$$\mathsf{U(y)} =_{\text{def}} \forall x \neg \mathsf{Gdl(x, y)}.$$

Suppose n is the g.n. of φ: then n satisfies $\mathsf{U(y)}$ just in case no super g.n. numbers a proof in PA of the diagonalization of φ.

For convenience, we'll further abbreviate $\mathsf{U(y)}$ simply as 'U' when we don't need to stress that it contains 'y' free. And now we diagonalize the wff U itself, to give

$$\mathsf{G} =_{\text{def}} \exists y(y = \ulcorner \mathsf{U} \urcorner \wedge \mathsf{U(y)}).$$

This is our first 'Gödel sentence' for PA.

Trivially, G is equivalent to $\mathsf{U}(\ulcorner \mathsf{U} \urcorner)$. Or unpacking that a bit,

G is equivalent to $\forall x \neg \mathsf{Gdl(x, \ulcorner U \urcorner)}$.

We then immediately have

Theorem 21.1 G *is true if and only if it is unprovable in* PA.

Proof By inspection. For consider what it takes for G to be true (on the interpretation built into L_A), given that the formal predicate Gdl expresses the numerical relation Gdl. By our equivalence, G is true if and only if there is no

153

number m such that $Gdl(m, \ulcorner U \urcorner)$. That is to say, given the definition of Gdl, G is true if and only if there is no number m such that m is the code number for a PA proof of the diagonalization of the wff with g.n. $\ulcorner U \urcorner$. But the wff with g.n. $\ulcorner U \urcorner$ is of course U; and its diagonalization is G. So, G is true if and only if there is no number m such that m is the code number for a PA proof of G. But if G is provable, some number would be the code number of a proof of it. Hence G is true if and only if it is unprovable in PA. Wonderful! ⊠

21.3 PA is incomplete: the semantic argument

The very simple proof that, if PA is sound, then PA is incomplete, now runs along the lines we sketched right back in Section 1.2.

Proof Suppose PA is a sound theory whose theorems are all true. If G (which is true if and only if it is *not* provable) could be proved in PA, the theory *would* prove a false theorem, contradicting our supposition. Hence, G is not provable in PA. Hence G is true. So ¬G is false. Hence ¬G cannot be proved in PA either. In Gödel's words, G is a 'formally undecidable' sentence of PA.   ⊠

But PA's logic is truth-preserving, so it will be sound so long as its (non-logical) axioms are true. So that establishes

> **Theorem 21.2** *If* PA *has true non-logical axioms, then there is an L_A-sentence φ such that neither* PA $\vdash \varphi$ *nor* PA $\vdash \neg\varphi$.

If we are happy with the semantic assumption that PA's axioms *are* true on interpretation and so PA *is* sound, the argument for incompleteness is as simple as that, once we have constructed G.

Now, for reasons that will become clearer when we consider Hilbert's Programme and related background in a later Interlude, it was very important to Gödel that incompleteness can *also* be proved *without* supposing that PA is sound: as he puts it, 'purely formal and much weaker assumptions' suffice. However, the overall argument that shows this is harder work. *So don't lose sight of Gödel's simpler 'semantic' argument for incompleteness.*

21.4 'There is an undecidable sentence of Goldbach type'

Before showing how much weaker assumptions suffice, let's note that, unlike Theorems 6.3 and 7.1, our new Theorem is proved constructively; in other words, our overall argument doesn't just make the bald existence claim that there is a formally undecidable sentence of PA, *it actually tells us how to construct one*: build the open wff Gdl which canonically captures Gdl, then form the corresponding sentence G.

Let's remark too that G, while horribly complex and ridiculously long (when spelt out in unabbreviated L_A), is in an important respect relatively simple. In the jargon of Section 11.5, G is a Π_1-equivalent, i.e. is equivalent by simple logic to a Π_1 sentence, i.e. is equivalent to a sentence 'of Goldbach type'.

Proof Gdl(x, y) is Σ_1. So Gdl(x, $\ulcorner U \urcorner$) is Σ_1. So its negation $\neg$Gdl(x, $\ulcorner U \urcorner$) is equivalent by elementary logic to a Π_1 wff $\psi(x)$. Hence $\forall x \neg$Gdl(x, $\ulcorner U \urcorner$) is equivalent to $\forall x \psi(x)$, which is of course still Π_1. But G is equivalent to $\forall x \neg$Gdl(x, $\ulcorner U \urcorner$), so it is – as claimed – still Π_1-equivalent. $\boxtimes$

Since G is undecidable given PA's soundness, so is any equivalent Π_1 sentence. So we can strengthen our statement of Theorem 21.2, to give us

Theorem 21.3 *If* PA *has true axioms, then there is an L_A-sentence φ of Goldbach type such that neither* PA $\vdash \varphi$ *nor* PA $\vdash \neg\varphi$.

21.5 Starting the syntactic argument for incompleteness

So far, we have only made use of the result that PA's language can *express* the relation *Gdl*. But in fact our chosen Gdl doesn't just express *Gdl* but (canonically) *captures* it. Using this fact about Gdl, we can again show that PA does not prove G, but this time *without* making the semantic assumption that PA is sound. We'll show first that

A. *If* PA *is consistent,* PA $\nvdash$ G.

Proof Suppose G *is* provable in PA. If G has a proof, then there is some super g.n. m that codes its proof. But by definition, G is the diagonalization of the wff U. Hence, by definition, $Gdl(m, \ulcorner U \urcorner)$.

Now we use the fact that Gdl captures the relation *Gdl*. That implies that, since $Gdl(m, \ulcorner U \urcorner)$, we have (i) PA $\vdash$ Gdl($\overline{m}, \ulcorner U \urcorner$).

But the assumption that G is provable is equivalent to: PA $\vdash \forall x \neg$Gdl(x, $\ulcorner U \urcorner$). Whence, by instantiating the quantifier, (ii) PA $\vdash \neg$Gdl($\overline{m}, \ulcorner U \urcorner$).

So, combining (i) and (ii), the assumption that G is provable in PA entails that PA is inconsistent. Therefore, if PA is consistent, there can be no PA proof of G. $\boxtimes$

21.6 ω-incompleteness, ω-inconsistency

Result (A) tells us that if PA is consistent then PA $\nvdash$ G. There is also a companion result (B), which tells us that if PA satisfies a rather stronger syntactic condition then PA $\nvdash \neg$G. This section explains the required stronger condition of 'ω-consistency'.

(a) But first, here's another standard definition we need:

An arithmetic theory T is *ω-incomplete* iff, for some open wff $\varphi(x)$, T can prove each $\varphi(\overline{m})$ but T can't go on to prove $\forall x \varphi(x)$.

155

So a theory is *ω-incomplete* if there are instances where it can prove, case by case, that each number satisfies some condition φ but it can't go the extra mile and prove, in one fell swoop, that all numbers satisfy φ.[1] It is an immediate corollary of our last proof that

Theorem 21.4 PA *is ω-incomplete if it is consistent.*

Proof   Assume PA's consistency. Then we've shown that

1. PA $\nvdash$ G, which is equivalent to PA $\nvdash \forall x \neg \mathsf{Gdl}(x, \ulcorner U \urcorner)$.

So no number is the super g.n. of a proof of G. That is to say, no number numbers a proof of the diagonalization of U. That is to say, for any particular m, it *isn't* the case that $Gdl(m, \ulcorner U \urcorner)$. Hence, again by the fact that Gdl captures Gdl, we have

2. For each m, PA $\vdash \neg \mathsf{Gdl}(\overline{m}, \ulcorner U \urcorner)$.

Putting $\varphi(\mathsf{x}) =_{\mathrm{def}} \neg \mathsf{Gdl}(\mathsf{x}, \ulcorner U \urcorner)$ in the definition above therefore shows that PA is ω-incomplete.   ⊠

In Section 10.5, we noted that Q exhibits a radical case of what we are now calling ω-incompleteness: although it can prove case-by-case all true equations involving numerals, Q can't prove many of their universal generalizations. For an extraordinarily simple example, put $\mathsf{Kx} =_{\mathrm{def}} (0 + \mathsf{x} = \mathsf{x})$; then for every n, Q $\vdash$ K$\overline{n}$, but we don't have Q $\vdash \forall \mathsf{x} \mathsf{Kx}$. In moving from Q to PA by adding the induction axioms, we vastly increase our ability to prove generalizations. But we now see that some ω-incompleteness remains even in PA.

(b)   And now here's the promised definition of the notion we need for part (B) of our syntactic incompleteness argument:

An arithmetic theory T is *ω-inconsistent* iff, for some open wff $\varphi(\mathsf{x})$, T can prove each $\varphi(\overline{m})$ and T can also prove $\neg \forall \mathsf{x} \varphi(\mathsf{x})$.

Or equivalently:

An arithmetic theory T is *ω-inconsistent* iff, for some open wff $\varphi'(\mathsf{x})$, $T \vdash \exists \mathsf{x} \varphi'(\mathsf{x})$, yet for each number m we have $T \vdash \neg \varphi'(\overline{m})$.

Compare and contrast. Suppose T can prove $\varphi(\overline{m})$ for each m. T is ω-incomplete if it can't also prove something we'd like it to prove, namely $\forall \mathsf{x} \varphi(\mathsf{x})$. While T is ω-inconsistent if it can actually prove the *negation* of what we'd like it to prove, i.e. it can prove $\neg \forall \mathsf{x} \varphi(\mathsf{x})$.

Note that ω-inconsistency, like ordinary inconsistency, is a syntactically defined property: it is characterized in terms of which wffs can be proved, not in

---

[1]To help explain choice of the terminology: 'ω' is the logicians' label for the natural numbers taken in their natural order. Relatedly, ω is the first infinite ordinal.

terms of what they mean. Note too that ω-consistency – defined of course as not being ω-inconsistent! – implies plain consistency. That's because T's being ω-consistent is a matter of its *not* being able to prove a certain combination of wffs, which entails that T can't be inconsistent and prove *all* wffs.

Now, ω-incompleteness in a theory of arithmetic is a regrettable weakness; but ω-inconsistency is a Very Bad Thing (not as bad as outright inconsistency, maybe, but still bad enough). For evidently, a theory that can prove each of $\varphi(\overline{m})$ and yet also prove $\neg\forall x\varphi(x)$ is just not going to be an acceptable candidate for regimenting arithmetic.

(c) That last observation can be made vivid if we bring semantic ideas temporarily back into play. Suppose the language of an arithmetic theory T is given a *standard* interpretation, by which we here mean just an interpretation whose domain comprises the natural numbers, and on which T's standard numerals denote the intended numbers (with the logical apparatus also being treated as normal, so that inferences in T are truth-preserving). And suppose further that on this interpretation, the axioms of T are all true. Then T's theorems will all be true too. So now imagine that, for some $\varphi(x)$, T does prove each of $\varphi(0)$, $\varphi(1)$, $\varphi(2)$, ... By hypothesis, these theorems will then be true on the given standard interpretation; so this means that every natural number must satisfy $\varphi(x)$; so $\forall x\varphi(x)$ is true since the domain contains only natural numbers. Hence $\neg\forall x\varphi(x)$ will have to be false on this standard interpretation. Therefore $\neg\forall x\varphi(x)$ can't be a theorem, and T must be ω-consistent.

Hence, contraposing, we have

> **Theorem 21.5** *If T is ω-inconsistent then T's axioms can't all be true on a standard arithmetic interpretation.*

Given that we want formal arithmetics to have axioms which *are* all true on a standard interpretation, we must therefore want ω-consistent arithmetics. And given that we think PA *is* sound on its standard interpretation, we are committed to thinking that it *is* ω-consistent.

(d) We now confirm that ω-consistency is a stronger requirement than plain consistency.

> **Theorem 21.6** *There are arithmetical theories which are consistent but not ω-consistent.*

Proof Consider the theory $\mathsf{PA}^\dagger = \mathsf{PA} + \neg\mathsf{G}$ (i.e. the theory you get by adding $\neg\mathsf{G}$ as a new axiom to PA).

Then if PA is consistent, so is $\mathsf{PA}^\dagger$. For if $\mathsf{PA} + \neg\mathsf{G}$ entailed a contradiction, we'd have $\mathsf{PA} \vdash \mathsf{G}$, which is impossible if PA is consistent.

But $\mathsf{PA}^\dagger$ trivially entails $\neg\mathsf{G}$, i.e. $\exists x\mathsf{Gdl}(x, \ulcorner \mathsf{U} \urcorner)$. And containing PA, it still entails $\neg\mathsf{Gdl}(\overline{m}, \ulcorner \mathsf{U} \urcorner)$ for each m (see the proof of Theorem 21.4). So $\mathsf{PA}^\dagger$ is ω-inconsistent. $\boxtimes$

21.7 Finishing the syntactic argument

(a)   We next show that PA can't prove the negation of G, again without assuming PA's soundness: this time we'll just make the syntactic assumption of ω-consistency.

B.   *If* PA *is* ω-*consistent,* PA $\nvdash \neg$G.

Proof   Suppose that PA is ω-consistent but $\neg$G is provable in PA. That's equivalent to assuming (i) PA $\vdash \exists x$Gdl(x, $\ulcorner$U$\urcorner$).

But if PA is ω-consistent, it is consistent. So if $\neg$G is provable, G is *not* provable. Hence for any m, m cannot code for a proof of G. But G is (again!) the wff you get by diagonalizing U. Therefore, by the definition of *Gdl*, our assumptions imply that $Gdl(m, \ulcorner$U$\urcorner)$ is false, for each m. So, by the requirement that Gdl captures *Gdl*, we have (ii) PA $\vdash \neg$Gdl($\overline{m}, \ulcorner$U$\urcorner$) for each m.

But (i) and (ii) together make PA ω-inconsistent after all, contrary to hypothesis. Hence, if PA is ω-consistent, $\neg$G is unprovable. $\boxtimes$

(b)   Now to put all the ingredients together. Recall that G is logically equivalent to a Π_1 sentence, i.e. to a sentence of Goldbach type. That observation, plus the result (A) from Section 21.5, plus the result (B) which we've just proved, gives us the classic syntactic incompleteness theorem for PA:

Theorem 21.7   *There is an* L_A-*sentence* φ *of Goldbach type such that, if* PA *is consistent then* PA $\nvdash \varphi$; *and if* PA *is* ω-*consistent then* PA $\nvdash \neg\varphi$.

Note too that – without further consideration of the interpretation of such a Π_1 undecidable sentence φ – it is immediate that it will be a true sentence of L_A. Why? PA $\nvdash \neg\varphi$, so φ is consistent with PA, hence φ is a Π_1 sentence consistent with Q, hence (by Theorem 11.6) φ is true.

21.8 Canonical Gödel sentences and what they say

So much for the two Gödelian proofs of the incompleteness of PA. We can either combine the bolder semantic assumption that PA is sound with the weak result that every p.r. function is expressible in L_A, or combine the more modest syntactic assumption that PA is ω-consistent (and so consistent) with the stronger and harder-to-prove result that every p.r. function is capturable in PA.[2] Either way, we can show that our G is 'formally undecidable' in PA.[3]

---

[2]In Section 25.3, we will see how to weaken the assumption needed for a syntactic proof of incompleteness to plain consistency; but that takes further work.

[3]Compare Remark A from Section 8.1 commenting on the different assumptions of our two informal arguments for incompleteness in Chapters 6 and 7.

The distinction between the 'semantic' and 'syntactic' routes to incompleteness is already there in Gödel's original 1931 paper. But the first commentator to highlight the point that

We might call our particular G a *canonical* Gödel sentence for three reasons: (i) it is defined in terms of a wff that we said captures *Gdl* in a canonical way by reflecting the full definition of that p.r. relation in a standard way; (ii) because it is the sort of sentence that Gödel himself constructed in his canonical paper; (iii) it is the kind of sentence most people standardly have in mind now when they talk of '*the*' Gödel sentence for PA (though since Gdl certainly isn't unique in canonically capturing *Gdl*, neither is G).

Now, it is often claimed that a Gödel sentence like G is not only true if and only if unprovable, but actually *says* of itself that it is unprovable: Gödel himself describes his original Gödel sentence this way (Gödel, 1931, p. 151). However, (i) this claim is never *strictly* true, though (ii) if we do restrict ourselves to canonical Gödel sentences,[4] then *these* can reasonably be said to *indirectly* say that they are unprovable.

(i)   First, let's stress that G (when fully unpacked) is just another sentence of PA's language L_A, the language of basic arithmetic. It is an enormously long wff involving the first-order quantifiers, the connectives, the identity symbol, and 'S', '+' and '×', which all have the standard interpretation built into L_A.[5] The semantics built into L_A therefore tells us that the various terms inside G (in particular, the numerals) have *numbers* as values – their values therefore aren't non-numerical items like, e.g., linguistic expressions. Because it is about numbers and not expressions, G can't strictly speaking refer to *itself* and so can't say that it is itself unprovable.

(ii)   That was trite, but important to stress. But now take the case where we are dealing with a canonical Gödel sentence like G. There is perhaps a reasonable sense in which this can be described as *indirectly* saying that it is unprovable.

Note, we are *not* going to be making play with some radical re-interpretation of G's symbols (for doing *that* would just make any claim about what G might say boringly trivial: if we are allowed radical re-interpretations – like spies choosing

---

there *are* two proofs seems to be Andrzej Mostowski in his (1952b) – see Section 23.1 below. Some other early writers in the intervening years do wobble over the point.

[4]Compare the wider class of undecidable Gödel sentences allowed in Section 24.6.

[5]The wff G embeds the standard numeral for U(y)'s Gödel number. Given our particular Gödel-numbering scheme, this number will be *huge*, vastly larger than the number of particles in the known universe. So writing out our G in pure L_A *without abbreviations* would not be a practical possibility.

How significant is this fact? Should we therefore try to find an alternative coding scheme that keeps the size of Gödel-numbers more manageable? Well, there are other coding schemes on the market which deliver much smaller, though still pretty large, Gödel-numbers. But writing out numerals SSS . . . S0 in unabbreviated L_A soon gets pretty impractical for any large numbers. So it just isn't worth fussing over the choice of coding scheme, at least on questions of size.

And after all, the situation here is actually not at all unusual in mathematics. Given our limited cognitive powers, we need to use chains of abbreviations all the time, and we often work with propositions whose official definitional unpacking into the relevant 'basic' terms would be far too long to grasp. This general phenomenon certainly isn't without its interest and its problematic aspects: see e.g. Isles (1992). However, the phenomenon occurs right across mathematics. So we aren't going to worry unduly that it crops up again here in our dealings with Gödel sentences.

to borrow ordinary words for use in a secret code – then any string of symbols can be made to say anything). No, it is because the symbols are still being given their *standard* interpretation that we can recognize that Gdl (when unpacked) will express *Gdl*, given the background framework of Gödel numbering which is involved in the definition of the relation *Gdl*. For remember, Gdl canonically captures *Gdl* in a way which perspicuously reveals which p.r. function it expresses. Therefore, given that coding scheme, we can recognize just from its construction that G will be true when no number m is such that $Gdl(m, \ulcorner U \urcorner)$, and so no number numbers a proof of G.

In short, given we know the coding scheme, *we can immediately see that G is constructed in such a way as to make it true just when it is unprovable*: we don't need further investigation. That is the rather limited sense in which we might claim that, via our Gödel coding, the canonical G already 'indirectly says' that it is unprovable.[6]

---

[6] Recall that if Gdl expresses *Gdl*, so does $Gdl^e(x, y) =_{def} Gdl(x, y) \wedge \theta$ for any free-loading true sentence θ (see Section 5.4(c)).

Suppose we don't actually know that θ is true. Then we won't be able just to read off from its construction that Gdl^e in fact expresses *Gdl*, in the way that we *can* read that off from the construction of Gdl. And now consider a corresponding G^e constructed out of Gdl^e as G was constructed from *Gdl*. Even when you know about the Gödel coding, you won't know that in fact G^e is true if and only if it is unprovable in PA – in contrast to the case of G, that won't be something that G^e shows on its face.

See also the discussion in Section 24.6.

22 Gödel's First Theorem

Back in Chapter 10, we introduced the weak arithmetic Q, and soon saw that it is boringly incomplete. In Chapter 12, the stronger arithmetic Δ_0 was defined, and this too can be seen to be incomplete without invoking Gödelian methods. Then in Chapter 13 we introduced the much stronger first-order theory PA, and remarked that we couldn't in the same easy way show that it fails to decide some elementary arithmetical claims. However, in the last chapter it has turned out that PA also remains incomplete.

Still, that result in itself isn't yet hugely exciting, even if it is perhaps rather unexpected (see Section 13.3). After all, just saying that a particular theory T is incomplete leaves wide open the possibility that we can patch things up by adding an axiom or two more, to get a complete theory T^+. As we said at the very outset, the real force of Gödel's arguments is that they illustrate *general* methods which can be applied to *any* theory satisfying modest conditions in order to show that it is incomplete. They reveal that a theory like PA is not only incomplete but in a good sense *incompletable*.

The present chapter explains these crucial points.

22.1 Generalizing the semantic argument

In Section 21.3, we showed that PA is incomplete on the semantic assumption that its axioms are true (given that its standard first-order logic is truth-preserving). In this section, we are going to extend the semantic argument for incompleteness to other theories.

(a) We said in Section 4.3 that a theory T is an effectively axiomatized formal theory iff (i) it is effectively decidable what counts as a T-wff/T-sentence, (ii) it is effectively decidable which wffs are T's (non-logical) axioms, (iii) T uses a proof system such that it is effectively decidable whether an array of T-wffs counts as a well-constructed derivation, and so (iv) it is effectively decidable which arrays of T-wffs count as proofs from T's axioms. We will now reflect this informal characterization in the following definition:

> A theory T is *p.r. axiomatized* iff (i′) the numerical properties of being the g.n. of a T-wff/T-sentence are primitive recursive, (ii′) the numerical property of being the g.n. of an axiom is p.r., likewise (iii′) the numerical property of being the super g.n. of a properly constructed proof is p.r., and therefore (iv′) the numerical property

of being the super g.n. of a properly constructed proof from T's axioms is p.r. too.[1]

Now, we know that there are effectively decidable properties which aren't primitive recursive: so this opens up the possibility that there are effectively axiomatized formal theories which aren't p.r. axiomatized. We will return to consider this possibility later. But for the moment, just note that any ordinarily axiomatized formal theory of a familiar kind will indeed be p.r. axiomatized (you won't have to do open-ended searches to check whether a wff is an axiom, for example).

(b)   Now let's turn to think again about how the semantic argument for PA's incompleteness worked. The key ingredient is the claim that

1. There is an open Σ_1 wff Gdl which *expresses* the relation *Gdl* – where, recall, $Gdl(m, n)$ iff m Gödel-numbers a proof in PA of the sentence which is the diagonalization of the wff Gödel-numbered by n.

And, let's not forget, we also need quantifiers and negation to form the wff G from Gdl.

Our demonstration that (1) holds for PA is underpinned by the facts that

2. The relation *Gdl* is primitive recursive (see Section 21.2).

3. PA's language L_A can express all p.r. functions – in fact, express them by Σ_1 wffs (see Sections 15.1–15.3).

And L_A of course provides negation and the numerical quantifiers needed for defining G.

Digging down a bit further, recall that *Gdl* can be defined in terms of *Prf* – where $Prf(m, n)$ iff m numbers a PA-proof of the sentence with number n – and the trivially p.r. *diag* function. And reviewing our proof that *Prf* is p.r. in Section 20.4, we see that this presupposes that

4. PA is p.r. axiomatized.

For the proof depends crucially on the facts that properties like *Wff* (of PA) and *Axiom* (of PA) are p.r., and that the relations that hold between the codes for the input and output of PA's inference rules are p.r. too.

And that's enough proof-mining to enable us to make a first generalization. Suppose that we are dealing with any other theory T such that

G1. T is p.r. axiomatized,

G2. T's language includes L_A,

---

[1] Recall, by the way, the argument of Section 19.2, where we explained why a numerical property like being the g.n. of a T-wff will be p.r. on any acceptable numbering scheme if it is p.r. on our default Gödel-style scheme. So the question whether a theory is p.r. axiomatized is not embarrassingly relative to our particular kind of numbering scheme.

where we say a language L includes L_A if all L_A-wffs are also L-wffs, and these shared wffs continue to receive the same interpretation (trivially, L_A counts as including itself).

Then (G1) gives us the analogue of (2), i.e. there will be a p.r. relation Gdl_T such that $Gdl_T(m, n)$ holds just so long as m numbers a T-proof of the sentence which is the diagonalization of the wff numbered by n. And (G2) gives us the analogue of (3): T's language, since it includes L_A, can express all p.r. functions by Σ_1 wffs.

There will therefore be, in particular, an open Σ_1 wff of T's language which expresses the p.r. relation Gdl_T. Let Gdl_T be such a wff. We then have numerical quantifiers and negation available to form a corresponding new Gödel sentence G_T in L_A, again trivially equivalent to some sentence φ of Goldbach type, which is true if and only if it is unprovable in T.[2]

From this point, we just use the same easy argument that we used to prove Theorem 21.2 in order to show that, if T is an *arithmetically sound* theory (i.e. if at least its L_A theorems are all true), then G_T is true-but-unprovable. Thus:

> **Theorem 22.1** *If a theory T, whose language includes L_A, is p.r. axiomatized and is arithmetically sound, then there is an L_A-sentence φ of Goldbach type such that $T \nvdash \varphi$ and $T \nvdash \neg\varphi$.*

22.2 Incompletability

Suppose T is a p.r. axiomatized, arithmetically sound theory, with a truth-preserving logic, whose language includes L_A; and suppose φ is one of the unde-cided arithmetical sentences such that neither $T \vdash \varphi$ nor $T \vdash \neg\varphi$. Either φ or $\neg\varphi$ is true. Consider the result of adding the true one to T as a new axiom. The new expanded theory T^+ is still arithmetically sound, and still a p.r. axiomatized theory, whose language still includes L_A. So, although T^+ by construction now decides φ the right way, T^+ *must still be incomplete*.

Take PA as an example. Assume it is sound. Then PA + G (the theory you get by adding G as a new axiom) is also sound. This new theory trivially entails G. But being sound, *its* canonical Gödel sentence $\mathsf{G}_{(\mathsf{PA+G})}$ is unprovable in the theory. The further augmented theory PA + G + $\mathsf{G}_{(\mathsf{PA+G})}$ proves it: but being sound, and still p.r. axiomatized, this theory too is incomplete. And so it goes.

Add in as many new true axioms to PA as you like, even augment the truth-preserving deductive apparatus, and the result will still be incomplete – unless

---

[2]Of course, when we move to consider a different theory T, the set of axioms and/or the set of rules of inference will change (and if T involves new symbols, then the scheme for Gödel-numbering will need to be extended). So the details of the corresponding relation Gdl_T will naturally change too. Hence the details of Gdl_T will change too, and likewise G_T. But still, the key point is that we can construct our new canonical Gödel sentence along exactly the same general lines as before in constructing the Gödel sentence G for PA, to get a sentence which 'indirectly says' that it is unprovable in T. In particular, we'll only need to use basic arithmetical vocabulary.

it ceases to be p.r. axiomatized. In a phrase, PA *is not just incomplete, but it is incompletable* (if we still want a sound, p.r. axiomatized theory).

22.3 Generalizing the syntactic argument

So far, so good. Now we turn to generalizing the syntactic argument for incompleteness. Looking at our proof for Theorem 21.7, we can see that the essential fact underpinning *this* incompleteness proof is:

1′. There is an open Σ_1 wff Gdl which *captures* the relation *Gdl*.

(Plus, again, we need to have quantification and negation available to form G.)
  Underpinning (1′) are the following facts:

2. The relation *Gdl* is primitive recursive.

3′. PA can capture all p.r. functions – in fact, capture them by Σ_1 wffs.

But, to repeat, (2) depends essentially on the fact that

4. PA is p.r. axiomatized.

And condition (3′) obtains because

5. PA extends Q,

which means we can establish (3′) by appeal to Theorem 17.2 which says that even Q can capture all p.r. functions using Σ_1 wffs.
  And that's enough proof-analysis for us to be able to generalize again. Suppose that we are dealing with any other theory T such that

G1. T is p.r. axiomatized.

G2′. T extends Q,

where, recall, T extends Q if all the wffs of L_A are wffs of T's language, and T can prove all Q-theorems. NB, as a limiting case, Q counts as 'extending' itself. Then (G1) gives us the analogue of (2) again: there is a p.r. relation Gdl_T such that $Gdl_T(m, n)$ holds just so long as m numbers a T-proof of the diagonalization of the wff numbered by n. (G2′) gives the analogue of (3′), so there will be a Σ_1 open wff Gdl$_T$ which captures the relation Gdl_T. Hence, using the now familiar construction, we can again form a corresponding Gödel sentence G$_T$.
  Here's a useful abbreviatory definition:

  We will say that a theory T is *nice* iff T is consistent, p.r. axiomatized, and extends Q.

There doesn't actually seem to be a standard label for this kind of theory: so 'nice' is our own local term.[3]

Then, once we have constructed the Gödel sentence G_T from Gdl_T, the line of argument will continue exactly as before (as in Sections 21.5 and 21.7). So we get

> **Theorem 22.2** *If T is a nice theory then there is an L_A-sentence φ of Goldbach type such that $T \nvdash \varphi$ and (if T is also ω-consistent) $T \nvdash \neg\varphi$.*

And, for just the same reasons that we noted after proving Theorem 21.7, such an undecidable Π_1 sentence of L_A must be true.

This result obviously gives us a corresponding fact about incompletability. Suppose we beef up some nice, ω-consistent, theory T by adding new axioms (e.g. by adding previously unprovable Gödel sentences). Then T will stay incomplete – unless it becomes ω-inconsistent (bad news) or stops being nice (even worse news).[4]

22.4   The First Theorem

Of course, Gödel in 1931 didn't know that Q is p.r. adequate (Q wasn't isolated as a minimal p.r. adequate arithmetic until twenty years later). So he didn't have the concept of 'niceness' and it would be anachronistic to identify our very neat Theorem 22.2 as *the* First Incompleteness Theorem. But it is a near miss.

Looking again at our analysis of the syntactic argument for incompleteness, we see that we are interested in theories which extend Q *because we are interested in p.r. adequate theories which can capture p.r. relations like Gdl.* So instead of mentioning Q, let's now instead explicitly write in the requirement of p.r. adequacy. Then we have, by just the same arguments,

> **Theorem 22.3** *If T is a p.r. adequate, p.r. axiomatized theory whose language includes L_A, then there is an L_A-sentence φ of Goldbach type such that, if T is consistent then $T \nvdash \varphi$, and if T is ω-consistent then $T \nvdash \neg\varphi$.*

---

[3]Perhaps 'minimally nice' or some such would have been better – since niceness in our sense is only a basic necessary condition for being nice in the everyday sense of being an attractively acceptable theory (for example, a nice theory in our limited sense could have lots of axioms beyond Q's which are *false* on the standard interpretation). But let's stick to the snappy shorthand.

[4]A reality check. Since PA doesn't decide every L_A sentence, assuming it is consistent, no *weaker* theory which proves less can decide every L_A sentence either! In other words, it *quite trivially* follows from PA's incompleteness that Q – and every other theory whose language is L_A and which is contained in PA – is incomplete. So the main interesting new content of our theorem is that theories *stronger* than PA must stay incomplete too, so long as they are sensibly axiomatized and remain consistent.

It is this rather less specific syntactic version of the incompleteness result which perhaps has as much historical right as any to be called Gödel's First Theorem. For in his 1931 paper, Gödel first proves his Theorem VI, which shows that the formal system P – which is his simplified version of the hierarchical type-theory of *Principia Mathematica* – has a formally undecidable sentence of Goldbach type. Then he immediately generalizes:

> In the proof of Theorem VI no properties of the system P were used besides the following:
>
> 1. The class of axioms and the rules of inference (that is, the relation 'immediate consequence') are [primitive] recursively definable (as soon as we replace the primitive signs in some way by the natural numbers).
>
> 2. Every [primitive] recursive relation is definable [i.e. is 'capturable'] in the system P.
>
> Therefore, in every formal system that satisfies the assumptions 1 and 2 and is ω-consistent, there are undecidable propositions of the form $[\forall x F(x)]$, where F is a [primitive] recursively defined property of natural numbers,[5] and likewise in every extension of such a system by a [primitive] recursively definable ω-consistent class of axioms. (Gödel, 1931, p. 181)

Putting that generalized version together with Gödel's Theorem VIII – which tells us that p.r. properties can be pinned down using only notions from basic arithmetic, i.e. from L_A – is tantamount to our Theorem 22.3.[6]

And so we have established the First Incompleteness Theorem at very long last. Let joy be unconfined!

---

[5] In our terms, F is $\neg\mathsf{Gdl}(\mathsf{x}, \ulcorner\mathsf{U}\urcorner)$ which expresses the p.r. property had by a number when it doesn't stand in the *Gdl* relation to the number $\ulcorner\mathsf{U}\urcorner$.

[6] 'Hold on! If *that's* the First Theorem, we didn't need to do all the hard work showing that **Q** and **PA** are p.r. adequate, did we?' Well, yes and no. No, proving Theorem 22.3 of course doesn't depend on proving that any particular theory can capture all p.r. functions. But yes, showing that this Theorem has real bite, showing that it applies to familiar arithmetics, does depend on proving the adequacy theorem.

By the way, Theorems 22.2 and 22.3 are not *quite* equivalent – in effect because there are some alternative very weak p.r. adequate arithmetics which are neither contained in nor contain **Q**. For some details, see e.g. Boolos et al. (2002, Sections 16.2, 16.4), comparing the theories which are there called **Q** and **R**. In fact, what's crucial for p.r. adequacy is that **Q** and **R** both deliver all the results that we listed in Section 11.8. But the differences here won't matter to us, and we'll continue to concentrate on our **Q** as the neatest weak p.r. adequate arithmetic.

23 Interlude: About the First Theorem

We have achieved our initial goal, namely to prove Gödel's First Incompleteness Theorem, and to do so in Gödelian style. Other proofs will come later. But it will do no harm to pause to survey what we've established so far and how we established it. Equally importantly, we should make it clear what we have *not* proved. The Theorem attracts serious misunderstandings: we will briefly block a few of these. Finally in this interlude, we will outline the next block of chapters which explore further around and about the First Theorem.

23.1 What we have proved

We begin, then, with the headlines about what we *have* proved (we are going to be repeating ourselves, but – let's hope! – in a good way).

Suppose we are trying to regiment the truths of basic arithmetic, meaning the truths expressible in terms of successor, addition, multiplication, together with the apparatus of first-order logic (so these are the truths that can be rendered in L_A). Ideally, we would like to construct a formally defined consistent theory T whose language includes L_A and which proves *all* the truths of L_A (and which entails no falsehoods). So we would like T to be negation-complete, at least for sentences of L_A. But the First Theorem tells us that, given some entirely natural assumptions, there just can't be such a negation-complete theory.

The first required assumption is that T should be set up so that it is effectively decidable whether a putative T-proof really *is* a well-constructed derivation from T's axioms. Indeed, we want rather more: we want it to be decidable what counts as a T-proof without needing open-ended search procedures (it would be an odd kind of theory of arithmetic where, e.g., checking whether some wff is an axiom takes an unbounded search). Sharpened up, this is the assumption that T should not just be an axiomatized formal theory but be a p.r. axiomatized theory.

But next, there's a fork in the path.

(a) Let's first assume that T is an *arithmetically sound* theory – after all, our original motivation for being interested in the issue of completeness was wanting a theory that regiments all the *truths* of L_A. Given the soundness assumption, we then have the following *semantic* argument for incompleteness (remember, we are assuming T's language includes L_A):

1. Every p.r. function and relation can be *expressed* in L_A, the language of basic arithmetic. (Theorem 15.1, proved in Sections 15.1 to 15.3 – the argument is quite elementary once we grasp the β-function trick.)

2. In particular, the relation $Prf_T(m, n)$, which holds when m codes for a T-proof of the sentence with code number n, is primitive recursive – on the assumptions that T is p.r. axiomatized and that we have a sensible coding system. (That's part of Theorem 19.1, for which we gave a quick and dirty but convincing argument in Section 19.4 and then a fuller proof-sketch in Section 20.4.) Likewise, the relation $Gdl_T(m, n)$, which holds when m codes for a T-proof of a sentence which is the diagonalization of the wff with code number n, is also primitive recursive (since Gdl_T is definable in terms of Prf_T and the trivially p.r. $diag_T$ function which maps the code for a wff to the code for its diagonalization).

3. Then, using the fact that Gdl_T must be expressible in L_A, and hence also expressible in the language of T, we can construct an arithmetic wff G_T which is true if and only if it is unprovable (see Section 21.2 for the construction which is ingenious but quite easy to understand). Moreover, we can do this in a way which makes G_T trivially equivalent to a Π_1 wff, i.e. to an L_A wff of Goldbach type.

4. Since G_T is true if and only if it is unprovable, then – given the assumption that T is arithmetically sound, i.e. proves no false L_A sentences – there's an entirely straightforward argument to the dual conclusions that $T \nvdash G_T$ and $T \nvdash \neg G_T$. The same applies, of course, to any Π_1 equivalent of G_T. (Use the argument which we first gave, in fact, in Section 1.2.)

In sum, then, Gödel's semantic argument shows that

Theorem 22.1 *If a theory T, whose language includes L_A, is p.r. axiomatized and is arithmetically sound, then there is an L_A-sentence φ of Goldbach type such that $T \nvdash \varphi$ and $T \nvdash \neg\varphi$.*

And let's now quickly compare this result with the upshot of the easy semantic incompleteness proof that we gave much earlier:

Theorem 6.3 *If T is a sound effectively axiomatized theory whose language is sufficiently expressive, then T is not negation-complete.*

What has all our hard work since Chapter 6 bought us?

One point of difference between the old and new results is that the old result applied to axiomatized theories generally, and the new result is so far only about p.r. axiomatized theories in particular. But that's minor. We've just noted that we are normally interested in axiomatized theories that are presented as being p.r. axiomatized: and, as we'll see in Section 26.1, *any* effectively axiomatized theory can in fact be recast as a p.r. axiomatized one.

So the key point of difference between the old and new results is this. The old theorem told us about theories with 'sufficiently expressive' languages (i.e. languages that can express any one-place computable function), but it didn't tell us anything at all about what such languages look like. The possibility was left open that sufficiently expressive languages would have to be rather rich.[1] And that in turn left open the possibility that – even though a theory with a richly expressive language can't be complete – weaker theories covering e.g. basic arithmetic could be complete.[2] Our new result closes off those possibilities: any sound theory of basic arithmetic must be incomplete.

(b) So much for Gödel's *semantic* route to his incompleteness result. Suppose, however, we now take the other fork in the path, and don't assume that T is sound. We make life a bit harder for ourselves. But we can still prove that T must be incomplete, given weaker assumptions. We can put one version of the *syntactic* argument as follows:

1'. Assume T contains at least as much arithmetic as Q. Then every p.r. function and relation can be *captured* in T (Theorem 17.1).

2'. As before, assuming that T is p.r. axiomatized, Prf_T and so $Gdl_T(m, n)$ are primitive recursive.

3'. Again, we can construct the wff G_T in the now familiar way, starting from a wff that captures Gdl_T.

4'. We can then show: if T is consistent, $T \nvdash G_T$; and if T is ω-consistent, $T \nvdash \neg G_T$. (Generalizing the arguments of Sections 21.5 and 21.7.)

Note that the stages $1'$ and $4'$ of the syntactic argument are a bit trickier than their counterparts in the semantic argument.[3] That's the price we pay for managing without the strong assumption of soundness. But we get a benefit, for we of course get an incompleteness result that now applies irrespective of the soundness of the theory in question:

---

[1] Though we did in fact say something to motivate the thought that sufficiently expressive languages don't have to be *very* rich in Section 6.1, fn. 1.

[2] Recall, Q is trivially incomplete, even though it extends the trivially complete BA: so we know that an incomplete theory in a richer language can indeed extend a complete sub-theory in a more restricted language.

[3] Trickier, yes. But still not *that* difficult. And it was very important to Gödel's overall conception of his enterprise that this was so. Georg Kreisel puts it like this in his memoir of Gödel:

> Without losing sight of the permanent interest of his work, Gödel repeatedly stressed – at least, during the second half of his life – how little novel mathematics was needed; only attention to some quite commonplace (philosophical) distinctions; in the case of his most famous result: between arithmetic truth on the one hand and derivability by (any given) formal rules on the other. Far from being uncomfortable about so to speak getting something for nothing, he saw his early successes as special cases of a fruitful general, but neglected, scheme.
>
> By attention to or, equivalently, analysis of suitable traditional philosophical notions and issues, adding possibly a touch of precision, one arrives painlessly at appropriate concepts, correct conjectures, and generally easy proofs. (Kreisel, 1980, p. 150)

169

Theorem 22.2 *If T is a nice theory – is a consistent, p.r. axiom-atized theory which extends Q – then there is an L_A-sentence φ of Goldbach type such that $T \nvdash \varphi$ and (assuming T is ω-consistent) $T \nvdash \neg\varphi$.*

Let's again quickly compare this result with the upshot of an earlier incompleteness argument, this time the 'syntactic' proof that:

Theorem 7.2 *If T is a consistent, sufficiently strong, effectively axiomatized theory of arithmetic, then T is not negation-complete.*

What has all our hard work since Chapter 7 bought us this time?

Well, one point of difference between the old and new results is again that the old one applied to axiomatized formal theories in general while the new result so far applies only to p.r. axiomatized theories: but, as promised, we will see how to bridge the apparent gap.

Another difference is that the old result required only the assumption that our theory T is consistent, while half of the new result requires the stronger assumption that T is ω-consistent. But in fact, this too is a difference which we can with some effort massage away. We can replace the Gödel sentence G_T with a cunningly designed Rosser sentence R_T, and replace the relatively easy argument for G_T's undecidability (assuming T's ω-consistency) with a correspondingly more complex argument for R_T's undecidability (assuming only that T is consistent). This is the so-called Gödel-Rosser theorem, which will enable us to show that nice theories are *always* incomplete, whether ω-consistent or not. (For more on this, see the Section 25.3.)

The critical difference between our old and new syntactic arguments for incompleteness is therefore this. The old theorem told us about theories which are 'sufficiently strong' (i.e. theories which can capture all decidable properties of numbers), but it didn't tell us what such theories look like. So the possibility was left open that such incomplete theories have to be immensely rich if they are to capture all decidable properties, leaving room for less rich but still powerful complete theories. Our new result has foreclosed that possibility too. It tells us that to get incompleteness via the syntactic route it is enough to be working with a theory which captures all p.r. properties and relations, and for that it suffices that the theory includes Q.

(c)   Let's summarize the summary! Beginning with the First Interlude, we have repeatedly stressed that there are *two* routes to incompleteness results: and now we have highlighted that each route can be followed either quickly-but-abstractly or slowly-but-more-concretely. The first route goes via the semantic assumption that we are dealing with sound theories, and otherwise uses a weak result about what certain theories can *express*: the second route goes via the syntactic assumptions of consistency/ω-consistency, and has to combine that with a stronger result about what theories can *prove*. And in each case, we can either give a fast argument that applies to an abstractly described class of theories (those with a 'sufficiently expressive' language, those which are 'sufficiently strong'), or we can do the work needed to show that the arguments apply to concretely described theories (theories whose language includes L_A, theories which extend Q).[4]

Now, it is true that in his 1931 paper Gödel rather downplays the semantic route to incompleteness, touching on it only in his informal introduction (in Chapter 37, we will say something about why he wants officially to avoid relying on the notion of truth in his paper and so highlights the syntactic route). But the distinction is very important none the less, and – as Andrzej Mostowski puts it –

> The different kinds of incompleteness proofs lead to different important corollaries. We obtain them when we investigate the problem of formalization of these proofs. It turns out that the semantical proof [for the incompleteness of theory S] is not formalizable within S itself. As a corollary we obtain the important theorem that the notion of "truth" for the system S is not definable within S. The syntactical proofs are on the contrary [normally] formalizable within S and studying carefully this fact we can recognize with Gödel that the consistency of S is not provable by means formalizable within S. (Mostowski, 1952b, p. 12)

Which is a nice observation: if we imagine setting out to formalize the semantic and syntactic incompleteness proofs, we will arrive at respectively Tarski's Theorem and Gödel's Second Theorem. More about this in due course.

23.2   Some ways to argue that G_T is true

The Gödelian arguments show that if a rich enough arithmetical theory T is sound, or indeed if it is just consistent, then there will be a canonical Gödel sentence G_T which is unprovable in T, and – because it indirectly 'says' it is unprovable – G_T will then be true.[5]

Believing on the basis of the Gödelian arguments that G_T really *is* true will therefore depend on believing that T is indeed sound, or at least is consistent. But note that our reasons for such a belief about T can be many and varied – and hence our reasons for accepting G_T as true can be equally varied. Let's take some examples.

(1)   Start with the most humdrum case. I believe that G_Q, the canonical Gödel sentence for Q, is true. Why? Well, I endorse Q as a good (though very partial) arithmetical theory; I accept its axioms as general truths of arithmetic, I recognize that first-order logic takes us from truths to truths, so I accept that the

---

[4]The more abstract our characterization of the target theories, the quicker our proof of incompleteness can be. For a particularly elegant, particularly abstract, fast-track proof, see Smullyan (1992, ch. 1).

[5]Reality check: why haven't we mentioned ω-consistency here?

theory is sound. The semantic argument for incompleteness then tells me that G_Q is unprovable and hence true.

(2)   A fancier case. Continue to assume that Q is sound, and now consider the theory Q_G that we introduced in Section 11.7, which is the theory we get by adding Goldbach's conjecture as an additional axiom to Q. Then Q_G is consistent if and only if Goldbach's conjecture is true of the natural numbers (the 'if' direction is trivial, for if the conjecture is true all Q_G axioms are true; the 'only if' direction is – as we noted – a corollary of Theorem 11.7). We should therefore accept the canonical Gödel sentence G_{Q_G} for Q_G as being true if and only if we in fact accept Goldbach's conjecture itself as true. For if the conjecture is *false*, Q_G is inconsistent, so proves everything including G_{Q_G} which is therefore false. So why might we tentatively accept Goldbach's conjecture, if we do? In part because of our failure ever to find any counterexamples to the conjecture. And in part, perhaps, because of some probabilistic considerations concerning the distribution of primes, and other 'almost-proofs' for the conjecture. It is those sorts of considerations we would have to appeal to in defending a claim that G_{Q_G} is true.

Here's a variant example (2'). This time take the theory Q_F which we get by adding a statement of Fermat's Last Theorem as an additional axiom to Q. Since we can (with some β-function trickery) regiment Fermat's Last Theorem as a Π_1 wff, then – by Theorem 11.7 again – Q_F is consistent if and only if Fermat's Last Theorem is true. We confidently believe *this* theory is consistent because we have Andrew Wiles's *proof* of the Theorem, although his proof involves some highly infinitary ideas going far beyond ordinary arithmetic. That highly sophisticated proof can then ground our belief that Q_F's canonical Gödel sentence must be true.

(3)   Consider next the case of the theory $\mathsf{PA}^\dagger = \mathsf{PA} + \neg G$ (which we met in proving Theorem 21.6). This theory is unsound on the interpretation built into its language L_A. But, as we saw, it must be consistent, assuming PA is. So, given we accept that PA *is* consistent and know about the syntactic incompleteness proof for PA, we'll accept that $\mathsf{PA}^\dagger$ is also consistent, and hence accept that *its* canonical Gödel sentence is true too.

(4)   For a final example, take the theory $Q^\dagger$ which we get by adding to Q the *negation* of $\forall x(0 + x = x)$ as an additional axiom. Again, $Q^\dagger$ is unsound. But in Section 10.5 we described a re-interpretation of Q's language in which we took the domain of quantification now to be the numbers plus a couple of rogue elements, and then we redefined the successor, addition and multiplication functions for this augmented domain. We can easily see that $Q^\dagger$'s axioms and hence all its theorems are true on this reinterpretation, and it follows that $Q^\dagger$ can't prove any contradiction (for contradictions would still be false on the new interpretation). In this case, therefore, we have an interpreted theory which speaks (falsely) about one structure; but we prove it consistent by reinterpreting

it as if it described another structure. And on that ground we'll accept $Q^\dagger$'s canonical Gödel sentence as true.[6]

Why note our different examples (1) to (4)? Two reasons. First, we want to drive home the message that, given a particular canonical Gödel sentence G_T, we might have *various* kinds of grounds for believing it true, because we have varying grounds for believing T is sound or consistent. But second, our examples also reveal that while our grounds for accepting Gödel sentences may be various, the reasons we adduce in cases like (1) to (4) are – so to speak – perfectly ordinary mathematical reasons. When we initially met the idea of incompleteness at the very outset, we wondered whether we must have some special, rule-transcending, cognitive grasp of the numbers underlying our ability to recognize Gödel sentences as correct arithmetical propositions (see Section 1.4). That speculation should now perhaps begin to seem unnecessarily fanciful.

23.3   What doesn't follow from incompleteness

What follows from our incompleteness theorems so far? In this section, we'll highlight a few claims that someone might carelessly think *are* justified by the theorems, but which certainly aren't. Thus consider

(a) 'For any nice theory T, T can't prove its Gödel sentences but we always can.'[7]

But not so, as the remarks in the last section should have made clear.

To hammer home the point, suppose G_T is a canonical Gödel-sentence for theory T. Then, true enough, we can show that, *if* T is indeed nice, then $T \nvdash G_T$. And hence, *if* we know that T is nice, we can conclude that G_T is true. But this line of thought, of course, doesn't give us a truth-establishing *proof* of G_T, unless we *do* know that T is nice: and that requires in particular knowing that T is consistent.[8] But very often, we *won't* be able to show that T is consistent.

Take, for example, the theory Q_C, which is Q augmented by some unproven Π_1 conjecture C (like Goldbach's conjecture again). By Theorem 11.7, to show

---

[6]It is worth commenting on the contrast between the cases of $PA + \neg G$ and $Q^\dagger$. We have a proof that the former is consistent if PA is, and we believe that PA *is* consistent: so, because it is a standard metatheorem of first-order logic that any consistent first-order theory has a model (i.e. an interpretation that makes all its theorems true – see Section 29.6), we will accept that there must exist a model of $PA + \neg G$. So in this case, at least in the first instance, we believe that there *is* such a model because we believe that the theory is consistent (it is quite a challenge to describe a suitable model). In the case of $Q^\dagger$ the inference goes exactly the other way about, from the existence of a model to the consistency of the theory – and hence to the truth of its Gödel sentence.

[7]See Section 37.6 for more on a sophisticated version of this line of thought.

[8]It is worth mentioning a famous historical episode here: towards the end of the first edition of his *Mathematical Logic* (1940), W. V. Quine proves Gödel's theorem for his proposed formal theory. Should we then conclude that the relevant Gödel sentence is true? We go wrong if we do: for it turns out that Quine's theory in that edition is inconsistent! Every sentence is provable so the system's Gödel sentence which 'says' it is itself unprovable is false.

Q_C is consistent is equivalent to proving C, which we may have no clue how to do. So we may have no clue at all whether Q_C's canonical Gödel sentence is true, even if in fact it is.

(b) 'There are truths of L_A – Gödel sentences – which are not provable in any (nice) axiomatized theory.'

But again, plainly not so. Take any arithmetical truth φ at all. Then this is trivially a theorem of the corresponding formal theory $Q + \varphi$, i.e. of the theory we get by adding φ as a new axiom to Q. (And, assuming Q is sound, $Q + \varphi$ is also sound: hence $Q + \varphi$ is a nice theory – for the theory is consistent because it is sound, p.r. axiomatized because it still has a finite list of axioms, and is p.r. adequate because it contains Q.)

There is a quantifier-shift fallacy lurking hereabouts: from the theorem that for every nice T there is a truth φ such that $T \nvdash \varphi$ it doesn't follow that there is a truth φ such that for every nice T, $T \nvdash \varphi$.

Of course, deriving a truth φ in some ad hoc formal theory in which we have adopted that very truth as an axiom is of no interest at all: what we want, if we are aiming to establish φ, are theories which we have *independent* reasons to accept as sound. So let's now bear that in mind, alongside the observation that we can't always ourselves prove the Gödel sentences for consistent theories. You might wonder if Gödel's First Theorem still gives us reason to accept something along the following lines:

(c) 'For some nice theory T, we can prove its Gödel sentence is true, even though this truth can't be derived in a formal theory we independently accept as sound.'

But once again not so. Take a case where we *can* show that G_T is true, by arguing outside the theory T that T is consistent. Then there's nothing at all in the First Theorem to stop us reflecting on the principles involved in this reasoning, and then constructing a richer formalized theory S (distinct from T, of course) which encapsulates these principles, a formal theory in which we prove G_T to be true. And we will presumably have good reason to accept *this* new formal theory S, since by hypothesis it just regiments some aspects of what we take to be sound methods of reasoning anyway.[9]

In sum, the claims (a), (b) and (c) are each fairly obviously wrong. Understand clearly *why* they are wrong, and you will already be inoculated against many of

---

[9] Careful! All that is being claimed is that there is nothing in the First Theorem that would prevent our regimenting the reasoning we use to show G_T is true – when we can show that – into some axiomatized system S. It doesn't follow, of course, that there could be some single master system S^* which wraps up any reasoning that we – or idealized versions of us – might use in proving true various G_T, a master system which is itself axiomatizable in a way we can survey. That would be another quantifier-shift fallacy.

the wilder claims about Gödel's First Incompleteness Theorem which are based on such misconceptions.[10]

23.4 What does follow from incompleteness?

So what *can* we infer from the theorem that if an arithmetic theory is sound/nice, it can't be complete? Well, let's here stress just one point about that simple but deep Gödelian fact: it does sabotage, once and for all, any project of trying to show that all basic arithmetical truths can be thought of as deducible, in some standard deductive system, from one unified set of fundamental axioms which can be specified in a tidy, p.r., way. In short, *arithmetical truth isn't provability in some single axiomatizable system.*

The First Theorem therefore sabotages the logicist project in particular, at least in its classic form of aiming to explain how we can deduce *all* arithmetic from a tidy set of principles which are either logical axioms or definitions.[11] Which is a major blow; for there surely remains something very deeply attractive about the thought we expressed at the very beginning of the book, i.e. the thought that in exploring the truths of basic arithmetic, we are just exploring the deductive consequences of a limited set of axiomatic truths which we grasp in grasping the very ideas of the natural number series and the operations of addition and multiplication.

If there is a silver lining to that cloud, it is this. The truths of basic arithmetic run beyond what is provable in any given formal system: even arithmetic is – so to speak – *inexhaustible*. Given any nice theory of arithmetic T which we accept as sound, we have to recognize that there are truths that T cannot prove (there's G_T for a start). So at least mathematicians are not going to run out of work, even at the level of arithmetic, as they develop ever richer formal frameworks in which to prove more truths.

23.5 What's next?

That's enough generalities for the moment: we need to move on. The next six chapters explore around and about The First Incompleteness theorem.

We haven't yet reached the strongest syntactic incompleteness theorem. We do this in two stages:

1. First we re-examine Gödel's construction in proving the syntactic First Theorem and show how the leading idea generalizes to yield the so-called Diagonalization Lemma. (Chapter 24)

---

[10]We haven't space to explode all the daft misconceptions about the Theorem: see Franzén (2005) for a wide-ranging demolition job. And also see our Section 37.6.

[11]For the philosophical prospects of non-classical forms of logicism, however, see e.g. Wright (1983), the essays on Frege in Boolos (1998), and Hale and Wright (2001).

2. Next, we note that Gödel's syntactic incompleteness proof needs somewhat less than the assumption of ω-consistency to make it fly: so-called '1-consistency' is enough. However, we can do even better. We can re-use the Diagonalization Lemma to prove the Gödel-Rosser Theorem, a syntactic argument for incompleteness that now relies only on the consistency (rather than 1-consistency or ω-consistency) of the theory in question. (Chapter 25)

At that point, we can usefully pause to look at two different ways of taking the proofs we already have and expanding the scope of their application.

3. We first show how to apply our results to formal theories which are in fact not p.r. axiomatized, and we derive an important corollary: the set of true L_A sentences can't be axiomatized. Then we show how the arguments can be applied as well to certain theories (like set theory) whose native languages don't yet include any arithmetical vocabulary, so aren't initially theories of arithmetic at all. (Chapter 26)

We then return to putting the Diagonalization Lemma to further use:

4. A second application of the Lemma gives us a version of Tarski's Theorem on the arithmetical indefinability of arithmetical truth. (Chapter 27)

5. For a third application, we use the Lemma in proving a so-called 'speed-up' theorem. For example, and rather surprisingly, if we add a Gödel sentence as a new axiom to PA, we can not only prove new theorems, but we can also get *vastly* shorter proofs of old theorems. (Chapter 28)

Finally, we return to a topic which we shelved earlier.

6. What does the incompleteness theorem tell us about *second-order* arithmetics? (Chapter 29)

So on we go . . . .

24 The Diagonalization Lemma

We first introduce the provability predicate Prov_T which applies to a number just when it numbers a T-theorem. We can then easily show that, if T is nice, $T \vdash \mathsf{G}_T \leftrightarrow \neg\mathsf{Prov}_T(\ulcorner\mathsf{G}_T\urcorner)$. This means that not only is T's canonical Gödel sentence true if and only if it is unprovable, but we can derive that claim inside T itself.

We then generalize the construction to establish that, for *any* $\varphi(\mathsf{x})$ with one free variable, if T is nice, there is some sentence γ such that $T \vdash \gamma \leftrightarrow \varphi(\ulcorner\gamma\urcorner)$. This is Gödel's Diagonalization Lemma (or Fixed Point Theorem). This Lemma will be used in later chapters to prove key theorems including the Gödel-Rosser Theorem and Tarski's Theorem.

24.1 Provability predicates

(a) Recall: $Prf(m, n)$ holds when m is the super g.n. for a PA-proof of the sentence with g.n. n. $Prov(n) =_{\mathrm{def}} \exists v\, Prf(v, n)$ will therefore hold when some number codes a proof of the sentence with g.n. n, so holds when n numbers a PA-theorem. (See Sections 19.4 and 20.4.)

Since the relation Prf is p.r., we know it can be canonically captured in PA by a Σ_1 wff (see Section 17.4). Let $\mathsf{Prf}(\mathsf{x}, \mathsf{y})$ stand in for a Σ_1 wff which canonically captures (and so expresses) Prf. Then we put

$$\mathsf{Prov}(\mathsf{x}) =_{\mathrm{def}} \exists\mathsf{v}\,\mathsf{Prf}(\mathsf{v}, \mathsf{x}).[1]$$

Evidently, this wff is also Σ_1.

We will call $\mathsf{Prov}(\mathsf{x})$ a *(canonical) provability predicate* for PA. It evidently *expresses* the provability property $Prov$. But be careful! – we mustn't assume that it *captures* that property: in fact, we'll soon prove that it doesn't (see Theorem 24.8).

(b) Next, we generalize in the now familiar way. For any theory T, there is similarly a relation $Prf_T(m, n)$ which holds when m is the super g.n. for a T-proof of the sentence with g.n. n.[2] If T is a nice theory, then Prf_T will again be a p.r. relation. Hence there will be a corresponding Σ_1 wff $\mathsf{Prf}_T(\mathsf{x}, \mathsf{y})$ which captures this relation in T (and again captures it in a perspicuous way, by recapitulating the p.r. definition of Prf_T).

---

[1] As usual, we will assume that we can quantify without clash of variables.

[2] Of course, we are now talking in the context of some appropriate scheme for Gödel-numbering expressions of T – see the second remark towards the end of Section 19.1.

So we can define $\mathsf{Prov}_T(\mathsf{x}) =_{\mathrm{def}} \exists \mathsf{v}\, \mathsf{Prf}_T(\mathsf{v}, \mathsf{x})$, where this new provability predicate expresses the property $Prov_T$ of numbering a T-theorem.

24.2   An easy theorem about provability predicates

Here's a straightforward result about provability predicates:

Theorem 24.1 *Let T be a nice theory. Then for any sentence φ,*

C1. *If $T \vdash \varphi$, then $T \vdash \mathsf{Prov}_T(\ulcorner\varphi\urcorner)$.*

Cω. *Suppose T is ω-consistent: then if $T \vdash \mathsf{Prov}_T(\ulcorner\varphi\urcorner)$,*
   $T \vdash \varphi$.

Proof for (C1) First assume $T \vdash \varphi$. Then there is a T proof of the sentence with g.n. $\ulcorner\varphi\urcorner$. Let this proof have the super g.n. m. Then, by definition, $Prf_T(m, \ulcorner\varphi\urcorner)$. Hence, since Prf_T is *captured* by Prf_T, it follows that $T \vdash \mathsf{Prf}_T(\overline{\mathsf{m}}, \ulcorner\varphi\urcorner)$. Hence $T \vdash \exists \mathsf{v}\, \mathsf{Prf}_T(\mathsf{v}, \ulcorner\varphi\urcorner)$, i.e. $T \vdash \mathsf{Prov}_T(\ulcorner\varphi\urcorner)$.   ⊠

An even quicker proof for (C1) If $T \vdash \varphi$, then $\mathsf{Prov}_T(\ulcorner\varphi\urcorner)$ will be true. But $\mathsf{Prov}_T(\ulcorner\varphi\urcorner)$ is Σ_1; hence, since Q proves all true Σ_1 sentences (by Theorem 11.5), $\mathsf{Q} \vdash \mathsf{Prov}_T(\ulcorner\varphi\urcorner)$. Hence $T \vdash \mathsf{Prov}_T(\ulcorner\varphi\urcorner)$.   ⊠

Proof for (Cω) Now assume T is ω-consistent and also that $T \vdash \mathsf{Prov}_T(\ulcorner\varphi\urcorner)$, i.e. $T \vdash \exists \mathsf{v}\, \mathsf{Prf}_T(\mathsf{v}, \ulcorner\varphi\urcorner)$. Suppose, for reductio, that $T \nvdash \varphi$. Then, for all m, it is not the case that $Prf_T(m, \ulcorner\varphi\urcorner)$. Therefore, since T is nice and captures Prf_T, for all m, $T \vdash \neg\mathsf{Prf}_T(\overline{\mathsf{m}}, \ulcorner\varphi\urcorner)$. But that makes T ω-inconsistent, contrary to hypothesis.   ⊠

Two quick comments on this. First, suppose that we are given only that T is nice, this time *without* the further assumption of something like ω-consistency. Then it *won't* always be the case that if $T \vdash \mathsf{Prov}_T(\ulcorner\varphi\urcorner)$ then $T \vdash \varphi$. Why? If T is ω-inconsistent, then T is not sound, and it has false theorems on the standard interpretation (see Section 21.6(c)). But if T is not sound, then among the things that T can get wrong are facts about what it can prove.

Second, with the additional assumption of ω-consistency back in place, it follows from (Cω) that if $T \nvdash \varphi$ then $T \nvdash \mathsf{Prov}_T(\ulcorner\varphi\urcorner)$. But this fact needs to be *very* sharply distinguished from the claim that, for any φ, if $T \nvdash \varphi$, then $T \vdash \neg\mathsf{Prov}_T(\ulcorner\varphi\urcorner)$. *That* second claim is in fact plain *false*.

For suppose otherwise. Then this supposition, combined with (C1), implies that Prov_T *captures* the provability property $Prov_T$. However, as we'll soon see in Section 24.7, *no* wff can do that if T is nice. So, even in nice theories, $T \vdash \neg\mathsf{Prov}_T(\ulcorner\varphi\urcorner)$ must at least *sometimes* fail to be true even when we have $T \nvdash \varphi$.

But there's more: Theorem 33.5 will in fact tell us that in typical nice theories, $T \vdash \neg\mathsf{Prov}_T(\ulcorner\varphi\urcorner)$ is *never* true. To put it vividly, such theories may know about what they *can* prove, but they know *nothing* about what they *can't* prove!

24.3   Proving G $\leftrightarrow$ ¬Prov(⌜G⌝)

We saw that our canonical Gödel sentence G for PA is true if and only if it is unprovable. Using the provability predicate Prov, we can now echo this claim about G in the language of PA itself, thus:

G $\leftrightarrow$ ¬Prov(⌜G⌝).

Indeed, we can do better; we can actually *prove* this very sentence inside PA.

To show this, let's start by thinking again about how our Gödel sentence for PA was constructed in Section 21.2.

We started by forming the wff

$$U(y) =_{\text{def}} \forall x \neg Gdl(x, y).$$

Here $Gdl(x, y)$ captures our old friend, the relation *Gdl*, where $Gdl(m, n)$ iff $Prf(m, diag(n))$. But the one-place p.r. function *diag* can be captured in Q and hence in PA by some open wff Diag(x, y). We can therefore give the following definition:

$$Gdl(x, y) =_{\text{def}} \exists z(Prf(x, z) \wedge Diag(y, z)).$$

And now let's do some elementary logical manipulations:

$$
\begin{aligned}
U(y) =_{\text{def}} \ & \forall x \neg \exists z(Prf(x, z) \wedge Diag(y, z)) && \text{(definition of Gdl)} \\
\leftrightarrow \ & \forall z \forall x \neg (Prf(x, z) \wedge Diag(y, z)) && \text{(quantifier logic)} \\
\leftrightarrow \ & \forall z(Diag(y, z) \rightarrow \neg \exists x\, Prf(x, z)) && \text{(rearranging after '}\forall z\text{')} \\
\leftrightarrow \ & \forall z(Diag(y, z) \rightarrow \neg \exists v\, Prf(v, z)) && \text{(changing variables)} \\
=_{\text{def}} \ & \forall z(Diag(y, z) \rightarrow \neg Prov(z)) && \text{(definition of Prov)}
\end{aligned}
$$

These manipulations can be done inside any theory in the language L_A which contains first-order logic (including Q and PA of course).

Now we form G by diagonalizing U. So trivially, we have

1. $\vdash$ G $\leftrightarrow$ U(⌜U⌝),

(the bare turnstile indicating a logical theorem). So therefore

2. $\vdash$ G $\leftrightarrow$ $\forall z(Diag(⌜U⌝, z) \rightarrow \neg Prov(z))$.

But diagonalizing U yields G. Hence, just by the definition of *diag*, we have $diag(⌜U⌝) = ⌜G⌝$. Since Diag captures *diag* in Q, it follows that

3. Q $\vdash$ $\forall z(Diag(⌜U⌝, z) \leftrightarrow z = ⌜G⌝)$.

(See Section 16.1(b).) Putting (2) and (3) together immediately implies

4. Q $\vdash$ G $\leftrightarrow$ $\forall z(z = ⌜G⌝ \rightarrow \neg Prov(z))$.

But the right-hand side of that biconditional is equivalent to ¬Prov(⌜G⌝). Whence

5. $Q \vdash G \leftrightarrow \neg\mathsf{Prov}(\ulcorner G \urcorner)$.

So that means we have, a fortiori,

Theorem 24.2 $\mathsf{PA} \vdash G \leftrightarrow \neg\mathsf{Prov}(\ulcorner G \urcorner)$.

And the same reasoning applies to other theories which contain first-order logic and which can capture p.r. functions and relations. In other words, if Prov_T is the provability predicate for T constructed analogously to the predicate Prov for PA, and if G_T is a Gödel sentence constructed analogously to G, then by exactly the same argument we have

Theorem 24.3 *If T is a nice theory,* $T \vdash G_T \leftrightarrow \neg\mathsf{Prov}_T(\ulcorner G_T \urcorner)$.

(though again, Q in fact suffices to prove the biconditional).

24.4   The Diagonalization Lemma

(a)   The construction in the last section generalizes. For a start, suppose L is a language which includes L_A, and let $\varphi(\mathsf{x})$ be *any* wff of its language with one free variable. Now consider the wff

$$\psi(\mathsf{y}) =_{\mathrm{def}} \forall \mathsf{z}(\mathsf{Diag}(\mathsf{y}, \mathsf{z}) \rightarrow \varphi(\mathsf{z}))$$

where – for the moment – Diag is a wff that *expresses* the diagonalization function *diag* which maps the g.n. of an L wff to the g.n. of its diagonalization (if L includes L_A, we know there must be such a wff).

Let γ be the diagonalization of ψ. So just from this definition, we know that

$$\gamma \leftrightarrow \forall \mathsf{z}(\mathsf{Diag}(\ulcorner \psi \urcorner, \mathsf{z}) \rightarrow \varphi(\mathsf{z}))$$

is true. But since by assumption Diag expresses *diag*, we also know the right-hand side will be true just when the g.n. of the diagonalization of the wff numbered $\ulcorner \psi \urcorner$ satisfies φ, i.e. just when the g.n. of γ satisfies φ. So that means

$$\gamma \leftrightarrow \varphi(\ulcorner \gamma \urcorner)$$

must be true. We might call this *Carnap's Equivalence* in honour of Rudolf Carnap.[3]

(b)   It is but a small step from noting the construction which gives us Carnap's Equivalence (a *semantic* claim about a certain sentence being true) to the so-called *Diagonalization Lemma* (which is the standard label for the following *syntactic* claim about a certain sentence being provable):

---

[3]In a footnote added to later reprintings of Gödel (1934), Gödel says that the general possibility of such a construction for any φ 'was first noted by Carnap [(1934)]'.

Theorem 24.4 *If T is a nice theory and $\varphi(\mathsf{x})$ is any wff of its language with one free variable, then there is a sentence γ such that $T \vdash \gamma \leftrightarrow \varphi(\ulcorner\gamma\urcorner)$.*

Proof We use the same basic proof idea as before: we do to the generic φ pretty much what our Gödelian construction in the last section did to '¬Prov'.

So, first step, put $\psi(\mathsf{y}) =_{\text{def}} \forall \mathsf{z}(\mathsf{Diag}_T(\mathsf{y}, \mathsf{z}) \to \varphi(\mathsf{z}))$ where now Diag_T is any wff which *captures* $diag_T$, the diagonalization function for T's language (and we know that if T is nice, there must be such a wff). Then construct γ, the diagonalization of ψ as before. Logic gives us

  i. $T \vdash \gamma \leftrightarrow \forall \mathsf{z}(\mathsf{Diag}_T(\ulcorner\psi\urcorner, \mathsf{z}) \to \varphi(\mathsf{z}))$.

Since $diag_T(\ulcorner\psi\urcorner) = \ulcorner\gamma\urcorner$ and Diag_T captures $diag_T$, it follows by the definition of capturing that

  ii. $T \vdash \forall \mathsf{z}(\mathsf{Diag}_T(\ulcorner\psi\urcorner, \mathsf{z}) \leftrightarrow \mathsf{z} = \ulcorner\gamma\urcorner)$.

(See Section 16.1(b) again.) From (i) and (ii) it easily follows that $T \vdash \gamma \leftrightarrow \forall \mathsf{z}(\mathsf{z} = \ulcorner\gamma\urcorner \to \varphi(\mathsf{z}))$. Whence, trivially,

  iii. $T \vdash \gamma \leftrightarrow \varphi(\ulcorner\gamma\urcorner)$. ☒

Alternative proof Reviewing that first proof, it is quite easy to spot that a variant construction is possible. So now put $\psi'(\mathsf{y}) =_{\text{def}} \exists \mathsf{z}(\mathsf{Diag}_T(\mathsf{y}, \mathsf{z}) \wedge \varphi(\mathsf{z}))$. Redefine γ to be the diagonalization of $\psi'(\mathsf{y})$. Then $T \vdash \gamma \leftrightarrow \exists \mathsf{z}(\mathsf{Diag}_T(\ulcorner\psi'\urcorner, \mathsf{z}) \wedge \varphi(\mathsf{z}))$. Since $diag_T(\ulcorner\psi'\urcorner) = \ulcorner\gamma\urcorner$, we have $T \vdash \forall \mathsf{z}(\mathsf{Diag}_T(\ulcorner\psi'\urcorner, \mathsf{z}) \leftrightarrow \mathsf{z} = \ulcorner\gamma\urcorner)$. Hence $T \vdash \gamma \leftrightarrow \exists \mathsf{z}(\mathsf{z} = \ulcorner\gamma\urcorner \wedge \varphi(\mathsf{z}))$. So again $T \vdash \gamma \leftrightarrow \varphi(\ulcorner\gamma\urcorner)$. ☒

(c) A quick remark about jargon. Suppose that the function f maps the argument a back to a itself, so that $f(a) = a$: then a is said to be a *fixed point* for f. And a theorem to the effect that, under certain conditions, there is a fixed point for f is a *fixed-point theorem*. By a somewhat strained analogy, the Diagonalization Lemma is also standardly referred to as *Gödel's Fixed Point Theorem*, with γ behaving like a 'fixed point' for the predicate $\varphi(\mathsf{x})$.

(d) And here's a quick little corollary of our first proof of the Lemma:

Theorem 24.5 *If T is nice and $\varphi(\mathsf{x})$ is Π_1-equivalent, then it has a fixed point in T which is Π_1-equivalent.*

Proof Choose a canonical Diag_T, which is Σ_1. Then $\forall \mathsf{z}(\mathsf{Diag}_T(\mathsf{y}, \mathsf{z}) \to \varphi(\mathsf{z}))$ is equivalent to a Π_1 wff and hence so is its diagonalization (since diagonalization is just equivalent to substituting a numeral for the free variable). ☒

24.5 Incompleteness again

Theorem 24.1 established that the following two conditions obtain for provability predicates for nice theories T:

C1. If $T \vdash \varphi$, then $T \vdash \mathsf{Prov}_T(\ulcorner\varphi\urcorner)$.

Cω. If T is ω-consistent then, if $T \vdash \mathsf{Prov}_T(\ulcorner\varphi\urcorner)$, $T \vdash \varphi$.

And these two principles immediately give us the following:

> **Theorem 24.6** *Let T be a nice theory, and let γ be any sentence such that $T \vdash \gamma \leftrightarrow \neg\mathsf{Prov}_T(\ulcorner\gamma\urcorner)$. Then $T \nvdash \gamma$; and if T is ω-consistent, then $T \nvdash \neg\gamma$.*

Proof By hypothesis, $T \vdash \gamma \leftrightarrow \neg\mathsf{Prov}_T(\ulcorner\gamma\urcorner)$. So if $T \vdash \gamma$ then $T \vdash \neg\mathsf{Prov}_T(\ulcorner\gamma\urcorner)$. But, by (C1), if $T \vdash \gamma$ then $T \vdash \mathsf{Prov}_T(\ulcorner\gamma\urcorner)$. So, if $T \vdash \gamma$, T would be inconsistent, contrary to hypothesis.

Now assume T is ω-consistent (so consistent), and suppose $T \vdash \neg\gamma$. Since $T \vdash \gamma \leftrightarrow \neg\mathsf{Prov}_T(\ulcorner\gamma\urcorner)$, it follows $T \vdash \mathsf{Prov}_T(\ulcorner\gamma\urcorner)$. Hence by (C$\omega$), $T \vdash \gamma$. So if $T \vdash \neg\gamma$, T would be plain inconsistent, contrary to hypothesis. $\boxtimes$

The general Diagonalization Lemma tells us that $\neg\mathsf{Prov}$ has fixed points, and since $\neg\mathsf{Prov}$ is a Π_1-equivalent wff, the little corollary Theorem 24.5 tells us that there are fixed points which are equivalent to Π_1 wffs. Apply the last theorem, then, and of course we immediately recover again our syntactic incompleteness theorem, Theorem 22.2.

24.6 'Gödel sentences' again

(a) Here is a corollary of our second proof of the Diagonalization Lemma, which is not always noted:

> **Theorem 24.7** *If T is a nice but* unsound *theory, and $\varphi(\mathsf{x})$ is any wff of its language with one free variable, then $\varphi(\mathsf{x})$ has a false fixed point.*

Proof Pick θ to be one of T's false theorems. Suppose $\mathsf{Diag}_T(\mathsf{x},\mathsf{y})$ captures $diag_T$. Then so does $\mathsf{Diag}'_T(\mathsf{x},\mathsf{y}) =_{\text{def}} [\mathsf{Diag}_T(\mathsf{x},\mathsf{y}) \wedge \theta]$, by the general point we noted in Section 5.5(b).

Now our proofs of the Diagonalization Lemma do not depend on which particular wff serves to capture $diag_T$. So Diag'_T will do just as well as Diag_T. In particular the alternative proof of Theorem 24.4 will go through with the fixed point γ identified as the diagonalization of $\exists\mathsf{z}(\mathsf{Diag}'_T(\mathsf{y},\mathsf{z}) \wedge \varphi(\mathsf{z}))$, i.e. the diagonalization of $\exists\mathsf{z}([\mathsf{Diag}_T(\mathsf{y},\mathsf{z}) \wedge \theta] \wedge \varphi(\mathsf{z}))$. But the diagonalization of that still has the false θ as a conjunct, so is false. $\boxtimes$

(b) What is the significance of that last little theorem? As we've just noticed, we can establish incompleteness by combining the claim that $\neg\mathsf{Prov}_T$ has fixed points, with the claim that any such fixed points must be undecidable. But if you do go via this route, you do need to be a bit careful about your *commentary*. In particular, you will get into trouble if you call *any* fixed point for $\neg\mathsf{Prov}_T$ (which is therefore undecidable) a 'Gödel sentence' for T, but you also repeat the usual claim that the Gödel sentences for T are true-if-and-only-if-unprovable-in-T, and hence are true (assuming consistency). For look again at our last theorem. That shows that if T is an *unsound* theory, there will be some *false* fixed points for $\neg\mathsf{Prov}_T$. So if T is consistent but unsound it will have *false* Gödel sentences in the wide sense.

Compare: *canonical* Gödel sentences for T – built up in something like Gödel's original way from a wff that perspicuously recapitulates a p.r. definition of the relation Gdl_T so indirectly say of themselves that they are unprovable – must be true if unprovable. More generally, any Π_1 undecidable sentence of a nice theory will be true (see the remarks after the proof of Theorem 21.7). But, once we move away from these cases and start using the idea of a Gödel sentence very generously for any fixed point of the relevant $\neg\mathsf{Prov}_T$ (as is often done), then we can't assume that such undecidable sentences always have to be true.[4]

24.7 Capturing provability?

Consider again the T-wff $\mathsf{Prov}_T(\mathsf{x})$ which *expresses* the property $Prov_T$ of being the g.n. of a T-theorem. The obvious next question to ask is: does this wff also case-by-case *capture* that property? The answer – as we've already indicated – is that it doesn't. In fact, we can prove something stronger:

> **Theorem 24.8** *No wff in a nice theory T can capture the numerical property $Prov_T$.*

Proof Suppose for reductio that $\mathsf{Pr}(\mathsf{x})$ abbreviates an open wff – not necessarily identical with $\mathsf{Prov}_T(\mathsf{x})$ – which captures $Prov_T$.

By the Diagonalization Lemma applied to $\neg\mathsf{Pr}(\mathsf{z})$, there is some wff γ such that

> 1. $T \vdash \gamma \leftrightarrow \neg\mathsf{Pr}(\ulcorner\gamma\urcorner)$.

By the assumption that Pr captures $Prov_T$, we have in particular

> 2. if $T \vdash \gamma$, i.e. $Prov_T(\ulcorner\gamma\urcorner)$, then $T \vdash \mathsf{Pr}(\ulcorner\gamma\urcorner)$,
> 3. if $T \nvdash \gamma$, i.e. not-$Prov_T(\ulcorner\gamma\urcorner)$, then $T \vdash \neg\mathsf{Pr}(\ulcorner\gamma\urcorner)$.

Contradiction quickly follows. By (3) and (1), if $T \nvdash \gamma$, then $T \vdash \gamma$. Hence $T \vdash \gamma$. So by (2) and (1) we have both $T \vdash \mathsf{Pr}(\ulcorner\gamma\urcorner)$ and $T \vdash \neg\mathsf{Pr}(\ulcorner\gamma\urcorner)$ making T inconsistent, contrary to hypothesis. $\boxtimes$

Hence $Prov_T$ cannot be captured in T: so – answering our original question – $\mathsf{Prov}_T(\mathsf{x})$ in particular doesn't capture that property.

Here is a quick corollary of our theorem. If T is nice, it includes Q and is p.r. adequate, and so can capture any p.r. property. Hence

---

[4]The need for care on this point has been vigorously pressed by Peter Milne (2007), who gives chapter and verse on the sins of various textbooks.

Theorem 24.9 *The provability property Prov$_T$ for any nice theory is not primitive recursive.*

Which is to say that there is no p.r. function of n which returns 0 if $Prov_T(n)$ is true, and 1 otherwise. (Later, we will prove that $Prov_T$ is not a decidable property at all – see Section 40.2; but of course we can't do that now, as we haven't yet got a general theory of what makes for a decidable property.)

Which all establishes two interesting general points:

1. There can be properties which are expressible in a given theory T but not capturable: $Prov_T$ is a case in point.

2. Although every p.r. property and relation can be expressed by an open Σ_1 wff, there are open Σ_1 wffs which don't express p.r. properties or relations: $\mathsf{Prov}_T(\mathsf{x})$ is a case in point, for any nice T.

25 Rosser's proof

One half of the First Theorem, recall, requires the assumption that we are dealing with a theory T which is not only nice but is ω-consistent. But we can improve on this in two different ways:

1. We can keep the *same* undecidable sentence G_T while invoking assumptions weaker than ω-consistency in showing that $T \nvdash \neg G_T$.

2. Following Barkley Rosser, we can construct a *different* and more complex sentence R_T such that we only need to assume T is plain consistent in order to show that R_T is formally undecidable.

Rosser's clever construction yields the stronger result, so that is our main topic. But we will begin with a quick nod to the first result (not because it is important for our purposes in this book – it is not – but because it might aid comparisons with some discussions elsewhere; you can skip the next section if you wish.)

25.1 Σ_1-soundness and 1-consistency

(a)   Assume that PA is consistent. Then we know from the first part of the syntactic incompleteness theorem that PA $\nvdash$ G, hence no number m numbers a proof of G, i.e. no number m numbers a proof of the diagonalization of U, hence

1. For all m, not-$Gdl(m, \ulcorner U \urcorner)$.

Since Gdl captures Gdl, that means

2. For all m, PA $\vdash \neg$Gdl$(\overline{m}, \ulcorner U \urcorner)$.

Now we want to show that PA doesn't prove $\neg$G. We suppose the opposite, which is equivalent to supposing

3. PA $\vdash \exists$xGdl$(x, \ulcorner U \urcorner)$,

and seek to derive a contradiction with something we might reasonably believe about PA.

Note first that, because of (1), the Σ_1 sentence $\exists$xGdl$(x, \ulcorner U \urcorner)$ must be false. So if (3) were true, PA would prove a false Σ_1 sentence. Hence one quick way of ruling out (3) is to assume that PA is Σ_1-sound, i.e proves no false Σ_1 sentence (see Section 11.6(a)). And it indeed is not uncommon for the incompleteness theorem for PA to be stated in this way: there is a Gödel sentence G such that if PA is consistent, PA $\nvdash$ G, and if PA is Σ_1-sound, PA $\nvdash$ ¬G. Similarly for nice theories more generally.

However this way of putting things is surely to be deprecated. It muddies the waters by mixing a syntactic assumption for one half of the theorem with a semantic assumption for the other half. So let's return to considering explicitly syntactic assumptions which, when added to (2), show PA $\nvdash$ ¬G.

(b)   If (2) and (3) were both true, that would make PA ω-inconsistent. Hence our result in Section 21.7(a): assuming PA is ω-consistent, it can't prove ¬G. Note, however, that Gdl is Σ_1. Suppose then that we say

> An arithmetic theory T is Σ_1-*inconsistent* iff, for some Σ_1 wff $\varphi(x)$,
> $T \vdash \exists x \varphi(x)$, yet for each number m we have $T \vdash \neg\varphi(\overline{m})$.

So Σ_1-inconsistency is just ω-consistency witnessed down at the level of Σ_1 wffs. And (2) and (3) together make PA not just ω-inconsistent, but Σ_1-inconsistent. Hence, the weaker assumption that PA is Σ_1-consistent (even if it isn't ω-consistent through and through) is enough to yield the conclusion that it can't prove ¬G.

(c)   That was an easy improvement, but we can in fact do even better. Let's say that

> An arithmetic theory T is *1-inconsistent* iff, for some Δ_0 wff $\varphi(x)$,
> $T \vdash \exists x \varphi(x)$, yet for each number m we have $T \vdash \neg\varphi(\overline{m})$,[1]

so 1-inconsistency is ω-inconsistency witnessed by a particularly elementary Σ_1 wff. Then we can show

Theorem 25.1 *If* PA *is 1-consistent, then* PA $\nvdash$ ¬G.

Proof   Gld$(x, \ulcorner U \urcorner)$ is Σ_1 so is provably equivalent to a wff of the form $\exists y \psi(x, y)$ where ψ is Δ_0 (by Theorem 12.4). So (2) and (3) are provably equivalent to

2′. For all m, PA $\vdash \neg \exists y \psi(\overline{m}, y)$,

3′. PA $\vdash \exists x \exists y \psi(x, y)$.

From (3′) by the proof of Theorem 12.2, we have

4. PA $\vdash \exists w (\exists x \leq w)(\exists y \leq w)\psi(x, y)$.

From (2′) we get

5. For all m, n, PA $\vdash \neg\psi(\overline{m}, \overline{n})$.

Using (O4) of Section 11.3, (5) easily implies

6. for all k, PA $\vdash (\forall x \leq \overline{k})(\forall y \leq \overline{k})\neg\psi(x, y)$,

---

[1] The definition of the notion goes back to Kreisel (1957).

which is equivalent to

7. for all k, $\mathsf{PA} \vdash \neg(\exists x \leq \overline{k})(\exists y \leq \overline{k})\psi(x, y)$.

But since $(\exists x \leq w)(\exists y \leq w)\psi(x, y)$ is Δ_0, the combination of (4) and (7) makes PA 1-inconsistent.

Hence, given that (2) is true, if PA is 1-consistent, (3) must be false, i.e. $\mathsf{PA} \nvdash \neg\mathsf{G}$. $\boxtimes$

We can therefore strengthen Theorem 21.7 to this:

Theorem 25.2 *There is an L_A-sentence φ of Goldbach type such that if* PA *is consistent then* $\mathsf{PA} \nvdash \varphi$, *and if* PA *is 1-consistent then* $\mathsf{PA} \nvdash \neg\varphi$.

There will be a parallel generalization to other nice theories T, so long as they have enough induction to prove Theorems 12.2 and 12.4. And 1-consistency is indeed the weakest condition for proving the original style of Gödel-sentence G_T is undecidable.

(d)   One last remark before leaving the idea of 1-consistency. Suppose we say

An arithmetical theory T is 1-sound if, for any Δ_0 wff $\varphi(x)$, if when $T \vdash \exists x\varphi(x)$, then $\exists x\varphi(x)$ is true.

So 1-soundness is Σ_1-soundness for particularly elementary Σ_1 wffs. Then it is easy to see that

Theorem 25.3 *A nice theory T is 1-consistent iff it is 1-sound.*[2]

Proof   Suppose T is 1-sound, yet is 1-inconsistent. 1-inconsistency means for some Δ_0 wff $\varphi(x)$, $T \vdash \exists x\varphi(x)$ but also for any m, we have $T \vdash \neg\varphi(\overline{m})$. By the soundness condition, $\exists x\varphi(x)$ is true, so for some m, $\varphi(\overline{m})$ must be true. Therefore, since nice theories correctly decide every Δ_0 wff (by Theorem 11.3), we have $T \vdash \varphi(\overline{m})$, which makes T inconsistent. Hence if T is nice (so consistent), if it is 1-sound, it must be 1-consistent.

For the converse, suppose T is 1-consistent and $T \vdash \exists x\varphi(x)$ where φ is Δ_0. By 1-consistency, T must fail to prove some wff of the kind $\neg\varphi(\overline{m})$; so this $\neg\varphi(\overline{m})$ can't be true (otherwise T would prove it, by Theorem 11.3 again). So this $\varphi(\overline{m})$ must be true, making $\exists x\varphi(x)$ true. So T is 1-sound. $\boxtimes$

25.2   Rosser's construction: the basic idea

We turn now to the main topic of this chapter, Rosser's 1936 construction of a new type of undecidable sentence.

---

[2]It is often said that 1-consistency is equivalent to Σ_1-soundness. But that doesn't quite work across the board with our official definition of Σ_1 wffs, as opposed to defining a Σ_1 wff as the single existential quantification of a Δ_0 wff. Cf. the final comment in Section 12.5(b).

Essentially, where Gödel constructs a sentence G_T which is true just when it is unprovable, Rosser constructs a sentence R_T which is true just when, if there is a proof of it, there is already a 'smaller' proof of its negation (i.e. if there is a proof of R_T with super g.n. n, then there is also proof of $\neg R_T$ super g.n. m, where $m < n$).

It is immediate that if T is a nice *sound* theory (and hence consistent), neither R_T nor $\neg R_T$ can be provable. For suppose R_T were a theorem. Then it would be true. In other words, 'if R_T is provable, $\neg R_T$ is already provable' would be true, and also this conditional would have a true antecedent. It would then follow that $\neg R_T$ is provable, making T inconsistent, contrary to hypothesis. Therefore R_T is unprovable. Which shows that the material conditional 'if R_T is provable, $\neg R_T$ is already provable' has a false antecedent, and hence is true. In other words, R_T is true. Hence its negation $\neg R_T$ is false, and is therefore also unprovable in a sound T.

However, that observation is not very exciting. What matters about Rosser's construction is that, in order to show that neither R_T nor $\neg R_T$ is provable, we do not need the semantic assumption that T is sound. The syntactic assumption of T's consistency is enough, as we now show.

25.3   The Gödel-Rosser Theorem

(a)   Define $\overline{Prf}_T(m, n)$ to hold when m numbers a T-proof of the *negation* of the sentence with number n. This relation is obviously p.r. given that Prf_T is; so assuming T is nice it can be canonically captured by a wff $\overline{\mathsf{Prf}}_T(\mathsf{x}, \mathsf{y})$.[3] Now we define the corresponding *Rosser provability predicate for* T:

$$\mathsf{RProv}_T(\mathsf{x}) =_{\mathrm{def}} \exists \mathsf{v}(\mathsf{Prf}_T(\mathsf{v}, \mathsf{x}) \wedge (\forall \mathsf{w} \leq \mathsf{v}) \neg \overline{\mathsf{Prf}}_T(\mathsf{w}, \mathsf{x})).$$

Then a sentence is Rosser-provable in T – its g.n. satisfies the Rosser provability predicate – if it has a proof (in the ordinary sense) and there is no 'smaller' proof of its negation.

(b)   Now we apply the Diagonalization Lemma, not to the negation of a regular provability predicate (which is what we just did to get Gödel's First Theorem again in Section 24.5), but to the negation of a Rosser provability predicate. The Lemma then tells us that there is a sentence R_T which is a fixed point for $\neg \mathsf{RProv}_T(\mathsf{x})$. That is to say, assuming T is nice,

---

[3]Consider the p.r. function defined by $neg(x) =_{\mathrm{def}} \ulcorner \neg \urcorner \star x$, where '$\star$' is the concatenation function from Section 20.1. We have $neg(\ulcorner \varphi \urcorner) = \ulcorner \neg \urcorner \star \ulcorner \varphi \urcorner = \ulcorner \neg \varphi \urcorner$. So neg takes the g.n. of a wff and returns the g.n. of its negation.

Suppose we introduce a function-symbol into T's language to capture this function (see Section 16.3(a)). We'll use the symbol '$\dot{\neg}$' to do the job. So now '$\neg$' has a double use: 'undotted' and attached to a wff, it is a truth-functional operator; 'dotted' and attached to a term, it expresses a corresponding numerical function. Such a dotting convention is useful to stop us getting confused.

With this neat new notation in the extended language, we could simply put $\overline{\mathsf{Prf}}_T(\mathsf{x}, \mathsf{y}) =_{\mathrm{def}} \mathsf{Prf}_T(\mathsf{x}, \dot{\neg}\mathsf{y})$.

$$T \vdash \mathsf{R}_T \leftrightarrow \neg\mathsf{RProv}_T(\ulcorner\mathsf{R}_T\urcorner).$$

Hence if T is sound and its theorems are true, then R_T will indeed be true just so long as it isn't Rosser-provable. In other words, R_T is true just when, if it's provable, there is already a proof of its negation. So by the argument we have already sketched, if T is sound, $T \nvdash \mathsf{R}_T$ and $T \nvdash \neg\mathsf{R}_T$.

(c) But now we want to show that we don't need the assumption of soundness: T's consistency is enough. We start with an analogue of Theorem 24.6:

> **Theorem 25.4** *Let T be a nice theory, and let γ be any fixed point for $\neg\mathsf{RProv}_T(\mathsf{x})$. Then $T \nvdash \gamma$ and $T \nvdash \neg\gamma$.*

Proof for first half Dropping subscripts for readability, we are given that γ is a fixed point for $\neg\mathsf{RProv}(\mathsf{x})$, i.e. $T \vdash \gamma \leftrightarrow \neg\mathsf{RProv}(\ulcorner\gamma\urcorner)$.

Suppose for reductio that $T \vdash \gamma$. Then (i) $T \vdash \neg\mathsf{RProv}(\ulcorner\gamma\urcorner)$.

But since γ is a theorem, for some m, $Prf(m, \ulcorner\gamma\urcorner)$. Since Prf captures Prf, $T \vdash \mathsf{Prf}(\overline{m}, \ulcorner\gamma\urcorner)$.

Also, since T is consistent, $\neg\gamma$ is unprovable, so for all n, not-$\overline{Prf}(n, \ulcorner\gamma\urcorner)$. Since $\overline{\mathsf{Prf}}$ captures $\overline{Prf}$, then for each $n \leq m$ in particular, $T \vdash \neg\overline{\mathsf{Prf}}(\overline{n}, \ulcorner\gamma\urcorner)$. Using the result (O4) of Section 11.3, that shows $T \vdash (\forall\mathsf{w} \leq \overline{m})\neg\overline{\mathsf{Prf}}(\mathsf{w}, \ulcorner\gamma\urcorner)$.

Putting these results together, $T \vdash \mathsf{Prf}(\overline{m}, \ulcorner\gamma\urcorner) \wedge (\forall\mathsf{w} \leq \overline{m})\neg\overline{\mathsf{Prf}}(\mathsf{w}, \ulcorner\gamma\urcorner)$. So, existentially quantifying, (ii) $T \vdash \mathsf{RProv}(\ulcorner\gamma\urcorner)$.

But (i) and (ii) contradict each other. So, given T's assumed consistency, γ is not provable: $T \nvdash \gamma$. ⊠

Proof for second half With the same γ, suppose that $\neg\gamma$ is a theorem. By the fixed point biconditional, (iii) $T \vdash \mathsf{RProv}(\ulcorner\gamma\urcorner)$.

Because $\neg\gamma$ is a theorem, for some m, $\overline{Prf}(m, \ulcorner\gamma\urcorner)$, so $T \vdash \overline{\mathsf{Prf}}(\overline{m}, \ulcorner\gamma\urcorner)$.

Also, since T is consistent, γ is unprovable, so for all n, not-$Prf(n, \ulcorner\gamma\urcorner)$. Hence, by a parallel argument to before, $T \vdash (\forall\mathsf{v} \leq \overline{m})\neg\mathsf{Prf}(\mathsf{v}, \ulcorner\gamma\urcorner)$. Elementary manipulation gives $T \vdash \forall\mathsf{v}(\mathsf{Prf}(\mathsf{v}, \ulcorner\gamma\urcorner) \to \neg\mathsf{v} \leq \overline{m})$. Now appeal to (O8) of Section 11.3, and that gives $T \vdash \forall\mathsf{v}(\mathsf{Prf}(\mathsf{v}, \ulcorner\gamma\urcorner) \to \overline{m} \leq \mathsf{v})$.

Combining these two results, it immediately follows that $T \vdash \forall\mathsf{v}(\mathsf{Prf}(\mathsf{v}, \ulcorner\gamma\urcorner) \to (\overline{m} \leq \mathsf{v} \wedge \overline{\mathsf{Prf}}(\overline{m}, \ulcorner\gamma\urcorner)))$. That implies $T \vdash \forall\mathsf{v}(\mathsf{Prf}(\mathsf{v}, \ulcorner\gamma\urcorner) \to (\exists\mathsf{w} \leq \mathsf{v})\overline{\mathsf{Prf}}(\mathsf{w}, \ulcorner\gamma\urcorner))$. So given our definition, (iv) $T \vdash \neg\mathsf{RProv}(\ulcorner\gamma\urcorner)$.

But (iii) and (iv) contradict each other. So the negation of the fixed point γ is not provable either: $T \nvdash \neg\gamma$. ⊠

(d) So we now know that any fixed point for $\neg\mathsf{RProv}_T$ must be formally undecidable in T. But the Diagonalization Lemma has already told us that there has to *be* such a fixed point R_T. Hence, assuming no more than T's niceness, it follows that T is negation-incomplete, with any R_T an undecidable sentence. Which yields the *Gödel-Rosser Theorem*:

> **Theorem 25.5** *If T is a nice theory, then there is an L_A-sentence φ such that neither $T \vdash \varphi$ nor $T \vdash \neg\varphi$.*

25.4   Improving the theorem

So we have got rid of the assumption of ω-consistency in the proof of the syntactic incompleteness theorem. But note that RProv_T is not straightforwardly equivalent to a Σ_1 wff (there are unbounded universal quantifiers buried in $\neg\overline{\mathsf{Prf}}$). So we can't use our proof for Theorem 25.4 as it stands to show that $\neg\mathsf{RProv}_T$ has a Π_1-equivalent fixed point, and hence to show that there is an undecidable sentence of Goldbach type. But can we do better?

Let's look at our proof and generalize the leading idea. Suppose, then, that instead of using the two-place predicates Prf and $\overline{\mathsf{Prf}}$ we use any other pair of two-place predicates P and $\overline{\mathsf{P}}$ which respectively 'enumerate' the positive and negative T-theorems, i.e. which satisfy the following conditions:

1. if $T \vdash \gamma$, then for some m, $T \vdash \mathsf{P}(\overline{\mathsf{m}}, \ulcorner\gamma\urcorner)$,
2. if $T \nvdash \gamma$, then for all n, $T \vdash \neg\mathsf{P}(\overline{\mathsf{n}}, \ulcorner\gamma\urcorner)$,
3. if $T \vdash \neg\gamma$, then for some m, $T \vdash \overline{\mathsf{P}}(\overline{\mathsf{m}}, \ulcorner\gamma\urcorner)$,
4. if $T \nvdash \neg\gamma$, then for all n, $T \vdash \neg\overline{\mathsf{P}}(\overline{\mathsf{n}}, \ulcorner\gamma\urcorner)$.

Now define $\mathsf{RP}_T(\mathsf{x}) =_{\mathrm{def}} \exists\mathsf{v}(\mathsf{P}(\mathsf{v},\mathsf{x}) \wedge (\forall\mathsf{w} \leq \mathsf{v})\neg\overline{\mathsf{P}}(\mathsf{w},\mathsf{x}))$. This gives us another Rosser-style predicate, and the argument will go through *exactly* as before: for a nice theory T, any fixed point of $\neg\mathsf{RP}_T(\mathsf{x})$ will be undecidable.

This tells us what we need to look for. Suppose we can find predicates P and $\overline{\mathsf{P}}$ which satisfy our four 'enumeration' conditions, but which are Δ_0 (i.e. lack unbounded quantifiers). Then the corresponding $\mathsf{RP}_T(\mathsf{x})$ will evidently be Σ_1: so its negation $\neg\mathsf{RP}_T(\mathsf{x})$ *will* be equivalent to a Π_1 wff, and will therefore have undecidable fixed points which are equivalent to Π_1 wffs, which will give us the Gödel-Rosser Theorem again but with an undecidable sentence of Goldbach type.

It just remains, then, to find a suitable pair of Δ_0 predicates P and $\overline{\mathsf{P}}$. Well, consider the Σ_1 formula $\mathsf{Prov}_T(\mathsf{x}) =_{\mathrm{def}} \exists\mathsf{vPrf}(\mathsf{v},\mathsf{x})$ which expresses the property $Prov_T$, i.e. the property of Gödel-numbering a T-theorem (see Section 24.1). Since it is Σ_1, $\mathsf{Prov}_T(\mathsf{x})$ is provably equivalent, given enough induction, to a wff of the form $\exists\mathsf{uP}(\mathsf{u},\mathsf{x})$ where P is Δ_0 (see Theorem 12.4).

But note that when γ is a theorem, $\exists\mathsf{uP}(\mathsf{u},\ulcorner\gamma\urcorner)$ is true, so for some m, $\mathsf{P}(\overline{\mathsf{m}},\ulcorner\gamma\urcorner)$ is true. So, being nice and hence Δ_0-complete, T proves that last wff. And if γ isn't a theorem, $\exists\mathsf{uP}(\mathsf{u},\ulcorner\gamma\urcorner)$ is false, so for every n, $\mathsf{P}(\overline{\mathsf{n}},\ulcorner\gamma\urcorner)$ is false, so each $\neg\mathsf{P}(\overline{\mathsf{n}},\ulcorner\gamma\urcorner)$ is true. Being Δ_0-complete, T proves all those latter wffs too. Hence P is Δ_0 and satisfies the 'enumerating' conditions (1) and (2).

We can similarly construct a Δ_0 wff $\overline{\mathsf{P}}$ from $\exists\mathsf{v}\overline{\mathsf{Prf}}(\mathsf{v},\mathsf{x})$. So, after quite a lot of work, we are done. We've shown that

> **Theorem 25.6** *If T is a nice theory which also has induction for at least Σ_1 wffs, then there is an L_A-sentence φ of Goldbach type such that neither $T \vdash \varphi$ nor $T \vdash \neg\varphi$.*

26 Broadening the scope

In this chapter, we note two different ways in which our incompleteness proofs can readily be extended in scope. First, the requirement for p.r. axiomatization can be weakened: the incompleteness theorems will apply to any effectively formalized theory. Second, we show how to get the theorems to apply to certain theories which aren't initially about arithmetic at all.

The second extension is much more important than the first. Formalized-but-not-p.r.-axiomatized theories are mostly strange beasts of little intrinsic interest, though we do extract a neat result by considering them. On the other hand, there is immediate interest in the claim that e.g. set theory – although not natively about numbers – must also be incomplete for Gödelian reasons.

26.1 Generalizing beyond p.r. axiomatized theories

(a)   Our intuitive characterization of a properly formalized theory T requires various properties like that of being an axiom of T to be effectively decidable. Given a sensible Gödel numbering scheme, that means that the characteristic functions of numerical properties like that of numbering a T-axiom should be effectively computable (see Sections 4.3 and 14.6). But we now know that not all computable functions are p.r. (Section 14.5). Hence we could in principle have an effectively axiomatized theory which isn't primitive-recursively axiomatized (in the sense of Section 22.1). Does this give us wriggle room to get around the Gödelian incompleteness theorems in the last chapter? Could there for example be a consistent effectively formalized theory of arithmetic containing Q which was complete because not p.r. axiomatized?

Well, as we noted in Section 23.1, a theory T that is effectively axiomatized but not p.r. axiomatized will be a rather peculiar beast; checking that a putative T-proof *is* a proof will then have to involve a non-p.r. open-ended search, which will make T *very* unlike the usual kind of axiomatized theory. Still, you might say, an oddly axiomatized arithmetic which is complete would still be a lot better than no complete formal theory at all. However, we can't get even that.

(b)   We will make use of the following result:

> **Theorem 26.1** *If Σ is a set of wffs which can be effectively enumerated, then there is a p.r. axiomatized theory whose theorems are exactly the members of Σ.*

Proof   Imagine running through some step-by-step algorithmic procedure Π which lists off Σ's members $\varphi_0, \varphi_1, \varphi_2, \ldots$ (by hypothesis, there is one). Count off the minimal computational steps as we go along. Most of these steps are interim computations; but very occasionally, a wff on the list will be printed out. Let s_j number the step at which φ_j is produced as we go through procedure Π.

Now for an ingenious trick due to William Craig (1953). Consider the theory T defined as follows: (i) for wff $\varphi_j \in \Sigma$, T has the axiom $\varphi_j \circ \varphi_j \circ \varphi_j \circ \ldots \circ \varphi_j$ with '$\circ$' a new symbol and s_j occurrences of φ_j; (ii) A T-proof is always a two-formula sequence of this form: $\varphi \circ \varphi \circ \varphi \circ \ldots \circ \varphi$ (for any number of repetitions of φ) followed by φ. Trivially, the only T-theorems – i.e. end-formulae of proofs – are the φ_j.

Given an arbitrary formula ψ, here is how to tell whether it is a T-axiom:

1. Read in ψ and test to see if it is $\varphi \circ \varphi \circ \varphi \circ \ldots \circ \varphi$ for some φ. If it passes this test, let n be the number of occurrences of φ.

2. Run through n steps of the procedure Π which lists off the T-theorems.

3. Check to see if at the end of these n steps, Π outputs the wff φ.

If the final check confirms that φ is indeed printed out, ψ is a T-axiom; otherwise ψ is not an axiom. No stage in this testing procedure requires an open-ended search. Hence we can decide what's a T-axiom by a p.r. procedure. So we can also decide what is a T-proof by a p.r. procedure.

Therefore, as required, T is a p.r. axiomatized theory whose theorems are the wffs in Σ. ☒

Recall next that if T is an effectively formalized theory, its theorems form an effectively enumerable set Σ (see Theorem 4.1). So as an immediate corollary of our last theorem we now have *Craig's Re-axiomatization Theorem* (Craig, 1953):

> **Theorem 26.2** *If T is an effectively axiomatized theory, then there is a p.r. axiomatized theory T' which has the same theorems.*

(c) Now we apply Craig's Theorem. Suppose that T is any sound axiomatized theory whose language includes L_A. Craig's Theorem tells us that there is a p.r. axiomatized theory T' which has the same theorems as T. But since T' shares the same theorems, this theory is sound too, so Theorem 22.1 applies. There is therefore an L_A-sentence φ (in fact of Goldbach type) such that $T' \nvdash \varphi$ and $T' \nvdash \neg\varphi$. Hence, again since T and T' share their theorems,

> **Theorem 26.3** *If an effective axiomatized theory T, whose language includes L_A, is sound, then there is an L_A-sentence φ of Goldbach type such that $T \nvdash \varphi$ and $T \nvdash \neg\varphi$.*

Suppose likewise that T is any consistent effectively axiomatized theory which extends Q. Then Craig's Theorem tells us that there is a p.r. axiomatized theory T' which has the same theorems, so is still consistent, still extends Q, and so is nice. So Theorem 25.5 applies, and there is an L_A-sentence such that $T' \nvdash \varphi$ and $T' \nvdash \neg\varphi$. Hence, since T and T' share their theorems,

> **Theorem 26.4** *If T is a consistent effectively axiomatized theory which extends Q, then there is an L_A-sentence φ such that $T \nvdash \varphi$ and $T \nvdash \neg\varphi$.*

In sum: given Craig's Theorem, Gödelian incompleteness infects any suitable sound/nice effectively axiomatized theory, whether or not it is p.r. axiomatized.

(d) To be sure, the proof of our informal version of Craig's Theorem here is rather quick-and-dirty. However we won't pause to tidy things up now. That's because in Chapter 40 – when we at last have a general account of computation and decidability to hand – we'll be able to prove more formal versions of Theorems 26.3 and 26.4 without going through Craig's Theorem.

Still, even before we dive into the general theory of computation, it has been worth noting that our Gödelian arguments which originally apply just to standardly presented (i.e. p.r. axiomatized) theories in fact apply more generally. And this yields an important corollary . . .

26.2   True Basic Arithmetic can't be axiomatized

Let True Basic Arithmetic be $\mathcal{T}_A$, the set of truths of L_A, i.e. the set of sentences which are true on the standard interpretation built into L_A. Then we have

> **Theorem 26.5** $\mathcal{T}_A$ *is not effectively axiomatizable.*

Proof Suppose otherwise, i.e. suppose T is an axiomatized formal theory which proves exactly the sentences in $\mathcal{T}_A$. Then T's theorems are all true, so T is sound; its language includes L_A; so Theorem 26.3 applies. Hence there is an L_A-sentence φ such that $T \nvdash \varphi$ and $T \nvdash \neg\varphi$. But one of φ and $\neg\varphi$ must be true, which means that T doesn't prove all the sentences in $\mathcal{T}_A$. Contradiction.   ⊠

Compare this result with Theorem 6.2, which tells us the truths of a 'sufficiently expressive' language are not effectively axiomatizable. It is plain what we have gained. The old theorem didn't tell us anything about what a sufficiently expressive language is like; it left open the possibility that the theorem only applied to *very* rich languages. Now we see that non-axiomatizability applies even down at the lowly level of the truths of L_A.

26.3   Generalizing beyond overtly arithmetic theories

(a) So far, our various versions of the first incompleteness theorem apply only to theories which can prove wffs in the language of basic arithmetic. However, the theorems can also be extended to apply to a theory T which isn't natively a

193

theory of arithmetic and whose language doesn't include L_A, if we can still 'do enough arithmetic' in T. In this section, then, we will show how to extend the scope of the syntactic proof of incompleteness, as this is the important case.

(b)   In general, a theory T may well be framed in a language $L =_{\text{def}} \langle \mathcal{L}, \mathcal{I} \rangle$ which is quite different from PA's now familiar language $L_A =_{\text{def}} \langle \mathcal{L}_A, \mathcal{I}_A \rangle$. L may initially contain no arithmetical vocabulary, and its native quantifiers – let's write them using the temporary notation '∀', '∃' – may or may not have numbers in their domain according to $\mathcal{I}$. For example, T might be couched in the language of set theory, whose only non-logical vocabulary is '∈' for set-membership, and whose quantifiers run over sets.

Still, even if the natural numbers aren't in the domain that T is talking about, there might be a 'copy' of them with the same structure lurking in that domain. For example, as we noted right back in Section 1.3, the universe of sets contains the infinite sequence of sets

$$\varnothing, \{\varnothing\}, \{\varnothing, \{\varnothing\}\}, \{\varnothing, \{\varnothing\}, \{\varnothing, \{\varnothing\}\}\}, \ldots$$

which is structured just like the numbers (each set in the sequence has a successor, different sets have different successors, the sequence never circles round on itself). So suppose we add definitions of arithmetical vocabulary to T so as to get $0, 1, 2, 3, \ldots$ to refer to elements of T's 'copy' of the numbers, and to get the numerical quantifiers to run over these elements. Then we might well still be able to apply the incompleteness proof.

That is the semantic motivation for the following syntactic construction. First we augment T's syntax as necessary with definitions of arithmetical vocabulary so that every string of symbols which is an $\mathcal{L}_A$ wff will now also count as a wff of T's definitionally extended language. This will take two stages:

1. We will need definitional rules for rendering the basic arithmetical vocabulary '0', 'S', '+' and '×' in terms of $\mathcal{L}$-expressions, if these aren't already present. For example, if T is a set theory, we might give the outright definition $0 =_{\text{def}} \varnothing$ and then have a rule for eliminating occurrences of 'S' using $\mathsf{S}\tau =_{\text{def}} \tau \cup \{\tau\}$.

2. If T doesn't already have numerical quantifiers, we will also need to specify a one-place $\mathcal{L}$-predicate 'Nat', where this is intended to be satisfied by the natural numbers (or by whatever plays the role of the numbers in T's native domain of quantification). Then we can re-introduce the $\mathcal{L}_A$ quantifiers thus: $\forall \xi \varphi =_{\text{def}} ∀\xi(\mathsf{Nat}(\xi) \to \varphi)$ and $\exists \xi \varphi =_{\text{def}} ∃\xi(\mathsf{Nat}(\xi) \land \varphi)$.

Suppose all that is done. Then we will now say

> T^A is an *A-extension* of T ('A' for 'arithmetical', of course!) if T^A is just T with its syntax augmented with arithmetical vocabulary and arithmetical quantifiers as described, and with new axioms giving the definitions of the new syntax in terms of the old, so

that for every L_A sentence φ^A of the extended syntax, there is a φ of the original language such that $T^A \vdash \varphi^A \leftrightarrow \varphi$.

Adding the new definitional axioms (instances of a few schemata) to a p.r. axiomatized T will give us a p.r. axiomatized T^A. And we assume that coherently framed definitions preserve consistency: i.e. if T is consistent, so is T^A.

(c)   Recalling the definition of 'T extends Q' in Section 11.7(b), we now say:

> T *extends** Q iff T has an A-extension which proves every Q-theorem.[1]

Then we can readily prove the key result

> **Theorem 26.6** *If T is a consistent, p.r. axiomatized theory which extends** Q, *then there is a sentence φ such that $T \nvdash \varphi$ and $T \nvdash \neg\varphi$.*

Proof By hypothesis, T has some A-extension T^A which is still a consistent, p.r. axiomatized theory, and which proves every Q-theorem (and hence proves all the wffs needed to capture p.r. functions, so is p.r. adequate). The Gödel-Rosser Theorem therefore applies to T^A. So there is an undecidable sentence φ^A of T^A. Take a sentence φ from the original language such that $T^A \vdash \varphi^A \leftrightarrow \varphi$ (we know there must be such a sentence). If T decided φ, then T^A would decide φ and hence decide φ^A after all. So T does not decide φ.   ⊠

The obvious application of this theorem is to set theory (in which we can 'do arithmetic' by defining the natural numbers in the way we indicated). Despite its enormous power, set theory is incomplete and can't even decide every sentence of basic arithmetic.

26.4   A word of warning

It is worth stressing, however, that even if a theory T talks about a domain which contains (an analogue of) the natural numbers, T *may* lack a way of picking them out, i.e. T may lack a way of defining a suitable predicate Nat. In such a case, T *won't* be able to prove or even express e.g. the analogues of Q's quantified theorems, and then our extended version of the incompleteness theorem won't get a grip.

Here's an illustration of the point (don't worry if the example is unfamiliar; for our purposes, it's only the general moral that matters).

Compare Q, an axiomatic theory of rational fields ('fractions'), and R, the textbook axiomatic theory of real closed fields ('real numbers'). Both theories are only true of domains big enough to contain a '0', a '1', and all their successors;

---

[1]We should perhaps echo fn. 9 from Section 11.7: like extending, extending* is a syntactic relation. What matters, if T is to extend* Q, is that T has an A-extension in which we can derive the right strings of symbols.

i.e. both theories are only true of domains which contain (something that looks like) the natural number series.

Now, Julia Robinson (1949) showed that you can indeed construct a Q-predicate Nat which is only true of the natural numbers – though doing this takes some rather sophisticated ideas.[2] Hence, you can construct a theory of the natural numbers inside Q: so Q extends Robinson Arithmetic Q and must be *incomplete* and incompletable too.

On the other hand, Alfred Tarski had earlier proved that R is a *complete* theory – which means that there can't be a predicate of R which picks out the natural numbers. Put it this way: while the real numbers contain (a copy of) the natural numbers, the pure theory of real numbers doesn't contain the theory of natural numbers.[3] So even if a theory T deals with a structure that has (a copy of) the natural numbers among its elements, the incompleteness of arithmetic need not entail the incompleteness of T.

---

[2] For accessible exposition, see Flath and Wagon (1991).
[3] For more on Tarski's theorem here, see e.g. Hodges (1997, Sections 2.7, 7.4).

27 Tarski's Theorem

We have seen two lines of argument for incompleteness: one assumes the relevant theory's soundness, the other its consistency. Our proofs, of course, have belonged to ordinary mathematics. But we might reasonably wonder: can these proofs themselves be coded up and 'done in arithmetic'?

As we will discover in Chapter 31, we can do this for the syntactic argument. Since a nice theory T can, via Gödel-coding, 'talk about' what it can prove, it can frame and prove a sentence that 'says' *If I am consistent, I can't prove* G_T. And this leads to Gödel's Second Theorem.

But in this chapter we will discover that there is a block to formalizing the semantic argument: nice arithmetics can't talk about their own soundness, because they can't talk about the truth of their own sentences. Or at least, that's the rough headline.[1]

27.1 Truth-predicates, truth-theories

Recall a very familiar thought: 'snow is white' is true iff snow is white. Likewise, 'The Higgs boson exists' is true iff the Higgs boson exists, '2 + 2 = 4' is true iff 2 + 2 = 4. And so on, for all other sensible instances of the schematic form

'S' is true iff S.

That's because of the meaning of the informal truth-predicate 'true'.[2]

We now consider the possibility of a formal arithmetical language's having a truth-predicate which satisfies analogous biconditionals. Of course, we don't have quotation marks in standard arithmetical languages; but we can use Gödel numbering instead to refer to sentences. So assume we have fixed on some scheme for numbering wffs of the interpreted arithmetical language L. Then we can define a corresponding numerical property $True_L$ as follows:

$True_L(n)$ iff n is the g.n. of a true sentence of L.

Because $True_L$ is a numerical property, we can now sensibly raise the question of whether it is expressible by a truth-predicate in the language L or in some richer language L^+ which includes L.[3]

---

[1]See again the quotation from Mostowski at the end of Section 23.1.

[2]A so-called deflationist about truth will say, indeed, that such biconditionals encapsulate more or less the *whole* truth about truth – see Stoljar and Damnjanovic (2012). Be that as it may. That such biconditionals must hold is certainly *part* of the truth about truth.

[3]As in Section 22.1(b), we say a language L^+ includes L if all L-wffs are also L^+-wffs, and these shared wffs continue to receive the same interpretation. Trivially, L includes itself.

So consider a language L^+ which includes L, and suppose that the open wff $\mathsf{T}(\mathsf{x})$ does indeed express in L^+ the numerical property $True_L$. Then, by definition, for any L-sentence φ,

'$\mathsf{T}(\ulcorner\varphi\urcorner)$' is true iff $True_L(\ulcorner\varphi\urcorner)$ iff 'φ' is true.

Hence, for any L-sentence φ, the corresponding L^+-sentence

$$\mathsf{T}(\ulcorner\varphi\urcorner) \leftrightarrow \varphi$$

will be true. This motivates our first main definition:

> An open L^+-wff $\mathsf{T}(\mathsf{x})$ is a *numerical truth-predicate for* L iff for every L-sentence φ, $\mathsf{T}(\ulcorner\varphi\urcorner) \leftrightarrow \varphi$ is true.

Here's a companion definition:

> A theory T (with language L^+ which includes L) is an *arithmetical truth-theory for* L iff there is some L^+-wff $\mathsf{T}(\mathsf{x})$ such that for every L-sentence φ, T proves $\mathsf{T}(\ulcorner\varphi\urcorner) \leftrightarrow \varphi$.

Quite often, such a truth-theory for L is called a 'definition of truth for L'.

27.2   The undefinability of truth

Suppose T is a nice arithmetical theory (with language L which includes L_A). One obvious question which arises is: could T be competent to define truth *for its own language* (i.e., can T include a truth-theory for L)? The answer is immediate:

Theorem 27.1 *No nice theory can define truth for its own language.*

Proof   Assume T defines truth for L, i.e. there is an L-predicate $\mathsf{T}(\mathsf{x})$ such that

1. $T \vdash \mathsf{T}(\ulcorner\varphi\urcorner) \leftrightarrow \varphi$ for every L-sentence φ.

Since T is nice, the Diagonalization Lemma applies, so applying the Lemma to $\neg\mathsf{T}(\mathsf{x})$ in particular, we know that there must be some sentence L such that

2. $T \vdash \mathsf{L} \leftrightarrow \neg\mathsf{T}(\ulcorner\mathsf{L}\urcorner)$.

But, from (1), we immediately have

3. $T \vdash \mathsf{T}(\ulcorner\mathsf{L}\urcorner) \leftrightarrow \mathsf{L}$.

So T is inconsistent and not nice, contrary to hypothesis. Hence our assumption must be wrong: T can't define truth for its own language. $\boxtimes$

(1) tells us that 'T' behaves like 'is true'. So in virtue of (2), L is provably equivalent to a sentence which 'says' that L isn't true. Hence L is like a Liar sentence. But note that L is produced by a simple diagonalization construction again; and the construction yields a theorem, not a paradox.

27.3 Tarski's Theorem: the inexpressibility of truth

Our syntactic Theorem 27.1 puts limits on what a nice theory can *prove* about truth. Our next theorem gives things a semantic twist by putting limits on what a theory's language can even *express* about truth.

Consider L_A for the moment, and suppose that there is an L_A truth-predicate T_A that expresses the corresponding numerical property $True_A$ (the property of numbering a true L_A sentence). Take the negation of that predicate and apply Carnap's Equivalence (the semantic result we noted in Section 24.4 en route to the Diagonalization Lemma). That tells us that for some L_A sentence L, the biconditional

    1. $\mathsf{L} \leftrightarrow \neg\mathsf{T}_A(\ulcorner\mathsf{L}\urcorner)$

will be a true L_A wff. But, by the assumption that T_A is a truth-predicate for L_A,

    2. $\mathsf{T}_A(\ulcorner\mathsf{L}\urcorner) \leftrightarrow \mathsf{L}$

must be true too. (2) and (3) immediately lead to contradiction again. Therefore our supposition that T_A is a truth-predicate has to be rejected. Hence no predicate of L_A can even *express* the numerical property $True_A$ (let alone serve to capture it in a sound theory built in L_A).

The argument evidently generalizes, to give us *Tarski's Theorem*:[4]

> **Theorem 27.2** *If L includes L_A, no L-predicate can express the numerical property* True$_L$ *(i.e. the property of numbering a truth of L).*

27.4 Capturing and expressing again

You can express typical *syntactic* properties of a sufficiently rich formal theory of arithmetic (like provability) inside the theory itself via Gödel numbering. By contrast, Tarski's Theorem tells us that you can't express some key *semantic* properties (like arithmetical truth) inside the theory.

So suppose T is a nice theory. Then

1. there are some numerical properties that T can both express and capture (the p.r. ones for a start, including those p.r. properties involved in the arithmetization of syntax);

2. there are some properties that T can express but not capture (for example $Prov_T$); and

---

[4] Alfred Tarski investigated these matters in his classic (1933); though Gödel had already noted the key point, e.g. in a letter to Zermelo written in October, 1931 (Gödel, 2003b, pp. 423–429). Also see the quotation at the end of this chapter.

3. there are some properties that T's language L cannot even express (for example $True_L$, the numerical property of numbering-a-true-L-wff).

It is not, we should hasten to add, that the property $True_L$ is mysteriously ineffable, and escapes all formal treatment. A richer theory T^+ with a richer language L^+ may perfectly well be able to express and capture $True_L$. But the point remains that, however rich a given theory of arithmetic is, there will be limitations, not only on what numerical properties it can capture but even on which numerical properties that particular theory's language can express.

27.5   The Master Argument?

Our results about the non-expressibility of truth give us a particularly illuminating take on the argument for incompleteness. Put it this way:

> Truth in L_A isn't provability in PA, because while PA-provability *is* expressible in L_A, truth-in-L_A *isn't*. So assuming that everything provable in PA is true (i.e. that PA is sound), this means that there must be truths of L_A which it can't prove. Similarly, of course, for other nice theories.

And we might well take this to be *the* Master Argument for incompleteness, revealing the roots of the phenomenon. Gödel himself wrote (in response to a query):

> I think the theorem of mine that von Neumann refers to is ... that a complete epistemological description of a language A cannot be given in the same language A, because the concept of truth of sentences in A cannot be defined [i.e., expressed] in A. *It is this theorem which is the true reason for the existence of undecidable propositions in the formal systems containing arithmetic.* I did not, however, formulate it explicitly in my paper of 1931 but only in my Princeton lectures of 1934. The same theorem was proved by Tarski in his paper on the concept of truth [Tarski (1933)].[5]

In sum, as we emphasized before, *arithmetical truth* and *provability-in-this-or-that-formal-system* must peel apart.

---

[5]The emphasis is mine. Gödel's letter is quoted in Feferman (1984), which also has a very interesting discussion of why Gödel chose not to highlight this line of argument for incompleteness in his original paper, a theme we'll return to in Chapter 37. The passage in the Princeton lectures to which Gödel refers is at (Gödel, 1934, p. 363).

28 Speed-up

I can't resist briefly noting here some rather striking results about the length of proofs. And it is indeed worth getting to understand what the theorems say, even if some of the proofs in this chapter are perhaps just for enthusiasts.

28.1 The length of proofs

We might expect that, as a general tendency, the longer a wff, the longer its proof (if it has one). But can there be any tidy order in this relationship in the case of nice theories?

We will say that a proof for φ is f-*bounded*, for a given function f, if the proof's g.n. is less than $f(\ulcorner\varphi\urcorner)$. Then it would indeed be rather tidy if, for some theory T, there were some corresponding p.r. function f_T which puts a general bound on the size of T-proofs – meaning that, for *any* φ, if it is a provable at all, it has an f_T-bounded proof. However, unfortunately,

> **Theorem 28.1** *If T is nice theory, then for any p.r. function f, there is a provable wff φ which has no f-bounded proof.*

Proof sketch Suppose the theorem is false. That is, suppose that there is a p.r. bounding function f_T such that for any φ, if it is T-provable at all, it has a f_T-bounded proof. Then there would be a p.r. procedure for testing whether φ has a proof in T. Just calculate $f_T(\ulcorner\varphi\urcorner)$, and do a bounded search using a 'for' loop to run through all the possible proofs up to that size to see if one of them is in fact a proof of φ. But Theorem 24.9 tells us that there can be no p.r. procedure for deciding whether φ is a T-theorem if T is nice. So there can be no such function f_T. So the theorem has to be true. ☒

But now note that some p.r. functions $f(n)$ grow fantastically fast (see e.g. Section 38.3 on 'super-duper-exponentiation') and soon have huge values even for low values of n. So our theorem means, roughly speaking, that there will inevitably be T-theorems which only have enormous proofs but which can, relative to the length of the proof, be stated pretty briefly.

28.2 The idea of speed-up

Our result that some relatively short wffs have enormous proofs is not too surprising. But our next result is perhaps more exciting.

We fix on an acceptable Gödel-numbering scheme. Let's say that

Theory T_2 *exhibits speed-up* over theory T_1 iff for *any* p.r. function f, there is some wff φ such that (i) both $T_1 \vdash \varphi$ and $T_2 \vdash \varphi$ but (ii) while there is a T_2-proof of φ with g.n. p, there is no T_1-proof with g.n. less than or equal to $f(p)$.

This means that, at least sometimes, T_2 gives much shorter proofs than T_1. Then we have the following:

Theorem 28.2 *If T is nice theory, and γ is some sentence of its language such that neither $T \vdash \gamma$ nor $T \vdash \neg\gamma$, then the theory $T+\gamma$ got by adding γ as a new axiom exhibits speed-up over T.*

In other words, adding a previously undecidable wff γ to a nice theory T as a new axiom not only enables us to prove *new* theorems (γ, for a start) but it also radically shortens the proofs of some *old* theorems. The general phenomenon was first noted in an abstract by Gödel (1936).[1]

Number theorists have long been familiar with cases where arithmetical theorems seem only to have complicated proofs in 'pure' arithmetic even though there are relatively nice proofs if we are allowed to extend our methods and appeal, for example, to the theory of complex analysis. What our theorem shows is that there is a kind of inevitability about this kind of situation: even without going outside the original language of T, adding (independent) new axioms can speed up proofs of old T-theorems.

Proof sketch Suppose the theorem is false. So there is a sentence γ which is undecided by T, but such that $T + \gamma$ does *not* exhibit speed-up over T. Then there is a p.r. function f such that for every T-theorem ψ, if ψ has a proof in $T + \gamma$ with g.n. p, then it has a proof in the original T with g.n. number no greater than $f(p)$.

Well, take any T-theorem φ. Then $(\gamma \vee \varphi)$ is also a T-theorem. But $(\gamma \vee \varphi)$ is trivially provable in $T+\gamma$. And there will be a very simple computation, with no open-ended searching, that takes us from the g.n. of φ to the g.n. of the trivial proof in $T + \gamma$ of $(\gamma \vee \varphi)$. In other words, the g.n. of the proof will be $h(\ulcorner\varphi\urcorner)$, for some p.r. function h.[2] So, by our supposition with $(\gamma \vee \varphi)$ for ψ, $(\gamma \vee \varphi)$ must have a proof in T with g.n. no greater than $f(h(\ulcorner\varphi\urcorner))$.

Next consider the theory $T + \neg\gamma$. Trivially again, for any φ, $T + \neg\gamma \vdash \varphi$ iff $T \vdash (\gamma \vee \varphi)$. So we have a p.r. decision procedure for telling whether an arbitrary φ is a theorem of $T+\neg\gamma$. Just run a 'for' loop examining in turn all the T-proofs with g.n. up to $f(h(\ulcorner\varphi\urcorner))$ and see if a proof of $(\gamma \vee \varphi)$ turns up.

---

[1] See also Section 29.9.

[2] To take the simplest case, imagine $T + \gamma$'s logical system is set up with the rule of $\vee$-introduction, so the little sequence γ, $(\gamma \vee \varphi)$ will serve as the needed proof. Then the super g.n. of this proof is
$$2^{\ulcorner\gamma\urcorner} \cdot 3^{\ulcorner(\gamma\vee\varphi)\urcorner} = 2^{\ulcorner\gamma\urcorner} \cdot 3^{\ulcorner(^\urcorner\star\ulcorner\gamma\urcorner\star\ulcorner\vee\urcorner\star\ulcorner\varphi\urcorner\star\ulcorner)\urcorner\urcorner}.$$
Evidently the proof's g.n. is then a p.r. function of $\ulcorner\varphi\urcorner$.

But $T + \neg\gamma$ is still a nice theory: it is consistent (else we'd have $T \vdash \gamma$, contrary to hypothesis), it is p.r. axiomatized, and it contains Q since T does. So there can't be a p.r. procedure for testing theoremhood in $T + \neg\gamma$ (by Theorem 24.9). Contradiction. $\boxtimes$

28.3 Long proofs, via diagonalization

The arguments above for our two theorems are fine as far as they go. However, there is some additional interest in noting that we can use the Diagonalization Lemma to prove the first of them 'constructively'. That is to say, we can actually give a recipe which takes an arbitrarily fast-growing p.r. function f, and constructs a wff φ which is provable but has no f-bounded proof.[3]

Suppose then that the *Prf* relation for T is captured by the open wff $\mathsf{Prf}(\mathsf{x}, \mathsf{y})$, and that our arbitrary p.r. function f is captured by $\mathsf{F}(\mathsf{x}, \mathsf{y})$. Now form the wff

$$\forall \mathsf{v}\{\mathsf{F}(\mathsf{z}, \mathsf{v}) \to (\forall \mathsf{u} \leq \mathsf{v})\neg\mathsf{Prf}(\mathsf{u}, \mathsf{z})\}.$$

By the Diagonalization Lemma, there is a wff φ such that

$$T \vdash \varphi \leftrightarrow \forall \mathsf{v}\{\mathsf{F}(\ulcorner\varphi\urcorner, \mathsf{v}) \to (\forall \mathsf{u} \leq \mathsf{v})\neg\mathsf{Prf}(\mathsf{u}, \ulcorner\varphi\urcorner)\}.$$

We now show that (i) $T \vdash \varphi$, but also (ii) the g.n. of the 'smallest' proof of φ is greater than $f(\ulcorner\varphi\urcorner)$.

Proof, for enthusiasts Suppose, for reductio, that there *is* a proof of φ with g.n. $p \leq f(\ulcorner\varphi\urcorner) = k$. Then we are assuming we have (1) $T \vdash \varphi$ but also (2) for some $p \leq k$, $Prf(p, \ulcorner\varphi\urcorner)$, so $T \vdash \mathsf{Prf}(\overline{\mathsf{p}}, \ulcorner\varphi\urcorner)$.

By (1), $T \vdash \forall \mathsf{v}\{\mathsf{F}(\ulcorner\varphi\urcorner, \mathsf{v}) \to (\forall \mathsf{u} \leq \mathsf{v})\neg\mathsf{Prf}(\mathsf{u}, \ulcorner\varphi\urcorner)\}$. So instantiating the quantifier we get $T \vdash \mathsf{F}(\ulcorner\varphi\urcorner, \overline{\mathsf{k}}) \to (\forall \mathsf{u} \leq \overline{\mathsf{k}})\neg\mathsf{Prf}(\mathsf{u}, \ulcorner\varphi\urcorner)$.

But since $f(\ulcorner\varphi\urcorner) = k$ and we're assuming F captures f, $T \vdash \mathsf{F}(\ulcorner\varphi\urcorner, \overline{\mathsf{k}})$. Hence $T \vdash (\forall \mathsf{u} \leq \overline{\mathsf{k}})\neg\mathsf{Prf}(\mathsf{u}, \ulcorner\varphi\urcorner)$. And now we can manipulate the result (O3) about Q (and hence T) from Section 11.3 to deduce that for any $p \leq k$, $T \vdash \neg\mathsf{Prf}(\overline{\mathsf{p}}, \ulcorner\varphi\urcorner)$. Which contradicts (2).

That establishes (ii), there is no proof of φ with g.n. $p \leq f(\ulcorner\varphi\urcorner) = k$. So, for each $p \leq k$, $Prf(p, \ulcorner\varphi\urcorner)$ is false, hence $T \vdash \neg\mathsf{Prf}(\overline{\mathsf{p}}, \ulcorner\varphi\urcorner)$. So by (O4) from Section 11.3, $T \vdash (\forall \mathsf{u} \leq \overline{\mathsf{k}})\neg\mathsf{Prf}(\mathsf{u}, \ulcorner\varphi\urcorner)$. But since F captures f, $T \vdash \forall \mathsf{v}(\mathsf{F}(\ulcorner\varphi\urcorner, \mathsf{v}) \leftrightarrow \mathsf{v} = \overline{\mathsf{k}})$, whence $T \vdash \forall \mathsf{v}\{\mathsf{F}(\ulcorner\varphi\urcorner, \mathsf{v}) \to (\forall \mathsf{u} \leq \mathsf{v})\neg\mathsf{Prf}(\mathsf{u}, \ulcorner\varphi\urcorner)\}$ and thus finally (i) $T \vdash \varphi$. $\boxtimes$

---

[3]To be quite honest, these details are more than you really need to know: but they are rather too pretty to miss out!

29 Second-order arithmetics

As we noted at the end of Chapter 9, it is rather natural to suggest that the intuitive principle of arithmetical induction should be regimented as a *second-order* principle that quantifies over numerical properties, and which therefore can't be directly expressed in a first-order theory that only quantifies over numbers. So why not work with a second-order theory, rather than hobble our formal arithmetic by forcing it into a first-order straightjacket?

True, we have discovered that – so long as it stays consistent and effectively axiomatized – *any* theory containing enough arithmetic will be incomplete. But still, we ought to say at least a little about second-order arithmetics, and this is as good a place as any. Indeed, if you have done a university mathematics course, you might very well be feeling rather puzzled by now. Typically, at some point, you are introduced to axioms for a version of 'Second-order Peano Arithmetic' and are given the elementary textbook proof that these axioms are *categorical*, i.e. pin down a unique type of structure. But if this second-order arithmetic *does* pin down the structure of the natural numbers, then – given that any arithmetic sentence makes a determinate claim about this structure – it apparently follows that this theory does enough to settle the truth-value of every arithmetic sentence. Which makes it sound as if there *can* after all be a (consistent) negation-complete axiomatic theory of arithmetic richer than first-order PA, flatly contradicting the Gödel-Rosser Theorem.

We need to sort things out; that's the business of this long chapter.[1]

29.1 Second-order syntax

As a preliminary, we need to characterize some suitable languages for second-order arithmetic, languages which extend the familiar first-order language L_A, i.e. $\langle \mathcal{L}_A, \mathcal{I}_A \rangle$. We want to add to our quantificational resources – which already allow us to say that all numbers satisfy some condition – a way of saying that all numerical properties satisfy some condition.

We begin by extending the syntax $\mathcal{L}_A$ by adding second-order quantifiers, to get the second-order syntax $\mathcal{L}_{2A}$. In $\mathcal{L}_A$, the *first-order variables* $x, y, z, \ldots$ can occupy the same positions as can be held by names. In $\mathcal{L}_{2A}$ we also have *second-order variables* $X, Y, Z, \ldots$, where these can replace predicates: we'll concentrate here on the case of second-order variables that act like monadic, i.e.

---

[1]It might help if you have already encountered a smidgen of second-order logic, as we must go quite briskly. Still, the aim is to sketch enough of the basics to enable you to understand what is going on.

one-place, predicates.[2] Both sorts of variable will have associated quantifiers, e.g. $\forall x, \exists y, \forall X, \exists Y, \ldots$.

In $\mathcal{L}_A$, there is just one type of atomic wff, of the form $\sigma = \tau$, where σ and τ are terms: wffs of this type are atomic wffs of $\mathcal{L}_{2A}$ too. But $\mathcal{L}_{2A}$ also has a second type of atomic wff. The new rule is:

> If Z stands in for some second-order variable and τ is a term, then
> $Z\tau$ is an atomic wff.[3]

For example,

$$X0, \ Xx, \ XSSS0, \ Y(S0 + z), \ Z((x + y) \times SS0)$$

are all atomic wffs of the new type. Very informally, as a heuristic prop, you can read these in Loglish[4] as '0 has property X', 'x has property X', 'SSS0 has property X' etc.

More complex wffs of $\mathcal{L}_A$ are built up from atomic wffs by using connectives and/or prefixing first-order quantifiers like $\forall x, \exists y$. And the sentences of $\mathcal{L}_A$ are the closed wffs, i.e. wffs without unquantified variables dangling free. Similarly, more complex wffs of $\mathcal{L}_{2A}$ are built up from atomic wffs by using connectives and/or prefixing first-order quantifiers and/or prefixing second-order quantifiers like $\forall X, \exists Y$. (Note then that every old $\mathcal{L}_A$ wff is also an $\mathcal{L}_{2A}$ wff.) And the sentences of $\mathcal{L}_{2A}$ are the closed wffs, i.e. wffs without unquantified variables of either type dangling free. For example, the following are sentences of $\mathcal{L}_{2A}$:

$$\exists X\,X0, \ \forall X(\forall x Xx \rightarrow X0), \ \exists X(\neg XS0 \wedge \forall x(Xx \rightarrow XSSx)).$$

Again in informal Loglish, we can read these as 'there is a property X such that 0 has property X', 'for all properties X, if every number has X then 0 has X', etc.

29.2 Second-order semantics

We now need to augment the official semantics $\mathcal{I}_A$ for first-order arithmetic to deal with the new quantifiers. Doing this in the most natural way gives us

---

[2]For more details see e.g. Shapiro (1991, ch. 3). If we were doing all this properly, we'd follow Shapiro by also including at the outset dyadic, triadic, etc. variables than can stand in for two-place, three-place, etc. relational predicates, with all *their* associated quantifiers too. Next we'd prove that even a very modest arithmetical theory can handle numerical codes for ordered pairs, triples etc. of numbers (cf. Section 2.4). We can then replace two-place predicates with one-place predicates that apply to codes-for-pairs, and n-place predicates with one-place predicates that apply to codes-for-n-tuples. So, in fact, we can then show that an arithmetic with the full apparatus of quantifiers running over n-place relations for any n is equivalent to one with just monadic second-order quantifiers. Which is why it is legitimate to cut corners here and concentrate on the monadic case, ignoring the rest.

[3]We are not departing from the policy announced in Section 5.1(b) of using Greek letters for metalinguistic variables. 'Z' is an upper case zeta!

[4]'Loglish' is that helpful unofficial logic/English mix which we cheerfully fall into in the logic classroom.

205

the 'full' second-order interpretation $\mathcal{I}_{2A}$, and we define L_{2A} to be the language $\langle \mathcal{L}_{2A}, \mathcal{I}_{2A} \rangle$. The full interpretation is so-called in contrast to e.g. the more modest interpretation $\mathcal{I}_{2a}$ which we will describe in a moment; and L_{2a} is the language with the same syntax but with this modest semantics, so is $\langle \mathcal{L}_{2A}, \mathcal{I}_{2a} \rangle$.

(a)   We have already hinted at an informal first pass at the semantics: let's start filling this out.

Here's one way of thinking of the *first-order* quantifiers. When ξ is a first-order variable, an $\mathcal{L}_{2A}$-sentence of the form $\forall \xi (\ldots \xi \ldots \xi \ldots)$ is true just when the corresponding sentence $(\ldots \tau \ldots \tau \ldots)$ is true whatever particular number is picked out by the closed term τ. Likewise for *second-order* quantifiers: an $\mathcal{L}_{2A}$-sentence of the form $\forall Z (\ldots Z \ldots Z \ldots)$ is true just when the corresponding sentence $(\ldots \varphi \ldots \varphi \ldots)$ is true whatever particular numerical property is picked out by the predicate φ. Similarly, of course, for the existential quantifiers: a sentence of the form $\exists Z (\ldots Z \ldots Z \ldots)$ is true just when the corresponding sentence $(\ldots \varphi \ldots \varphi \ldots)$ is true on some way of assigning a particular numerical property as interpretation to the predicate φ.

Those readings of the quantifiers ensure that, as the Loglish rendering suggests, $\exists X \, X0$ is true just when at least one property is had by the number zero. $\forall X (\forall x \, Xx \to X0)$ will be true just when, for any property, if every number has it, then zero in particular has it. And $\exists X (\neg XS0 \land \forall x (Xx \to XSSx))$ is true just when there is a property lacked by 1 such that, if any number n has it, so does $n + 2$ (which makes this third sentence true like the others, since the property of *being even* is an obvious instance).

(b)   However, that first pass at the semantics appeals to the somewhat murky notion of a *property*. But recall how things go for the standard official semantics of first-order languages. We deal with predicates there by assigning them (not properties but) satisfaction conditions (see Section 4.2(c)). Or what comes to the same thing, we fix the predicate's *extension* (cf. Section 3.2(c)) – for an object satisfies the predicate φ just in case it falls into the predicate's extension. Correspondingly, then, we will officially treat second-order quantifiers as ranging over (not properties but their) extensions, where these are certain subsets of the domain of objects, i.e. in the present case certain sets of natural numbers.

So, as a second pass at the semantics, we will say that a sentence of the form $\forall Z (\ldots Z \ldots Z \ldots)$ is true just when the corresponding sentence $(\ldots \varphi \ldots \varphi \ldots)$ is true, whatever suitable set of numbers is assigned as extension to the predicate φ. So, for example, $\forall X (\forall x \, Xx \to X0)$ is true just when any suitable set of numbers satisfies the condition that, if all numbers are in it, then zero in particular is. Similarly for existentially quantified sentences.

(c)   Talking about extensions – that is to say, sets – rather than properties is certainly a nice gain in clarity.[5] But there still remains a question: which

---

[5] If you *aren't* already rather puzzled by the notion of a property, then try reading the editors' introduction and some of the papers in Mellor and Oliver (1997)!

extensions should we recognize as being relevantly available for us to quantify over when we use second-order numerical quantifiers? In other words, which are the suitable subsets of $\mathbb{N}$ that we want to be in the range of our second-order variables?

Well, as we hinted before in Section 9.4, the obvious first thought is: *we want to quantify over the full collection of every arbitrary finite or infinite subset of the numbers*. And that indeed is the answer which will be built into the natural, 'full', interpretation $\mathcal{I}_{2A}$.

Yet, on second thoughts, we might hesitate over that idea of arbitrary infinite sets of numbers – sets which are supposedly perfectly determinate but are in the general case beyond any possibility of our specifying their members. We can readily make sense of the membership of a set of numbers being determined by possession of some characterizing property which gives a recipe for picking out the numbers; and we can readily make sense of the membership being merely stipulated (more or less arbitrarily). However, the first idea gives us infinite sets but not arbitrary ones; and the second idea may give us arbitrary sets (whose members share nothing but the gerrymandered property of having been selected for membership) but not infinite ones – unless we are prepared to conceive of a completed infinite series of arbitrary choices. Neither initial way of thinking of sets immediately makes sense of the classical idea of arbitrary infinite sets of numbers.

Now, in pointing this out, I'm not arguing that we should be sceptical about the classical idea we've just built into $\mathcal{I}_{2A}$; I am not saying that we really can make so sense of the idea. Rather, for present purposes, the moral is simply this: *if we do interpret the second-order quantifiers as ranging over all arbitrary subsets of the domain of numbers, then this commits us to making sense of a clearly infinitary conception*, one that arguably goes beyond anything that is given just in our ordinary understanding of elementary arithmetic (even when combined with the logical idea of sets-as-extensions-of-concepts).

But let's suppose we are happy to go along with this new infinitary idea. Then, in sum, the corresponding 'full' interpretation $\mathcal{I}_{2A}$ for our second-order arithmetical syntax $\mathcal{L}_{2A}$ can be summed up like this:

> $\mathcal{I}_{2A}$ agrees with $\mathcal{I}_A$ that the domain of (first-order) quantification remains the natural numbers $\mathbb{N}$, and '0', 'S', '+' and '×' get the same interpretation as before. The connectives and first-order quantifiers are dealt with in the usual way. But we now add the crucial stipulation that *the second-order quantifiers run over the full collection of all arbitrary subsets of the domain, i.e. over all subsets of $\mathbb{N}$.*

(d)   If the interpretation $\mathcal{I}_{2A}$ makes play with a distinctively infinitary idea of arbitrary subsets of the numbers, is there some alternative, less infinitary, way of understanding the second-order quantifiers?

Well, if we want to stick more closely to the conceptual resources of ordinary elementary arithmetic, one way to go would be to adopt what we'll label $\mathcal{I}_{2a}$, which interprets the second-order quantifiers as ranging only over 'arithmetical' sets. These are the sets of numbers that can be explicitly characterized in purely arithmetical terms, i.e. as the extensions of the monadic L_A predicates (predicates which lack second-order quantifiers).

Note that the monadic predicates of L_A can be enumerated (as we remarked in the proof of Theorem 7.1), so the arithmetical sets can be enumerated too. But arbitrary sets of numbers can't be enumerated (see the proof of Theorem 2.2, for the power set of the numbers just is the set of all arbitrary sets of numbers). So the interpretation $\mathcal{I}_{2a}$ does indeed interpret the second-order quantifiers more restrictedly than the interpretation $\mathcal{I}_{2A}$. We won't say much more here, however, about this modest sort of interpretation. We mention it just as a gesture towards the important point that there *are* choices to be made in how we interpret the second-order quantifiers, even if $\mathcal{I}_{2A}$ includes by far the most usual one.

(e)   What about *semantic entailments* between sentences of L_{2A}?

Recall the general definition of semantic entailment. For sentences in language L, $\Sigma \vDash \varphi$ (the sentences in Σ entail φ) iff every admissible (re)interpretation of the syntax which makes all the sentences in Σ true makes φ true too. Here, the admissible interpretations are those which keep fixed the meanings of the logical apparatus of L, while varying the non-logical aspects of the interpretation.

So, applied to the present case, $\Sigma \vDash_2 \varphi$ (i.e., interpreted as L_{2A} sentences, the sentences in Σ entail φ) iff every admissible interpretation which makes all the sentences in Σ true makes φ true too. As we spin the interpretations, moving from $\mathcal{I}_{2A}$ to other admissible (re)interpretations of L_{2A}, we vary the domain of quantification and/or the interpretations of the non-logical symbols but keep fixed the interpretations of the connectives, first-order quantifiers and the identity sign. And then, crucially, we *also* keep fixed $\mathcal{I}_{2A}$'s requirement that the second-order quantifiers run over the full collection of *all* subsets of the domain, whatever the domain now is.

(Entailment for L_{2a} sentences is defined similarly, except this time we keep fixed $\mathcal{I}_{2a}$'s requirement that the second-order quantifiers run over the sets which are the extensions of open wffs which lack second-order quantifiers.)

29.3   The Induction Axiom again

Once we are using a formal language with second-order quantifiers, we can render the informal second-order induction axiom we met in Section 9.4 as follows:

Induction Axiom $\forall X((X0 \land \forall x(Xx \rightarrow XSx)) \rightarrow \forall xXx)$.

But – to take up again the point we've just highlighted – note that a sentence like this does *not* wear its interpretation on its face.

There's an often-quoted remark due to Georg Kreisel:

A moment's reflection shows that the evidence of the first-order schema derives from the second-order [axiom]. (Kreisel, 1967, p. 148)

Well, *this* much may be right: we are prepared to give a blanket endorsement to all the instances of the first-order Schema because we accept the intuitive thought that if zero has some property, and if that property is passed from one number to the next, then all numbers have that property, whatever property that is (cf. our introduction of the idea of induction in Section 9.1).

However, there is no obvious and immediate reason to suppose that our intuitive thought here aims to generalize over more than some 'natural' class of arithmetical properties (e.g. those with arithmetically definable sets as extensions), or that it already involves the extended conception of properties whose extensions are quite arbitrary subsets of the numbers.[6] In other words, the intuitive thought arguably falls well short of the content of the formal Induction Axiom *when interpreted as a sentence of* L_{2A}, i.e. when interpreted as quantifying over the full collection of arbitrary sets of numbers (as opposed to when interpreted, for example, as a sentence of L_{2a}).

So we shouldn't slide *too* readily from the intuitive induction principle to the formal Induction Axiom interpreted in the strong way.

29.4 'Neat' second-order arithmetics

(a)   Let's put questions of semantics on hold for the moment, and consider next what happens if we start from first-order Robinson Arithmetic Q and add the second-order Induction Axiom to start building an axiomatized second-order arithmetic whose syntax is $\mathcal{L}_{2A}$.

Now of course, we'll need to add to Q's deductive apparatus if we are to derive anything from our added Axiom! So what additional principles should we adopt?

You might suppose it should just be a question of adding to the existing rules for dealing with *first*-order quantifiers some analogous rules for the new *second*-order quantifiers. So, for example, just as from $\forall x\, Sx \neq x$ we can infer $S\tau \neq \tau$ for any term τ, so from $\forall X(\forall x Xx \to X0)$ we can infer $\forall x \varphi(x) \to \varphi(0)$ for any one-place predicate φ.

But hold on! This assumes – to fall harmlessly back into using property-talk – that any one-place $\mathcal{L}_{2A}$-predicate φ, however complex, does indeed specify one of the properties which the second-order quantifiers run over (it assumes that there are more genuine properties than, for example, are allowed by the modest interpretation $\mathcal{I}_{2a}$). And *that* assumption shouldn't be just slipped in under the radar. So let's proceed more carefully, in two steps. At step (1), we give neutral,

---

[6]Careful mathematicians are alert to this point. For example, in his classic survey of modern mathematics, Saunders Mac Lane quite explicitly first presents the second-order principle of induction as an induction over the natural arithmetical properties that can be identified by first-order formulae. And he then distinguishes this intuitive 'induction over properties' (as he calls it) from the stronger 'induction over [arbitrary] sets'. See Mac Lane (1986, p. 44).

purely logical, rules for dealing with the second-order quantifiers, for moving between wffs with bound variables and wffs with free variables/parameters. Then, step (2), we make a quite explicit assumption about which predicates specify properties within the range of the second-order quantifiers.

So, for step (1), given that we officially are using a Hilbert-type proof system, we can give the following rules:

1. From $\forall Z\psi(Z)$, infer $\psi(Z)$ for any second-order variable Z.

2. From $\varphi \to \psi(Z)$, infer $\varphi \to \forall Z\psi(Z)$, assuming Z doesn't occur free in φ or in any premiss of the deduction.

3. $\exists Z\psi(Z) =_{\mathrm{def}} \neg\forall Z\neg\psi(Z)$.

Then, at step (2) we specify a class of wffs C and say that (the closure of) every instance of the

Comprehension Schema $\exists X\forall x(Xx \leftrightarrow \varphi(x))$,

is to be an axiom, where φ is a C-predicate not containing 'X' free (if we want to end up with a properly axiomatized theory, we will want it to be decidable which wffs belong to class C). This tells us that the range of the second-order quantifiers contains at least a property corresponding to every C-predicate. If we *are* feeling generous, we can allow class C to include any one-place $\mathcal{L}_{2A}$-predicate at all; while if we want to be more modest we might e.g. restrict C to the class of arithmetic $\mathcal{L}_A$-predicates which don't embed second-order quantifiers.

(b)   Given those brisk remarks by way of motivation, let's now give a summary definition of what we will call *neat* second-order arithmetics.[7] These are arithmetics whose syntax is $\mathcal{L}_{2A}$ and which have a standard first-order logic, augmented with the 'neutral' rules given above for the second-order quantifiers (or some equivalent). Their axioms are

1. the non-logical axioms of Q;

2. the second-order Induction Axiom;

3. and in addition, for some class of wffs C (whose membership is decidable), the universal closures of all instances of the Comprehension Schema with φ a wff in C.[8]

---

[7]'Neat' is our informal term: there isn't a standard one. We should note, by the way, that not every interesting formal theory of second-order arithmetic discussed in the literature is neat in our sense. Still, these neat theories – ordered by the increasing generosity of C, the class of wffs we can put into the Comprehension Schema – do form a spine running through the class of second-order arithmetics. For an encyclopedic survey of theories of second-order arithmetic see Simpson (1991), whose first chapter gives a wonderfully helpful overview.

[8]There is some redundancy here. For a start, we can drop the third axiom of Q, just as we did in presenting first-order PA.

29.5   Introducing PA$_2$

(a)   We now introduce *Second-Order Peano Arithmetic*, PA$_2$, which is syntactically the *strongest* neat arithmetic: it allows *any* open wff at all to appear in an instance of the Comprehension Schema.[9] But how are we to understand this theory? What semantics should we assign its wffs?

You might wonder for a moment whether there is a coherent semantics to be had here. An instance of the Comprehension Schema aims to define a numerical property in the range of the second-order quantifiers. Being maximally generous, PA$_2$'s comprehension principle allows us to define a numerical property using a wff which itself embeds second-order variables which range over properties including therefore the one we are currently defining! So, on the one hand, we need to ensure that the newly defined property really is one of those the second-order variables are already ranging over. Yet, on the other hand, how can this be legitimate? How can we define something in terms of a collection which is supposed already to include the very item we are trying to define?

Bertrand Russell famously thought that this involves a vicious circle.[10] The obvious response, however, is Gödel's, which echoes an earlier discussion by Frank Ramsey. This sort of apparently circular definition is in fact unproblematic, if we assume that 'the totality of all properties [of numbers] exists somehow independently of our knowledge and our definitions, and that our definitions merely serve to pick out certain of these previously existing properties'.[11] Putting it in terms of property-extensions: if we assume that there already exists a fixed totality of relevant subsets of the domain, there is no problem about the idea that we might locate a particular member of that totality by reference to a quantification over it – any more than there is a problem about the idea that we might locate a particular man as the tallest of all the men in the room (including him!).

Now, the semantic interpretation $\mathcal{I}_{2A}$ accords with this Ramsey/Gödel line. For $\mathcal{I}_{2A}$ assumes that there is already 'out there' a fixed totality of *all* the properties of numbers – or rather a fixed totality of their extensions, *all* arbitrary sets of numbers. Given this assumption, an expression $\varphi(\mathsf{x})$ that embeds second-order quantifiers running over all these extensions does have a determinate sense, and will determine a definite set of numbers as its extension. And this extension will be one of those already in the range of the second-order quantifiers, as on $\mathcal{I}_{2A}$ all sets of numbers are in the range.

So, in sum, $\mathcal{I}_{2A}$ seems to be a natural interpretation which would warrant PA$_2$'s generous comprehension principle. But the other axioms are also intuitively true on this interpretation. So $\mathcal{I}_{2A}$ *intuitively makes all* PA$_2$*'s theorems true.*

---

[9]Incidentally, this theory is at least equally often called Z$_2$, the name given it in Hilbert and Bernays (1934).

[10]Thus Russell (1908, p. 63): 'Whatever involves *all* of a collection must not be one of the collection.' For discussion see Chihara (1973, esp. ch. 1) and Potter (2000, esp. ch. 5).

[11]The quotation is from Gödel (1933, p. 50); compare the remark on the supposed vicious circle in Ramsey (1925, p. 204).

Henceforth then, we'll assume $\mathcal{I}_{2A}$ is indeed PA$_2$'s intended interpretation (so $L_{2A} =_{\text{def}} \langle \mathcal{L}_{2A}, \mathcal{I}_{2A} \rangle$ is the theory's native language).

(b)   Since PA$_2$ can quantify over arbitrary numerical sets, it can (with some coding tricks) handle the standard construction of the real numbers in terms of suitable infinite sets of numbers. We can therefore do a great deal of classical analysis in this theory, which is why this rich theory is sometimes itself just called *analysis* (a fact that points up again the thought that here we are pushing beyond the boundaries of what we need to grasp in grasping arithmetic).

But let that pass. Here we are going to be particularly interested in the *arithmetical* strength of PA$_2$, i.e. in the question of what it can prove by way of first-order arithmetical truths (true sentences of L_A). And the important headline news is predictably this:

> **Theorem 29.1** PA$_2$ *can prove every* L_A *wff that* PA *can prove, and also some that* PA *can't prove.*

The first half of that theorem is, of course, trivial (because PA$_2$ includes Q, and PA$_2$'s Induction Axiom together with its comprehension principle gives us all the first-order instances of induction, i.e. gives us the other axioms of PA).

As for the second half, we here just report that PA$_2$ can formalize the intuitive reasoning for PA's consistency (by defining a notion of truth-in-PA and showing that the axioms of PA are true in this sense and its inference-rules are truth-preserving). It can also formalize our informal reasoning in proving Gödel's theorem that shows that if PA is consistent, then PA's canonical Gödel sentence G is unprovable and hence true. So, by modus ponens, PA$_2$ can prove G, which PA can't do (assuming its consistency).

Of course, PA$_2$ is not only 'neat' but 'nice' (assuming it too is consistent). Hence, while PA$_2$ *can* prove PA's Gödel sentence G, it *can't* prove its own canonical Gödel sentence G$_2$.[12]

29.6   Categoricity

PA and PA$_2$ are both intuitively sound theories (if we buy the idea of arbitrary infinite sets of numbers), and both incomplete. There are, however, deep differences between these two theories which we now need to highlight over the next two sections.

To introduce them, we need some preliminary definitions (probably entirely familiar ones if you have done any serious logic at all).

  i. Suppose T is a theory built in the language $L =_{\text{def}} \langle \mathcal{L}, \mathcal{I} \rangle$. Echoing what we just said in Section 29.2, an admissible (re)interpretation $\mathcal{J}$ for theory

---

[12]Note, by the way, that like any Gödel sentence built in the now familiar way, G$_2$ is a purely arithmetic Π_1-equivalent sentence that lacks second-order quantifiers.

T is an interpretation of the syntax $\mathcal{L}$ which keeps the same interpretation of the logical vocabulary as is given by $\mathcal{I}$ but perhaps varies the interpretation of the non-logical vocabulary. If $\mathcal{J}$ makes all the axioms of a theory T true (and hence all T's theorems true too, assuming T's logic is a sound one), then we'll say that $\mathcal{J}$ is a *model* for T.

ii. Two admissible interpretations are *isomorphic* iff they structurally look exactly the same.

More carefully: $\mathcal{I}$ and $\mathcal{J}$ are isomorphic iff (i) there's a one-to-one correspondence f between the domain of $\mathcal{I}$ and the domain of $\mathcal{J}$;[13] (ii) if $\mathcal{I}$ assigns the object o as referent to the term τ, then $\mathcal{J}$ assigns τ the corresponding object $f(o)$; (iii) if $\mathcal{I}$ assigns the set E as the extension to the predicate φ, then $\mathcal{J}$ assigns that predicate the corresponding extension $f(E) =_{\text{def}} \{f(o) \mid o \in E\}$; and so on, and so forth. This means that the atomic wff $\varphi(\tau)$ is true on $\mathcal{I}$ iff o is in E, which holds iff $f(o)$ is in $f(E)$ (since f is a one-one correspondence), and that holds just if $\varphi(\tau)$ is true on $\mathcal{J}$. And the point obviously generalizes to more complex wffs. In other words, more generally, ψ is true on interpretation $\mathcal{I}$ if and only if it is also true on any isomorphic interpretation $\mathcal{J}$.

iii. A theory is *categorical* iff it has models but all its models are isomorphic – i.e. the theory has just one model 'up to isomorphism', as they say.

And with that jargon to hand, we can state the first key contrast between PA and PA₂ like this:

Theorem 29.2 *Assuming both theories do have models,* PA *isn't categorical, but* PA₂ *is.*

Proof sketch: PA *isn't categorical*   Assume PA has a model and so is consistent. We then know that both of the expanded theories PA + R and PA + ¬R must also be consistent (where R is the undecidable Rosser sentence). Hence, by the familiar theorem that any consistent first-order theory has a model, indeed a model with an enumerable domain, both these expanded theories have models.[14]

---

[13]Note the cheerful quantification over functions here, including functions we may not be able to finitely specify. Should we now pause to furrow our brows over this infinitary idea, in the way we earlier worried about the idea of arbitrary infinite sets of numbers? Well, the situation is different. Earlier the question was whether PA₂, interpreted as quantifying over arbitrary sets of numbers, overshoots as a regimentation of what we grasp *in understanding ordinary arithmetic* (see also the following Interlude on this). But now we are touching on a bit of elementary *model theory*, and here we inevitably tangle with infinitary ideas.

[14]For 'the familiar theorem', see any standard logic text, e.g. Mendelson (1997, p. 90, Prop. 2.17). But note also that, in order to show that PA isn't categorical, we don't in fact need to appeal to Gödelian incompleteness: the so-called 'Upward Löwenheim-Skolem Theorem' already tells us that any first-order theory like PA which has a model whose domain is the size of the natural numbers ℕ has other, differently structured, models with bigger, indenumerable, domains. What our Gödelian argument adds is that there are also non-isomorphic models of PA with enumerable domains. However, we haven't space in this book to pursue issues in

Since one of these models makes R true and the other makes ¬R true, these models can't be isomorphic replicas. But any model for PA + R and equally any model for PA + ¬R is a fortiori a model for plain PA. So PA has non-isomorphic models (with enumerable domains). In other words, PA is not categorical. ☒

Proof sketch: PA₂ is categorical Suppose $\mathcal{M}$ is a model for PA_2. $\mathcal{M}$ will need to pick out some object o to be the reference of '0', and to pick out some function ς which satisfies the successor axioms for 'S' to denote. And then, according to $\mathcal{M}$, the numerals $0, S0, SS0, SSS0, \ldots$ must pick out the objects $o, \varsigma o, \varsigma\varsigma o, \varsigma\varsigma\varsigma o, \ldots$. Let's abbreviate '$\varsigma\varsigma \ldots \varsigma o$' with m occurrences of 'ς' by '$\widetilde{m}$'. Then we can put it this way: according to $\mathcal{M}$, the numeral '$\overline{m}$' picks out $\widetilde{m}$.

We'll denote the set of all the elements $\widetilde{m}$ by $\mathbb{M}$. And let's say that a model $\mathcal{M}$ is *slim* iff its domain is just the corresponding set $\mathbb{M}$ – i.e. iff its domain contains just the 'zero' element o, the 'successors' of this 'zero', *and nothing else*.

We now argue in two stages. (i) We'll show that any two *slim* models of PA_2 are isomorphic. Then (ii) we'll show that PA_2 can *only* have slim models. It will then follow that, if PA_2 has a model at all, it is a slim one, unique 'up to isomorphism' – a result essentially first shown by Richard Dedekind (1888).[15]

Stage (i) Suppose $\mathcal{M}$ is a slim model. If $m \neq n$, then even $BA \vdash \overline{m} \neq \overline{n}$, and hence $PA_2 \vdash \overline{m} \neq \overline{n}$. But $\mathcal{M}$ is a model and so makes all PA_2's theorems true. So if $m \neq n$, then $\mathcal{M}$ must make $\overline{m} \neq \overline{n}$ true, and that requires $\widetilde{m} \neq \widetilde{n}$. Since m and n were arbitrary, it immediately follows that the objects $\widetilde{0}, \widetilde{1}, \widetilde{2}, \ldots$ must all be different. So $\mathcal{M}$'s domain $\mathbb{M}$ contains (nothing but) a sequence of objects arranged by the ς function which looks exactly like a copy of the natural numbers $\mathbb{N}$ arranged by the successor function.

$\mathcal{M}$ will also pick out some function $+_{\mathcal{M}}$ for '+' to denote. It is easy to show that $+_{\mathcal{M}}$, when applied to the objects $\widetilde{m}$ and $\widetilde{n}$, also behaves just like addition on the numbers. For, if $m + n = k$, then $PA_2 \vdash \overline{m} + \overline{n} = \overline{k}$, so to make that true on $\mathcal{M}$ we must have $\widetilde{m} +_{\mathcal{M}} \widetilde{n} = \widetilde{k}$. Similarly for multiplication. So again, there's only one way that 'addition' and 'multiplication' in a slim model can work: just like addition and multiplication of natural numbers.

Hence all the slim models must 'look the same', i.e. be isomorphic.

Stage (ii) Now consider the wff

$$\sigma(x) =_{\text{def}} \forall Z\{(Z0 \land \forall y(Zy \to ZSy)) \to Zx\}.$$

---

model theory, so we can't here pause over the intriguing question of the fine structure of all the whacky, quite unintended, models that PA inevitably has.

For the general Löwenheim-Skolem Theorem, see e.g. Mendelson (1997, p. 128), or perhaps more accessibly Bridge (1977, ch. 4). For a wonderful exploration of the non-standard-but-enumerable models of arithmetic see the classic Kaye (1991).

[15] For a full-dress modern proof see e.g. Shapiro (1991, pp. 82–83). A technical note: it is crucial for Stage (ii) that the second-order quantifiers are required to run over the full collection of subsets of the domain. Versions of second-order arithmetic which lift that requirement – e.g. the theories which are in effect built in a two-sorted *first*-order language, explored in Simpson (1991) – aren't categorical.

so $\sigma(\bar{n})$ says, informally, that n has any hereditary property of zero (see Section 18.1). Or more officially, it says that n belongs to any set of numbers which contains zero and which, if it contains a number, contains its successor.

PA$_2$'s generous comprehension principle will apply to $\sigma(x)$ in particular. In other words, PA$_2$ treats $\sigma(x)$ as a predicate expression that genuinely picks out a real property. So this property can figure in an instance of induction, and we have the theorem

$$\{\sigma(0) \wedge \forall x(\sigma(x) \to \sigma(Sx))\} \to \forall x \sigma(x).$$

But $\sigma(0)$ and $\forall x(\sigma(x) \to \sigma(Sx))$ hold trivially (check that!). So it is a PA$_2$ theorem that $\forall x \sigma(x)$.

From this it follows that any model of PA$_2$, i.e. any interpretation $\mathcal{M}$ that makes all its theorems true, must make $\forall x \sigma(x)$ true in particular. But, by definition, something satisfies $\sigma(x)$ on a given interpretation $\mathcal{M}$ just if it belongs to every subset Z of the domain which contains o and which if it contains x contains ςx (where o and ς are the zero and the successor function according to $\mathcal{M}$). Hence, in particular, something satisfies $\sigma(x)$ just if it belongs to the *smallest* subset of the domain which contains o and which, if it contains x, contains ςx. Hence something satisfies $\sigma(x)$ if it is the zero or one of its successors. So given $\mathcal{M}$ makes $\forall x \sigma(x)$ true, every element of this model's domain must be either the zero or one of its successors. Therefore the model $\mathcal{M}$ is slim. $\boxtimes$

The paradigm slim model for PA$_2$ is, of course, $\mathcal{I}_{2A}$ (assuming that that interpretation *is* a model).

Two final comments on this. First, note that the Stage (i) argument equally well shows that every slim model of first-order PA is isomorphic to the standard model $\mathcal{I}_A$. But we can't then run the Stage (ii) argument to show that first-order PA is categorical and *only* has slim models; that's because the property of being a-zero-or-one-of-its-successors can't be expressed in L_A, and PA's Induction Schema can only deal with properties that can be expressed by L_A predicates.

Second, note that all the real work in the categoricity argument is done by the successor axioms (to show that slim models comprise a copy of the natural numbers) plus an instance of induction involving the successor function (to show that PA$_2$ only has slim models). So we could have proceeded by first considering a cut-down theory with just the successor axioms and the induction axiom (such a theory is in fact traditionally called Peano Arithmetic), shown that this theory is already categorical, and only then moved on to showing that addition and multiplication can be added to its models in just one way.

29.7 Incompleteness and categoricity

(a) The fact that PA$_2$ is categorical and PA isn't entails another key difference between the first-order and second-order theory. Starting with the first-order case, suppose φ is some L_A truth that PA can't prove, e.g. a Gödel sentence like

G or a Rosser sentence like R. Then the axioms of PA don't semantically entail φ either. In standard symbols, if PA $\nvdash \varphi$, then PA $\nvDash \varphi$. That's because the built-in first-order deductive system is sound and complete: PA $\vdash \varphi$ iff PA $\vDash \varphi$.

Now for the contrasting second-order case: suppose φ is some L_{2A} truth which PA$_2$ can't prove, e.g. its true canonical Gödel sentence G$_2$ or Rosser sentence R$_2$. This time, however, PA$_2$ *will* still semantically entail φ. That's because we have

Theorem 29.3 *Assuming $\mathcal{I}_{2A}$ is in fact a model for* PA$_2$, PA$_2$ *semantically entails all L_{2A}-truths.*

Proof   PA$_2$ semantically entails φ – in symbols, PA$_2 \vDash_2 \varphi$ – if any admissible interpretation which makes the axioms of PA$_2$ true makes φ true. In other words, any model for PA$_2$ makes φ true.

However, we've just seen that PA$_2$ is a categorical theory. So given $\mathcal{I}_{2A}$ is a model for PA$_2$, *all* its models are isomorphic to $\mathcal{I}_{2A}$. But, trivially, if φ is true on $\mathcal{I}_{2A}$ (i.e. is an L_{2A}-truth), it will be true on all the isomorphic interpretations. It immediately follows that *all* interpretations which make the axioms of PA$_2$ true make φ true. So PA$_2 \vDash_2 \varphi$ for any wff φ that is true on $\mathcal{I}_{2A}$.   ☒

Hence, in particular, although PA$_2 \nvdash$ G$_2$, PA$_2 \vDash_2$ G$_2$. Which immediately reveals that PA$_2$'s *deductive system isn't complete* – there are semantic entailments which it can't prove. And this isn't because we have rather dimly forgotten to give PA$_2$ enough logical deduction rules: the incompleteness can't be repaired, so long as we keep to an effectively axiomatized logical system. For adding new logical axioms and/or inference rules to PA$_2$ will – so long as we stay consistent and decidably axiomatized – just give us a new theory which is still categorical and still subject to Gödel's Theorem. Whence

Theorem 29.4 *There can be no sound and complete effectively axiomatized logical deductive system for second-order logical consequence (when second-order quantifiers are interpreted as quantifying over the full collection of arbitrary subsets of the domain).*

(b)   In summary, then, while the claim 'PA settles all arithmetical truths' is false however we interpret it, the situation with the corresponding claim 'PA$_2$ settles all arithmetical truths' is more complex. (Here, then, we are at last answering the puzzle raised in the preamble at the beginning of the chapter.)

Assuming $\mathcal{I}_{2A}$ is a model for PA$_2$ (so the theory is consistent), the axioms of PA$_2$ are enough to *semantically entail* all true sentences of L_{2A}. But the Gödel-Rosser Theorem tells us this formal deductive theory is not strong enough to *prove* all true L_{2A} sentences – and we can't expand PA$_2$ so as to prove them all either, so long as the expanded theory remains consistent and properly axiomatized. Make the distinction between what is *semantically entailed* and what is *deductively proved*, and we reconcile the apparent conflict between the implication of Dedekind's categoricity result ('PA$_2$ settles all the truths') and Gödelian incompleteness ('PA$_2$ leaves some truths undecided').

216

For vividness, it may well help to put that symbolically. We'll use $\{PA, \vdash\}$ to denote the set of theorems that follow in PA's formal proof system, and $\{PA, \vDash\}$ to mean the set of sentences semantically entailed by PA's axioms (given the standard semantics of L_A). Similarly, we'll use $\{PA_2, \vdash\}$ to mean the set of theorems that follow in PA_2's formal proof system, and $\{PA_2, \vDash_2\}$ to mean the set of sentences semantically entailed by PA_2's axioms (given the 'full' semantics for the quantifiers built into L_{2A}). Finally – as before, in Section 26.2 – we'll use $\mathcal{T}_A$ to denote the set of truths of L_A (True Basic Arithmetic); and we'll now use $\mathcal{T}_{2A}$, the set of truths of L_{2A} (True Second-Order Arithmetic). Then we can very perspicuously display the relations between these sets as follows, using '$\subset$' to indicate strict containment (i.e. $X \subset Y$ when Y contains every member of X and more besides):

$$\{PA, \vdash\} = \{PA, \vDash\} \subset \mathcal{T}_A$$
$$\{PA_2, \vdash\} \subset \{PA_2, \vDash_2\} = \mathcal{T}_{2A}.$$

The completeness of first-order logic which yields $\{PA, \vdash\} = \{PA, \vDash\}$ is, though so familiar, a remarkable mathematical fact which is enormously useful in understanding PA and its models. This has led some people to think the strict containment $\{PA_2, \vdash\} \subset \{PA_2, \vDash_2\}$ is a sufficient reason to be unhappy with second-order logic. Others think that the categoricity of theories like second-order arithmetic, and the second-order definability of key mathematical notions like finiteness, suffices to make second-order logic the natural logic for mathematics – see Shapiro (1991) for explanations and a classic defence. Apart from a few more remarks in Section 30.6, we will have to pass by this intriguing issue.

29.8 Another arithmetic

As we saw, understanding the semantics $\mathcal{I}_{2A}$ built into PA_2 requires us to make sense of the infinitary concept of quantifying over all arbitrary sets of numbers. Suppose we want to restrict ourselves instead to the conceptual resources of non-infinitary pure arithmetic itself, and so adopt e.g. the interpretation $\mathcal{I}_{2a}$ (see the end of Section 29.2). Then we'll be interested in the corresponding weaker 'neat' arithmetic whose quantifiers in effect run over only those sets of numbers that we can pick out arithmetically. Putting that formally, this involves requiring that any wff $\varphi(x)$ which we substitute into the Comprehension Scheme must lack second-order quantifiers. The resulting neat axiomatized theory with this arithmetical comprehension principle is known in the trade as ACA_0.

Arguably, *this* theory doesn't go beyond what is given by our understanding of basic arithmetic (together with general logical ideas) – though despite its apparently unambitious character, we can *still* reconstruct a surprising amount of analysis and other applicable mathematics inside ACA_0. However, this time – by contrast with PA_2 – we get no new purely arithmetical truths:

Theorem 29.5 ACA_0 *is conservative over first-order* PA.

In other words, for any L_A sentence φ such that $\mathsf{ACA_0} \vdash \varphi$, it is already the case that $\mathsf{PA} \vdash \varphi$.[16]

So we have the following suggestive contrast. We can derive more L_A sentences in $\mathsf{PA_2}$ than in PA. But if we are to accept these formal derivations as *proofs* which give us reason to accept their conclusions, then we will need to accept the axioms of $\mathsf{PA_2}$ as true. And to accept the axioms as true – including all those instances of Comprehension for formulae φ which quantify over subsets of the domain – will involve accepting infinitary ideas that go beyond those essential to a grasp of elementary arithmetic. By contrast, accepting the weaker formal theory $\mathsf{ACA_0}$ arguably doesn't involve more than a grasp of arithmetic together with some very general purely logical ideas; but then the theory doesn't give us any more basic arithmetic than PA. We'll return to this point in the next chapter.

29.9   Speed-up again

Finally, just for dessert after the main course, let's very briefly make a connection with Chapter 28, on the general topic of the length of proofs. We've said that $\mathsf{ACA_0}$ proves just the same L_A wffs as PA; but interestingly it does *massively* speed up some proofs. Define

$$2 \star k = 2^{2^{2^{\cdot^{\cdot^{2}}}}} \quad \text{where the stack of 2s is } k \text{ high.}$$

Then Robert Solovay has shown that there is a family of L_A wffs $\varphi_0, \varphi_1, \varphi_2, \ldots$, such that (i) PA proves each φ_n; (ii) the PA proof of φ_n requires at least n bits, so in particular the PA proof of $\varphi_{2 \star k}$ is at least $2 \star k$ bits long; but (iii) there is a constant c such that, for any k, there is an $\mathsf{ACA_0}$ proof of $\varphi_{2 \star k}$ less than ck^2 bits long. Hence, for some values of n, there is an $\mathsf{ACA_0}$ proof of φ_n which is *radically* shorter than the shortest proof for φ_n in PA.[17] And since any $\mathsf{ACA_0}$ proof is also a $\mathsf{PA_2}$ proof, the result carries over. There are PA theorems which have radically shorter proofs in $\mathsf{PA_2}$.

Gödel himself noted this phenomenon long ago:

> [P]assing to the logic of the next higher order [e.g. moving from a first-order to a second-order setting] has the effect, not only of making provable certain propositions that were not provable before, but also of making it possible to shorten, by an extraordinary amount, infinitely many of the proofs already available. (Gödel, 1936, p. 397)

But a full proof of this Gödelian speed-up claim wasn't in fact published until Buss (1994).[18]

---

[16] For the proof, see Simpson (1991, Sec. IX.1).
[17] Solovay notes his result in http://www.cs.nyu.edu/pipermail/fom/2002-July/005680.html.
[18] For more, see Pudlák (1998).

30 Interlude: Incompleteness and Isaacson's Thesis

This Interlude discusses a couple of further questions about incompleteness that might well have occurred to you as you have been reading through recent chapters. But, given that the chapters since the last Interlude have been really rather densely packed, we should probably begin with a quick review of where we have been.

30.1 Taking stock

Here, then, is some headline news which is worth highlighting again:

1. Our first pivotal result was Theorem 24.4, the Diagonalization Lemma: if T is a nice theory and $\varphi(\mathsf{x})$ is *any* wff of its language with one free variable, then there is a 'fixed point' γ such that $T \vdash \gamma \leftrightarrow \varphi(\ulcorner\gamma\urcorner)$. And further, if $\varphi(\mathsf{x})$ is Π_1-equivalent, then it has a Π_1-equivalent fixed point. (Section 24.4)

2. $\mathsf{Prov}_T(\overline{\mathsf{n}})$ says that n numbers a proof of a T-theorem. We next proved the rather easy Theorem 24.6: if γ is *any* fixed point for $\neg\mathsf{Prov}_T(\mathsf{x})$, then, if T is nice, $T \nvdash \gamma$, and if T is also ω-consistent, then $T \nvdash \neg\gamma$. Since the Diagonalization Lemma tells us that there *is* a fixed point for $\neg\mathsf{Prov}_T(\mathsf{x})$, that gives us the standard incompleteness theorem again. (Section 24.5)

3. We then proved Theorem 25.4: if T is a nice theory, and γ is *any* fixed point for $\neg\mathsf{RProv}_T(\mathsf{x})$, then $T \nvdash \gamma$ and $T \nvdash \neg\gamma$ – where $\mathsf{RProv}_T(\overline{\mathsf{n}})$ says 'there's a proof of the sentence with g.n. n, and no "smaller" proof of its negation'. Since the Diagonalization Lemma also tells us that there is a fixed point for $\neg\mathsf{RProv}_T(\mathsf{x})$, we now get incompleteness without assuming ω-consistency. A bit more work gives us a Π_1-equivalent undecidable sentence, proving the full Gödel-Rosser Theorem: if T is a nice theory, then there is an L_A-sentence φ of Goldbach type such that neither $T \vdash \varphi$ nor $T \vdash \neg\varphi$. (Sections 25.3, 25.4)

4. Then we showed that the restriction of our incompleteness theorem to p.r. axiomatized theories is in fact no real restriction. Appealing to a version of Craig's Theorem, we saw that the incompleteness result applies equally to any consistent effectively axiomatized theory which contains Q. (Section 26.1)

5. We also saw how to apply the incompleteness theorem to non-arithmetical theories, so long as they can be definitionally extended to 'do enough arithmetic'. (Section 26.3)

This all reinforces our earlier remarks on incompletability. Suppose T is a nice theory. It is incomplete. Throw in some unprovable sentences as new axioms. Then, by the Gödel-Rosser Theorem the resulting T^+ will still be incomplete, unless it stops being nice. But adding new axioms can't make a p.r. adequate theory any less adequate. So now we know the full price of T's becoming complete. Either (i) our theory ceases to be p.r. axiomatized because you've added too disordered a heap of new axioms, or (ii) it becomes flatly inconsistent. Outcome (i) is bad (for we now know that retreating to an axiomatized-but-not-p.r.-axiomatized theory won't let us escape the incompleteness results: a complete theory will have to stop being effectively axiomatized at all). Outcome (ii) is worse. Keep avoiding those bad outcomes, and T is incompletable.

6. We then moved on to use the Diagonalization Lemma to prove Tarski's Theorem that, if the theory's language L is arithmetically rich enough to formulate a theory like Q, then L can't even *express* the numerical property of numbering a truth of L. (Section 27.3)

7. We also proved some results about the *length* of proofs. For example, if T is nice, then there are always some wffs which have relatively enormous proofs. Take any p.r. function f at all, as fast growing as you like: there will be always be some wff with g.n. n whose proof has a g.n. greater than $f(n)$. (Section 28.1)

8. Finally, we explained how the incompleteness of even a nice theory like PA$_2$ is compatible with its categoricity (even though categoricity implies that – in one good sense – PA$_2$ *does* settle all arithmetical truths). In the second-order case, we crucially need to distinguish questions about what is *provable* in a theory from questions about what is *semantically entailed* by its axioms. (Section 29.6)

But, despite all that, our technical elaborations around and about the First Incompleteness Theorem still leave a number of further questions unanswered. Here are two:

1. Back in Chapter 13, we said that PA is the benchmark first-order theory of basic arithmetic, and that – unlike Q, for example – it is not *obviously* incomplete. Since then, PA has turned out to be in fact incomplete, perhaps contrary to expectation. But our only specific examples of basic arithmetical truths that are unprovable-in-PA (such as G and R) are constructed using coding tricks; and, as we noted before, spelt out in all their detail, with all the abbreviations unpacked, these will be quite horribly long and messy sentences. Looked at purely as sentences of arithmetic,

they have no *intrinsic* mathematical interest. Only someone who already knows about a particular, quite arbitrary, Gödel coding scheme will be in a position to recognize G as 'saying' that it is unprovable. Which raises the question: are there true sentences of basic arithmetic which *are* of intrinsic mathematical interest but which are not derivable in PA? Or are the unprovable L_A-truths just Gödelian oddities?

2. If the ideas regimented by PA don't suffice to pin down the structure of the natural numbers, how come we all do seem to arrive at a shared understanding of that structure (and arguably without making play with the infinitary ideas encapsulated in PA$_2$)? What further idea have we all got our heads around which pins down that structure?

In this Interlude, we will say something briefly about these questions. However, the results we mention aren't taken up much in later chapters; so don't get bogged down, and do feel free to skim, or even to skip straight on to the discussion of the Second Theorem in the next chapter.

30.2 The unprovability-in-PA of Goodstein's Theorem

Let us start with a rather lovely example of a natural, non-Gödelian, arithmetical statement which is true, statable in the language of basic arithmetic, yet demonstrably not provable in PA. The argument is due to Jeff Paris and Laurence Kirby (1982), but it really has its roots in Gerhard Gentzen's wonderful (1936).

(a) To set things up, we need some initial definitions. First,

i. The *pure base k representation* of n is the result of writing n as a sum of powers of k, then rewriting the various exponents of k themselves as sums of powers of k, then rewriting these new exponents as sums of powers of k, etc., ... (writing k^0 as simply 1).

For example,

$$266 = 2^8 + 2^3 + 2^1.$$

So the pure base 2 representation of 266 is

$$266 = 2^{2^{(2^1+1)}} + 2^{(2^1+1)} + 2^1.$$

Similarly,

$$266 = 3^5 + 3^2 + 3^2 + 3^1 + 1 + 1.$$

And its pure base 3 representation is

$$266 = 3^{3^1+1+1} + 3^{1+1} + 3^{1+1} + 3^1 + 1 + 1.$$

Now for our second definition:

ii. To evaluate the *bump function* $B_k(n)$, take the pure base k representation of n; bump up every k to $k+1$; and then subtract 1 from the resulting number.

Let's calculate, for example, $B_2(19)$:

Start with the pure base 2 representation of 19: $2^{2^{2^1}} + 2^1 + 1$;

Bump up the base: $3^{3^{3^1}} + 3^1 + 1$;

Subtract 1 to get $B_2(19) = 3^{3^{3^1}} + 3^1 = 7625597484990$.

We need one more definition:

iii. *The Goodstein sequence* for n is: n, $B_2(n)$, $B_3(B_2(n))$, $B_4(B_3(B_2(n)))$, etc.

In other words, start with n; then keep applying the next bump function to the last term in the sequence.

Let's give a couple of examples, the Goodstein sequences starting with 3 and 19 respectively. For brevity, we'll denote the k-th term of the sequence starting with n by 'n_k'.

$3_1 = 3$: i.e. $2^1 + 1$
$3_2 = B_2(3_1) = 3^1 + 1 - 1 = 3^1$
$3_3 = B_3(3_2) = 4^1 - 1 = 1 + 1 + 1$
$3_4 = B_4(3_3) = 1 + 1 + 1 - 1 = 1 + 1$
$3_5 = B_5(3_4) = 1 + 1 - 1 = 1$
$3_6 = B_6(3_5) = 1 - 1 = 0.$

$19_1 = 19 = 2^{2^{2^1}} + 2^1 + 1$
$19_2 = B_2(19_1) = 3^{3^{3^1}} + 3^1 + 1 - 1 = 3^{3^{3^1}} + 3^1 \ (\approx 7 \cdot 10^{13})$
$19_3 = B_3(19_2) = 4^{4^{4^1}} + 4^1 - 1 = 4^{4^{4^1}} + 1 + 1 + 1 \ (\approx 1.3 \cdot 10^{154})$
$19_4 = B_4(19_3) = 5^{5^{5^1}} + 1 + 1 \ (\approx 2 \cdot 10^{2184})$
$19_5 = B_5(19_4) = 6^{6^{6^1}} + 1 \ (\approx 2.6 \cdot 10^{36305})$
$19_6 = B_6(19_5) = 7^{7^{7^1}}.$

Which no doubt suggests that, while the Goodstein sequence for n might eventually hit zero for *very* small n, for later values of n the sequence, which evidently starts off wildly inflationary, must run away for ever.

(b)   *But not so!* In his (1944), R. L. Goodstein showed that

Theorem 30.1 *For all n, the Goodstein sequence eventually terminates at zero.*

And very surprisingly, the proof is actually quite straightforward – or at least it is straightforward if you know just a little of the theory of ordinal numbers. Let's quickly outline the proof:

Sketch of a proof sketch   Take the Goodstein sequence for n. Render its k-th term into its pure base $k + 1$ representation as in our examples above (with each sum presented in descending order of exponents). Now consider the parallel sequence that you get by going through and replacing each base number by ω (the first infinite ordinal). For example, the parallel sequence to the Goodstein sequence for 19 starts

$$\omega^{\omega^{\omega^1}} + \omega^1 + 1$$
$$\omega^{\omega^{\omega^1}} + \omega^1$$
$$\omega^{\omega^{\omega^1}} + 1 + 1 + 1$$
$$\omega^{\omega^{\omega^1}} + 1 + 1$$
$$\omega^{\omega^{\omega^1}} + 1$$
$$\vdots$$

It isn't hard to show that this parallel sequence of ordinals will in every case be strictly decreasing.

But there just cannot be an infinite descending chain of such ordinals – that is a quite fundamental theorem about ordinals. Hence the ordinal sequence must terminate. And therefore the parallel Goodstein sequence for n must terminate too![1]   ⊠

Don't worry at all, however, if you find that proof-sketch baffling. All you really need to take away is the idea that Goodstein's Theorem *can* easily be proved, *if* we invoke ideas from the theory of infinite ordinal numbers, i.e. if we invoke ideas that go beyond the basic arithmetic of finite numbers.

(c)   Note next that the sequence-computing function $g(n, k) = n_k$ is evidently primitive recursive (to calculate the k-th value of the Goodstein sequence starting at n, we can write a program just using 'for' loops). Hence this two-place function is expressible in L_A by a three-place Σ_1 wff $S(x, y, z)$.

Goodstein's theorem is therefore itself *expressible* in PA by the corresponding Π_2 sentence $\forall x \exists y\, S(x, y, 0)$. It is certainly not obvious, however, how we might go about *proving* Goodstein's theorem in PA, using only the arithmetic of finite numbers. And indeed it can't by done, by the Kirby-Paris Theorem:[2]

Theorem 30.2 *Goodstein's Theorem is not provable in* PA *(assuming* PA *is consistent).*

---

[1] For a relatively gentle introduction to the ordinals, see e.g. Goldrei (1996). For a modern presentation of the theorem and its set-theoretic proof, see Potter (2004, pp. 212–218).

[2] For the original proof, see Kirby and Paris (1982); and for elaboration of the background see Kaye (1991, ch. 14). For a different method of proof, see Fairtlough and Wainer (1998).

Which gives us an answer to the first question we posed at the beginning of this Interlude: there are indeed some non-Gödelian truths of basic arithmetic which have at least *some* intrinsic mathematical interest but which provably can't be derived in PA.

(d)   Recall our result in Section 12.4 which showed that the exponential 2^n grows too fast for the function to be proved in $I\Delta_0$ to be total, i.e. to have a value for every argument. In symbols, $I\Delta_0 \nvdash \forall x \exists y \varepsilon(x,y)$. (But increase our inductive power by going up to $I\Sigma_1$, and then we *can* prove that sentence.)

Here is a comparable way of looking at the Kirby-Paris result. Consider the two-variable wff $\widetilde{S}(x,y) =_{\mathrm{def}} S(x,y,0) \wedge (\forall z \le y)(S(x,z,0) \to z = y)$.[3] Then $\widetilde{S}(\overline{n}, \overline{k})$ is true just when k numbers the first step when the Goodstein sequence for n hits zero. Hence $\widetilde{S}(x,y)$ expresses the one-place function $lg(n)$ which returns the number of steps in the Goodstein sequence for n before it gets to zero. This is an *immensely* fast-growing function of n. And the fact that PA $\nvdash \forall x \exists y\, S(x,y,0)$ – which implies PA $\nvdash \forall x \exists y\, \widetilde{S}(x,y)$ – comes to this: PA can't prove that the ultra-fast-growing function lg is total, i.e. takes a value for every input.

The observation that arithmetics of increasing strengths will show ever more fast-growing functions to be total, but that even PA has its limitations in this respect, goes back ultimately to Kreisel in the 1950s.

30.3   An aside on proving the Kirby-Paris Theorem

For those who *do* know just a bit about ordinals, here is a brief gesture at a proof of the Kirby-Paris Theorem, since it links to a couple of key results which we'll be discussing later. Others can skip.

Goodstein's Theorem, we have already noted, depends on the fundamental fact that there can't be an infinite decreasing chain of ordinals which are sums of powers of ω, i.e. there can't be an infinite decreasing chain of ordinals less than ε_0, the first ordinal that comes after all the sums of powers of ω. Proving that fundamental result is equivalent to showing that *transfinite induction* up to ε_0 is sound.[4]

However, there are natural Gödel-numberings for the ordinals which are sums of powers of ω; so we can transmute claims about these ordinals into arithmetical claims about their numerical codes. So being able to prove Goodstein's theorem

---

[3]Compare a similar construction in Section 16.2.

[4]"Transfinite induction up to ε_0'? Ordinary course-of-values induction amounts to this principle: suppose (i) 0 is F; and suppose (ii) if all numbers less than n are F, then n is F; then (iii) all numbers are F. It is easily seen to be equivalent to the Least Number Principle which is that, given any set of natural numbers, one of them is the least. Transfinite induction up to ε_0 is the parallel principle: suppose (i) 0 is F; and suppose (ii) if all ordinals less than α are F, then α is F (where α is a sum of powers of ω); then (iii) all ordinals which are sums of powers of ω are F. That is similarly equivalent to the principle, given any set of ordinals which are sums of powers of ω, one of *them* is the least.

inside PA would be tantamount to PA's being able to handle (via our codings) transfinite induction up to ε_0.

And now we appeal to two future results. First, this kind of transfinite induction is in fact strong enough to prove the consistency of PA by Gentzen's argument (see Section 32.4). Hence, if PA could prove Goodstein's theorem, it could also prove its own consistency. But second, PA can't prove its own consistency, by Gödel's Second Theorem (see Section 31.3). So PA can't prove Goodstein's theorem.

30.4   Isaacson's Thesis

The next key point to note is that the other known cases of mathematically interesting L_A truths which are provably independent of PA share an important feature with Goodstein's Theorem. *The demonstrations that they are L_A truths likewise use conceptual resources which go beyond those which are required for understanding the basic arithmetic of finite natural numbers.*

For example, proving the so-called Paris-Harrington theorem – which gives another arithmetical truth that is unprovable-in-PA – requires König's Lemma, which says that an infinite tree that only branches finitely at any point must have an infinite path through it.[5]

And – in a rather different way – appreciating the truth of undecidable Gödel sentences for PA also seems to involve conceptual abilities that go beyond a grasp of elementary operations on the finite numbers. Maybe in this case we don't need to invoke infinitary ideas like transfinite induction; but we surely have to be able to reflect on our own arithmetical theorizing in order to recognize e.g. that canonical Gödel sentences are true (see Section 36.5). We have to be able to make the move from (i) implicitly assuming in our unreflective mathematical practice that (say) every natural number has a unique successor to (ii) explicitly accepting that a certain theory which has that proposition as an axiom is sound/consistent. And this *is* a move, because knowing your way around the numbers doesn't in itself entail the capacity to be able to reflect on that ability.

Putting these points about the Gödelian and non-Gödelian cases together suggests an interesting speculation:

> *Isaacson's Thesis*   If we are to give a proof of *any* true sentence of L_A which is independent of PA, then we will need to appeal to ideas that go beyond those which are constitutive of our understanding of basic arithmetic.[6]

---

[5] For details, see Kaye (1991, ch. 14) again; the proofs of the variant Ramsey theorem and of its independence from PA were first given in Paris and Harrington (1977). For a gentle introduction see Kolata (1982) which also touches on Harvey Friedman's finite version of Kruskal's theorem, another truth which can be expressed in PA but is independent of it. Friedman's result is also discussed in Smoryński (1982) and Gallier (1991).

[6] Compare Daniel Isaacson's (1987), where he suggests that the truths of L_A that can't

If that's right, then PA in fact reaches as far into the truths of basic arithmetic as any properly axiomatized theory can reach, at least if it aims to encapsulate no more than what follows from our purely arithmetical knowledge.

But *is* the thesis right? It isn't exactly clear what is involved in 'purely arithmetical' knowledge. But even so – at least before our discussions in the last chapter – we might well have thought that there is a way of going beyond first-order PA while keeping within the confines of what is given to us in our understanding of elementary arithmetic, namely by exploiting our informal understanding of induction which arguably seems to involve grasp of a *second-order* principle.

However, what we have discovered about second-order arithmetics is in fact entirely in conformity with Isaacson's Thesis. To repeat, there are indeed L_A sentences which we can derive in PA_2 but which aren't derivable in PA. But if we are to accept these formal derivations as genuine *proofs*, i.e. chains of reasoning which do give us grounds to accept their conclusions, then we must endorse PA_2's generous treatment of the Comprehension Schema. And that involves something like making sense of the non-arithmetic infinitary idea of quantifying over arbitrary subsets of $\mathbb{N}$. By contrast, accepting the weaker formal theory ACA_0, for example, doesn't seem to involve more than a grasp of arithmetic together with some very general logical ideas; but this theory doesn't give us any more basic arithmetic than PA does.[7]

30.5   Ever upwards

Going *second-order*, we've noted, enables us to prove new *first-order* arithmetical sentences that we couldn't prove before. And we see the same phenomenon appearing again as we push on further up the hierarchy, and allow ourselves to talk not just of numbers and sets of numbers, but of sets of sets, sets of sets of sets, and so on. Embracing each new level as we go up the hierarchy of types of sets allows us to prove some truths at lower levels which we couldn't prove

---

be proved in PA 'are such that there is no way that their truth can be perceived in purely arithmetical terms' (p. 203).

However, Isaacson goes further, adding – by way of explanation? – that 'via the phenomenon of coding [such truths] contain essentially hidden higher-order, or infinitary, concepts' (pp. 203–204). But that seems certainly wrong for the non-Gödelian cases. Take the unprovable wff $\forall x \exists y\, S(x, y, 0)$, where S is as before the Σ_1 wff that expresses the p.r. function which computes the k-th member of the Goodstein sequence starting from m. In so far as there is any coding associated with this wff, it is the coding of the steps in an entirely finitary arithmetical computation. So although the *proof* of Goodstein's theorem involves infinitary concepts, the *content* of the theorem doesn't. But we won't pursue this point: for we can cleanly separate Isaacson's interesting conjecture from the rather contentious additional gloss he puts on it.

[7]There is, to be sure, a whole range of neat (and not-so-neat) theories of intermediate strength worth considering, with comprehension axioms stronger than ACA_0's and weaker than PA_2's – and there is a lot of interesting work to be done to see, for example, exactly what strength of comprehension principle is required to prove various results like Goodstein's Theorem or the Gödel sentence for PA. However, the headline news is that these additional technicalities don't seem to change the basic picture as far as Isaacson's Thesis is concerned.

before, including more ground-level L_A truths: richer theories can prove Gödel sentences for weaker theories (as PA_2 proves the canonical Gödel sentence for PA).

Gödel himself remarked on this phenomenon even in his original paper:

> [I]t can be shown that the undecidable propositions constructed here become decidable whenever appropriate higher types are added. (Gödel, 1931, p. 181, fn. 48^a)[8]

And what about non-Gödelian cases? Are there intrinsically interesting arithmetical truths which are formally undecided by PA, also formally undecided by PA_2, and essentially *require* higher-order theories to prove them?

Well, there is extensive work by Harvey Friedman which aims to produce 'natural' arithmetical statements whose proof is supposed essentially to require the existence of 'large cardinals' – i.e. the existence of sets very high up the transfinite hierarchy. The jury is still out, however, on the real significance of his results.[9]

30.6   Ancestral arithmetic

(a)   We have said something in response to the first question we posed at the end of Section 30.1. But our discussion raises further issues: in particular, there's our second question, about the business of understanding arithmetic.

As we said at the very outset (Section 1.1), it seems that anyone who comes to understand basic arithmetic gets to share an informal yet quite determinate conception of the structure of the natural number series ordered by the successor relation, and has an equally determinate conception of the operations of addition and multiplication for numbers. And in virtue of this shared grasp of these intuitive conceptions, it seems that we have a shared understanding of the idea of the standard, intended, interpretation for first-order PA. (We hope and believe, of course, that this standard interpretation is a model!)

However, assuming it is consistent, PA has non-isomorphic models. Which means that accepting its axioms as correct is quite compatible with understanding the theory as talking about some deviant, unintended model. So what more, beyond accepting PA's axioms as correct, *is* involved in understanding the stan-

---

[8]If you know any set theory at all, then you'll know that the hierarchy of types of sets continues into the transfinite (so after all the finite types at levels $1, 2, 3, \ldots$, we take their union at level ω and keep on going through levels $\omega + 1$, $\omega + 2$ and up through the ordinals). And as we go further up, yet more ground-level arithmetical sentences become provable. Indeed, in his rather enigmatic footnote, Gödel suggests that the fact that 'the formation of ever higher types can be continued into the transfinite' – which means that there is an ever growing list of new arithmetic truths that become provable as we go up the hierarchy – is 'the true reason' for the incompleteness in any particular formal theory of arithmetic.

[9]For a programmatic outline of Friedman's overall approach, see his (2000). For a critical response, see Feferman (2000).

dard structure of the natural numbers: what more is involved in fixing on the intended 'slim' model (whose domain comprises just a zero and its successors)?

It might have been tempting to conclude – given that the second-order Peano Arithmetic PA$_2$ *is* categorical and *does* pin down a unique structure (up to isomorphism) – that our everyday understanding must involve the sort of second-order ideas built into PA$_2$. But our discussion hasn't really favoured this suggestion, for it amounts to something like the idea that we need to understand the infinitary idea of quantifying over *all* possible subsets of the numbers in order to understand elementary arithmetic. And that arguably overshoots. However, if we deploy instead some weaker and more tractable second-order ideas, like that of quantifying just over arithmetical sets, we end up with a theory which in crucial ways is no stronger than PA. So what to do?

(b)   To pin down the intended 'slim' model of PA, we need the idea that the numbers comprise zero, the successors of zero (all different, and never circling round to zero again), *and nothing else*. In other words, the numbers are what you can reach by repeated moves from a number to its successor.

Here's a similar idea at work. We familiarly define the class of wffs of a formal language by specifying a class of atomic wffs, and then saying e.g. that any wffs you can build up from these using connectives and quantifiers are wffs, and *nothing else* is a wff. So in this case we are saying that the wffs are what you can get by repeated applications of the wff-building operations.

Or take an even simpler case, the notion of an ancestor. An ancestor is someone you can reach by repeated moves back from a person to one of their parents. Now, to be sure, we *can* define the idea of ancestor from the idea of a parent by invoking second-order ideas (i.e., in this case, by talking not just of people but of sets of people). Thus someone is one of my ancestors if they belong to every set of people which includes me and which, if it includes N, also includes N's parents. Further, we can't define the idea of an ancestor from the idea of a parent using just first-order quantifiers, identity and the connectives.[10] However, although we *can* define the idea of an ancestor from the idea of a parent in second-order terms, and *can't* do it in first-order terms, it surely doesn't follow that the child who so easily grasps the concept *ancestor* must already in effect be quantifying over arbitrary sets of people.

Likewise we can define the idea of a wff in second-order terms. For example, an expression is a wff of L_A iff it belongs to every collection of expressions which contains certain atoms, and which, if it contains φ and ψ, contains $\neg\varphi$, $(\varphi \land \psi)$, $\forall\xi\varphi$, etc. But the availability of this sort of definition again doesn't show that the humdrum notion of an L_A-wff – something that we need to grasp to understand the Induction Schema of first-order PA – is already essentially second-order.

---

[10]For a proper proof, see Shapiro (1991, pp. 99–100). But the following thought is suggestive. If R stands for the relation of *being a parent to*, then we could express 'm is an ancestor of n' by Rmn $\lor$ $\exists$x(Rmx $\land$ Rxn) $\lor$ $\exists$x$\exists$y(Rmx $\land$ Rxy $\land$ Ryn) $\lor$ $\exists$x$\exists$y$\exists$z(Rmx $\land$ Rxy $\land$ Ryz $\land$ Rzn) $\lor$ . . ., if we were allowed infinitely long sentences. But a first-order language doesn't allow unending disjunctions, nor in general does it allow us to construct equivalent wffs.

Here's a comparison. We can define identity in second-order terms by putting $x = y =_{\text{def}} \forall X(Xx \leftrightarrow Xy)$; but that doesn't show that understanding identity is to be explicated in terms of understanding quantification over arbitrary subsets of the domain. It can't show that, because understanding the idea of arbitrary sets of objects of some kind presupposes an understanding of what makes for an object, and that involves an understanding of what makes candidates the same or different objects, i.e. it already involves an understanding of identity.

So, in sum, to pin down the intended domain of numbers (to keep it slim, banning extraneous objects) requires deploying the general idea of something being one of the things we can reach by repeating a certain operation indefinitely often, the same idea that is involved in specifying what it is to be a wff or an ancestor. And the suggestion is that, although this idea isn't definable in first-order terms, it is *not* second-order in the full sense either.

(c) So the child learns that her parents have parents, and that they have parents too. In sepia tones, her great-grandparents have parents in their turn. And she learns that other children have parents and grandparents too. The penny drops: she realizes that in each case you can keep on going. And the child gets the idea of an ancestor, i.e. the idea of being someone who turns up eventually as you trace people back through their parents. Similarly, she learns to count; the hundreds are followed by the thousands, and the tens of thousands, and the hundreds of thousands and the millions. Again the penny drops: you can keep on going. And she gets the idea of being a natural number, i.e. of turning up eventually as you keep on moving to the next number.

Generalizing, then, these are cases where the child moves from a grasp of a relation to a grasp of the *ancestral* of that relation – where the ancestral R^* of a relation R is that relation such that R^*ab holds just when there is an indefinitely long but finite chain of R-related things between a and b. Now, we don't want to be multiplying conceptual primitives unnecessarily. But the concept of the ancestral of a relation doesn't seem a bad candidate for being a fundamental logical idea; grasping this concept seems to involve a distinctive level of cognitive achievement. So what happens if we try to build *this* idea into a formal theory of arithmetic which is otherwise basically first order? – which is an old question that goes back at least to R. M. Martin and John Myhill over fifty years ago.[11]

(d) To begin, we need to expand our first-order logical vocabulary with an operator – we'll symbolize it with a star – which attaches to two-place expressions $\varphi(x, y)$: $\varphi^*(x, y)$ is to be interpreted as expressing the ancestral of the original relation expressed by $\varphi(x, y)$.[12]

Suppose that we augment the language L_A with such an operator to get the language L^*. We will write the result of applying the star operator to $Sx = y$ as

---

[11]See Martin (1943) and the follow-up note (1949) where Martin urges that his construction is 'nominalistic', i.e. doesn't commit you to the existence of sets. This work was then developed in the rather more accessible Myhill (1952).

[12]We are being forgivably careless about the syntactic details here.

S*xy. So consider then what happens when you take the familiar axioms of PA but add in the new axiom

S. $\forall x(x = 0 \lor S^*0x)$

which says – as we want – that every number is zero or one of its successors. Any interpretation of this expanded set of axioms which respects the fixed logical meaning of the ancestral-forming operator evidently must be slim. So, by the argument for Theorem 29.2, a theory with these axioms will be categorical. Let's define, then, the related semantic entailment relation $\Gamma \vDash^* \varphi$, which obtains if every interpretation which makes all of Γ true makes φ true – where we are now generalizing just over interpretations which give the star operator its intended fixed logical meaning (and otherwise treats the logical vocabulary standardly). Then, because of categoricity, our expanded axioms semantically entail any true sentence of the expanded language L^*, and hence entail any true L_A sentence.

There can't, however, be a complete effective axiomatization of this 'ancestral arithmetic' in the language L^*. The argument is as before. Take any effectively axiomatized theory A* which extends PA plus (S) by adding rules for the star operator. The incompleteness theorem still applies (assuming consistency), so there will be an unprovable-yet-true L_A sentence G* for this theory. In other words, $A^* \nvdash G^*$, although $A^* \vDash^* G^*$. So our theory's deductive system can't be complete.

However, there can of course be *partial* axiomatizations of ancestral arithmetic. We can lay down various rules for handling the ancestral operator. Suppose, for brevity, that we write $H(\psi, \varphi)$ as short for $\forall x \forall y((\psi(x) \land \varphi(x, y)) \to \psi(y))$ – i.e. the property expressed by ψ is *hereditary* with respect to the relation φ (i.e. is passed down a chain linked by φ). Then Myhill's proposed axioms for the star operator are tantamount to the following schematic rules:

From $\varphi(a, b)$ infer $\varphi^*(a, b)$.
From $\varphi^*(a, b)$, $\varphi(b, c)$ infer $\varphi^*(a, c)$.
From $H(\psi, \varphi)$, infer $H(\psi, \varphi^*)$.

These first two rules are 'introduction' rules for the star operator; and the third rule is easily seen to be equivalent to the following 'elimination' rule:

From $\varphi^*(a, b)$ infer $\{\psi(a) \land H(\psi, \varphi)\} \to \psi(b)$.

This is a kind of generalized inductive principle which says that given that b is a φ-descendant of a then, if a has some property which is passed down from one thing to another if they are φ-related, then b has that property too. And taking the particular case where $\varphi(x, y)$ is $Sx = y$, having this rule will enable us to derive all the instances of the familiar first-order induction schema.

So let's now briefly consider the formal system PA* which extends Q by adding our new axiom (S) plus Myhill's rules for handling the ancestral operator. The obvious next question is: what is the deductive power of this system? Is this another case like PA_2 where we also only had a partial axiomatization of the relevant semantic relation, though PA_2 could prove more L_A sentences than PA? Or is PA* another extension of PA like ACA_0, which is conservative over PA?

If the first case held, then we'd have a very interesting challenge to Isaacson's Thesis. For the ancestral arithmetic PA* is arguably within the conceptual reach of someone who has fully understood basic arithmetic; and so, if we could use it to prove new sentences of basic arithmetic not provable in PA, then Isaacson's Thesis would fall. But in fact, the issue doesn't arise:

Theorem 30.3 PA* *is conservative over* PA *for* L_A *sentences.*

In other words, for any L_A sentence ψ such that PA* $\vdash \psi$, it is already the case that PA $\vdash \psi$.

Proof sketch Recall that we can express facts about sequences of numbers in PA by using a β-function (see Section 15.2). So suppose R is some relation. Then

A. R^*ab is true just so long as, for some number x, there is a sequence of numbers $k_0, k_1, \ldots, k_x$ such that: $k_0 = a$, and if $u < x$ then Rk_uk_{Su}, and $k_x = b$.

Using a three-place β-function, that means

B. R^*ab is true iff for some x, there is a pair c, d such that: $\beta(c, d, 0) = a$, $\beta(c, d, x) = b$, and if $u < x$ then $R(\beta(c, d, u), \beta(c, d, Su))$.

So consider the following definition:

C. $\varphi^{**}(a, b) =_{\text{def}} \exists x \exists c \exists d \{ B(c, d, 0, a) \wedge B(c, d, x, b) \wedge$
$(\forall u \leq x)[u \neq x \rightarrow \exists v \exists w \{ (B(c, d, u, v) \wedge B(c, d, Su, w)) \wedge \varphi(v, w) \}] \}$

where B captures the now familiar three-place Gödelian β-function.

It is easy to check that the Myhill inference rules for single-starred φ^* apply equally to our defined double-starred construct φ^{**} in PA (that's essentially because the moves are then valid semantic entailments within PA, and the theory's deductive system is complete). And the double-starred analogue of axiom (S) is also a theorem of PA. So corresponding to any proof involving starred wffs in PA* there is an exactly parallel proof in plain PA involving double-starred wffs. Hence in particular, any proof using starred wffs whose conclusion is a pure (unstarred) L_A wff in PA* will have a parallel proof in plain PA going via double-starred wffs. Which establishes what we needed to show.[13] ⊠

(e) Let's summarize this last section. Our everyday understanding of basic arithmetic pins down a unique structure for the natural numbers, at least up to isomorphism. Hence our grasp of basic arithmetic involves more than is captured

---

[13]Thanks to Andreas Blass and Aatu Koskensilta for discussion of this. The argument would seem to generalize to any natural first-order variant of PA*. For something on richer, second-order, ancestral logics, see Heck (2007).

by first-order PA. But what more? It is enough that we have the idea that the natural numbers are zero and its successors and nothing else. And getting our head round this idea, we suggested, involves the general idea of the ancestral; the numbers are what stand in the ancestral of the successor relation to zero.

Now, the ancestral of a relation can be defined in second-order terms, but it seems overkill to suppose that our understanding of ordinary school arithmetic is essentially second-order. Why not treat the operation that takes a relation to its ancestral to be a (relatively unmysterious, even though not purely first-order) logical primitive? If we do, we can construct a theory PA* which naturally extends PA in a way that arguably still reflects our everyday understanding of arithmetic. And PA* has the semantic property of categoricity – it pins down the structure of the natural numbers in exactly the way we want.

Since PA* is still effectively axiomatized, however, we know that it will be incomplete (assuming it is consistent). But we might have suspected that it would at least have proved more than PA: but not so. PA* is deductively conservative over PA for L_A sentences: so we can't in fact use this expanded theory to deduce new truths of basic arithmetic that are left unsettled by PA.

Hence, to put it the other way around, it seems that if we are to come up with proofs of L_A truths unsettled by PA, and to do this in an effectively axiomatized system, then we'll have to deploy additional premises and/or logical apparatus that go beyond PA* or simple variants of it. Which suggests that we'll need to invoke new ideas which go beyond those essential to our ordinary understanding of basic arithmetic. For PA*'s guiding idea that all the numbers can be reached from zero by repeatedly adding one, i.e. the idea that all numbers are related to zero by the ancestral of the successor relation, is very plausibly at the limit of what is *necessary* to ground a grasp of the natural number structure and arithmetic concepts which can be defined over it. Which gives us Isaacson's Thesis again.

31 Gödel's Second Theorem for PA

We now, at long last, turn to considering the *Second* Incompleteness Theorem for PA.

We worked up to the First Theorem very slowly, spending a number of chapters proving various preliminary technical results before eventually taking the wraps off the main proofs in Chapters 21 and 22. But things go rather more smoothly and accessibly if we approach the Second Theorem the other way about, working backwards from the vicinity of the target Theorem to uncover the underlying technical results needed to demonstrate it. So in this chapter, we will simply *assume* a technical result about PA which we will call the 'Formalized First Theorem': we then show that it very quickly yields the Second Theorem for PA, and derive some corollaries.

In Chapter 33, we then show that the Formalized First Theorem and hence the Second Theorem can similarly be derived in any nice arithmetic theory T when certain 'derivability conditions' hold for the provability predicate Prov_T. In Chapter 35 we finally dig down to discover what it takes for those derivability conditions to obtain.

31.1 Defining Con

We begin with four reminders. Fix a scheme for Gödel numbering. Then:

1. $Prf(m, n)$ holds when m is the super g.n. of a PA-proof of the sentence with g.n. n. And we defined $Prov(n)$ to be true just when n is the g.n. of a PA theorem, i.e. just when $\exists m\, Prf(m, n)$. Thus, $Prov(\ulcorner\varphi\urcorner)$ iff PA $\vdash \varphi$. (Section 20.4)

2. We introduced $\mathsf{Prf}(\mathsf{x}, \mathsf{y})$ as an abbreviation for a Σ_1 wff of PA's language that canonically captures the relation Prf by perspicuously recapitulating its p.r. definition. (Section 24.1)

3. We went on to define $\mathsf{Prov}(\mathsf{x}) =_{\mathrm{def}} \exists \mathsf{v}\, \mathsf{Prf}(\mathsf{v}, \mathsf{x})$. This complex predicate is also Σ_1, and it expresses the numerical property $Prov$. Hence $\mathsf{Prov}(\ulcorner\varphi\urcorner)$ is true iff PA $\vdash \varphi$. (Section 24.1)

4. Recall, finally, that we showed that PA $\vdash \mathsf{G} \leftrightarrow \neg\mathsf{Prov}(\ulcorner\mathsf{G}\urcorner)$. (That was Theorem 24.2.)

Now for two new definitions:

5. Suppose that we have set up the first-order logic of PA using the familiar absurdity constant '$\bot$'. Then, of course, PA is consistent if and only if PA $\nvdash \bot$.

Suppose on the other hand that the absurdity constant isn't built into PA's official logical system. We can introduce it by definition in various ways. Here is one. PA's Axiom 1 immediately proves the wff $0 \neq 1$; so if PA *also* proved $0 = 1$, it would be inconsistent. And conversely, since PA has a classical logic, if it is inconsistent we can derive anything, including $0 = 1$. So, when '$\bot$' isn't built in, we can put $\bot =_{\text{def}} 0 = 1$. Again, PA is consistent if and only if PA $\nvdash \bot$.

Henceforth, then, we'll take the absurdity constant '$\bot$' to be available in PA, either built in or by definition (and similarly in other theories).

6. PA is consistent, i.e. PA $\nvdash \bot$, iff it isn't the case that $Prov(\ulcorner\bot\urcorner)$, i.e. just when $\neg\mathsf{Prov}(\ulcorner\bot\urcorner)$ is true. So that motivates the following abbreviation:

$$\mathsf{Con} =_{\text{def}} \neg\mathsf{Prov}(\ulcorner\bot\urcorner).$$

Being the negation of a Σ_1 sentence, Con is a Π_1-equivalent wff. And given the background Gödel coding scheme in play, we can immediately see, without need for any further argument, that Con is constructed in such a way as to make it true if and only if PA is consistent. So, in the spirit of Section 21.8, we might comment that Con *indirectly says* that PA is consistent.

31.2 The Formalized First Theorem in PA

In Section 21.5 we proved the following (it's the easier half of the First Theorem):

If PA is consistent, then G is not provable in PA.

We now know that one way of representing the antecedent of this conditional in L_A is by the formal wff we are abbreviating as Con, while the consequent can of course be represented by $\neg\mathsf{Prov}(\ulcorner\mathsf{G}\urcorner)$. So, in sum, the wff

$$\mathsf{Con} \rightarrow \neg\mathsf{Prov}(\ulcorner\mathsf{G}\urcorner)$$

expresses one half of the incompleteness theorem for PA, and does so *inside L_A, the language of* PA *itself.*

But that point by itself isn't particularly exciting. The novel and interesting claim is this next one, call it the *Formalized First Theorem*: PA can itself *prove* that formal conditional.

Of course, we've seen this sort of thing before. Remember how we initially constructed G in Section 21.2, and then immediately noted that G is true if and only if it is unprovable. That informally derived biconditional (which we reached while looking at PA 'from the outside', so to speak) can readily be expressed in the language of PA, by the formal wff G $\leftrightarrow$ $\neg$Prov($\ulcorner$G$\urcorner$). And Theorem 24.2 tells us that this diagonalization biconditional can be formally *proved* inside PA:

D.   PA $\vdash$ G $\leftrightarrow$ $\neg$Prov($\ulcorner$G$\urcorner$).

We now have a very similar situation. Informal reasoning (looking at PA 'from the outside') leads to Gödel's result that PA's consistency entails G's unprovability. And we are saying that this result can also be formally derived within PA itself:

F.   PA $\vdash$ Con $\rightarrow$ $\neg$Prov($\ulcorner$G$\urcorner$).

Gödel doesn't actually *prove* (F) or anything like it in his 1931 paper.[1] He just invites us to observe the following:

> All notions defined or statements proved [in establishing the First Theorem] are also expressible or provable in P [the formal system Gödel is working with]. For throughout, we have used only the methods of definition and proof that are customary in classical mathematics, as they are formalizable in P. (Gödel, 1931, p. 193)

Gödel could very confidently assert this because P, recall, is his version of Russell and Whitehead's theory of types; it is a higher-order theory which is a lot richer than either first-order PA or second-order PA_2, and which was already well known to be sufficient to formalize great swathes of mathematics. So, agreed, it is entirely plausible that P has the resources to formalize the very straightforward mathematical reasoning that leads to the First Theorem for P.

It isn't so obvious, however, that all the reasoning needed for the proof of the First Theorem for PA can be formally reflected in a relatively low-power theory like PA itself. Checking that we can formalize the proof inside PA (indeed, inside weak subsystems of PA) requires some hard work. But let's assume for the moment that the work has been done: that's the big technical assumption on which the rest of this chapter proceeds. Then we will have arrived at the result (F), the Formalized First Theorem.

31.3   The Second Theorem for PA

Suppose now (for reductio) that

1.   PA $\vdash$ Con.

Then, given the Formalized First Theorem (F), modus ponens yields

2.   PA $\vdash$ $\neg$Prov($\ulcorner$G$\urcorner$).

But (D) tells us that $\neg$Prov($\ulcorner$G$\urcorner$) and G are provably equivalent in PA. Whence

---

[1]It seems that he intended to make good the deficit in a never-written Part II of his paper.

3. PA ⊢ G.

However, that flatly contradicts the First Theorem, assuming PA is consistent. Therefore our supposition (1) must be false, unless PA is inconsistent.

Hence, assuming we *can* establish the Formalized First Theorem, that shows

Theorem 31.1 *If* PA *is consistent,* PA ⊬ Con.

Call this Gödel's *Second Incompleteness Theorem* for PA. Suppose that the axioms of PA are true on the standard interpretation and hence all its theorems are true, so PA *is* consistent. Then Con will be another true-but-unprovable wff.

31.4 On ω-incompleteness and ω-consistency again

Assume PA *is* consistent. Then ⊥ isn't a theorem. Therefore no number is the super g.n. of a proof of ⊥ – i.e. for all n, it isn't the case that $Prf(n, \ulcorner \bot \urcorner)$. Since Prf captures *Prf*, we therefore have

1. for any n, PA ⊢ ¬Prf($\bar{n}, \ulcorner \bot \urcorner$).

Now, if we *could* prove Con then – unpacking the abbreviation – we'd have

2. PA ⊢ ∀v¬Prf(v, $\ulcorner \bot \urcorner$).

The unprovability of Con means, however, that we *can't* get from (1) to (2). So this is another example of PA's initially surprising ω-incompleteness (see Section 21.6).

Now suppose for a moment that PA ⊢ ¬Con. In other words, suppose that PA ⊢ ∃vPrf(v, $\ulcorner \bot \urcorner$). This, given (1) above, would make PA ω-inconsistent. Which immediately gives us the following very easy companion result to Theorem 31.1 (this one *doesn't* depend on the Formalized First Theorem):

Theorem 31.2 *If* PA *is* ω-*consistent, then* PA ⊬ ¬Con.

31.5 So near, yet so far

To repeat, for any n, PA can prove ¬Prf($\bar{n}, \ulcorner \bot \urcorner$). So, a proof of ∀v¬Prf(v, $\ulcorner \bot \urcorner$), i.e. of Con, is – as it were – only just beyond PA's grasp.

Here is another way in which PA comes so near to proving its own consistency. Suppose T is a finitely-axiomatized sub-theory of PA. That is to say, suppose T has a finite number of arithmetical axioms (where those axioms are provable in PA). Since T's axioms are finite in number, it is effectively decidable whether a sentence is an axiom, and so the relation $Prf_T(m, n)$ which holds when m numbers a T-proof of the sentence number n is primitive recursive, and can be canonically captured in PA by a wff $Prf_T(m, n)$. Construct Con_T from Prf_T in the way that Con is constructed from Prf, so $Con_T =_{\text{def}} ¬∃v\,Prf_T(v, \ulcorner \bot \urcorner)$. Then

Mostowski (1952a) proved that PA ⊢ Con$_T$. In a slogan, PA *can* prove each of its finitely axiomatized sub-theories to be consistent. It just can't go the extra step of proving consistent the result of putting those sub-theories together into one big theory (even though any inconsistency in PA, being provable in a finite number of steps, would have to show up in some finitely-axiomatized sub-theory).[2]

Incidentally, Mostowski's theorem – which we won't prove here – entails the following significant corollary:

Theorem 31.3 PA *cannot be finitely axiomatized.*[3]

We presented PA as having an infinite number of axioms, including (the closures of) all the possible instances of the Induction Schema. We now see that this was inevitable. For suppose we could re-package PA as the finitely axiomatized theory T. Then by Mostowski's theorem, we'd have PA ⊢ Con$_T$, i.e. T ⊢ Con$_T$. But since by hypothesis T just *is* equivalent to PA, we'd have Gödel's Second Theorem for PA again in the form T ⊬ Con$_T$. Contradiction.

31.6 How should we interpret the Second Theorem?

So much for some initial technical results. We end the chapter with some introductory remarks about how we should *interpret* the Second Incompleteness Theorem for PA.

(a) Looked at one way, all we have done in this chapter is find another wff which exemplifies the *first* incompleteness theorem for PA: we have seen that, as with G, if PA is consistent then PA ⊬ Con, and if PA is ω-consistent then PA ⊬ ¬Con. So what is the special significance of finding this new undecidable sentence?

Here is how Gödel interprets his generalized version of the Second Theorem:

> [For a suitable consistent theory κ] the sentential formula stating that κ is consistent is not κ-provable; in particular, the consistency of P is not provable in P, provided P is consistent. (Gödel, 1931, p. 193)

So the analogous gloss on our Second Theorem for PA would be: *the consistency of* PA *is not* PA-*provable, assuming* PA *is consistent.* And that's how we put things in the last section.

(b) But is that quite right? We might remark that Con is *one* natural way of 'indirectly saying' that PA is consistent. But we haven't shown that it is the *only*

---

[2]There's a related theorem: PA can also prove the consistency of any theory whose instances of the Induction Schema are restricted to wffs up to a certain level of quantifier complexity, Σ_n for some n (see Section 11.5(e) for the idea of a Σ_n wff, for which the paradigm is a wff $\exists x \forall y \exists z \forall w \ldots \varphi$ with n initial alternating quantifiers and φ a Δ_0 wff). For much more on this sort of result, see Hájek and Pudlák (1993, ch. 1, §4).

[3]For another line of proof, see (Kaye, 1991, p. 132).

way of expressing PA's consistency in L_A.[4] So although PA can't prove Con, the question remains: couldn't it still prove some *other* sentence which states that PA is consistent?

Here are two obvious alternative ways of expressing PA's consistency. First, we could put

$$\mathsf{Con'} =_{\mathrm{def}} \neg\exists x(\mathsf{Prov}(x) \land \mathsf{Prov}(\dot{\neg}x)),$$

where the dotted negation added to L_A is such that '$\dot{\neg}x$' expresses the p.r. function that maps the g.n. of a wff φ to the g.n. of its negation $\neg\varphi$.[5] Second, sticking to pure L_A and relying on the fact that inconsistent classical theories 'prove' every sentence, we could define

$$\mathsf{Con''} =_{\mathrm{def}} \exists x(\mathsf{Sent}(x) \land \neg\mathsf{Prov}(x)),$$

where $\mathsf{Sent}(x)$ captures the p.r. property of being a closed wff of PA's language (this happens to be Gödel's own preferred type of consistency sentence).

Still, as you'd probably expect, we can show without too much effort that PA $\vdash$ Con $\leftrightarrow$ Con', and PA $\vdash$ Con $\leftrightarrow$ Con''. And being provably equivalent to Con, these alternative consistency sentences are of course all equally unprovable in PA. It is therefore reasonable enough to think of Con as a *canonical* consistency statement for PA.

(c)   But still we can press the question: could there perhaps be other, more oblique and less obvious, ways of expressing PA's consistency in L_A? Well, there is a genuine issue here. However, let's not get tangled up in this further question now: instead, we will return to the topic in Section 36.1. For the moment we'll just announce that the headline news is this: *if we care about informative consistency proofs, the canonical consistency statement* Con *(or an equivalent) is what we'll want to prove.* And while we might well have expected an arithmetically rich consistent theory like PA to 'know about' its own consistency (when we code up that claim), it can't. In fact – as Theorem 33.3, the generalized version of the Second Theorem, shows – if *any* rich enough theory 'proves' its own canonical consistency sentence, it must be *inconsistent*.

---

[4]Of course, Con, like G, unpacks into pure L_A as just another horribly complicated arithmetical sentence, whose cumbersome details will be contingent on the vagaries of our choice of coding. This unpacked L_A sentence will have no more intrinsic mathematical interest than the unpacked version of G. It is only when we look at it through the lens of coding that Con becomes interesting, and 'indirectly says' that PA is consistent.

[5]See Section 25.3, fn. 3 for more explanation of our dotting notation.

32 On the 'unprovability of consistency'

So what can we learn from the fact that PA cannot prove its own consistency? What constraints does the result put on the possibility of informative consistency proofs for PA?

32.1 Three corollaries

You might initially think: 'OK, so we can't derive Con in PA. But that fact is of course no evidence at all *against* PA's consistency, since we already know from the First Theorem that *lots* of true claims about underivability are themselves underivable in PA: for example, PA can't prove the true claim ¬Prov(G). While if, *per impossibile*, we could have given a PA proof of Con, that wouldn't have given us any special evidence *for* PA's consistency either – because even if PA were inconsistent we'd still be able to derive Con, since we can derive *anything* in an inconsistent theory! Hence the derivability or otherwise of a canonical statement of PA's consistency inside PA itself can't show us a great deal.'

That's correct – except for the final conclusion which doesn't follow. For the Second Theorem *does* yield three plainly important and substantial corollaries:

(a) First, the Theorem tells us that *even* PA isn't enough to derive the consistency of PA, so we certainly can't derive the consistency of PA using a *weaker* theory.

(b) The Theorem also tells us that PA isn't enough to derive the consistency of *even* PA, so we certainly can't use PA to demonstrate the consistency of a *stronger* theory.

(c) The Theorem further implies that if we *are* going to produce any interestingly informative consistency proof for PA then, since a weaker theory isn't up to the job and using a stronger theory which fully subsumes PA would presumably be question-begging, we'll need to use a theory which is weaker in some respects and stronger in others.

We develop the first two points over the next two sections. Then we comment on the third point by briefly describing Gentzen's consistency proof for PA.

32.2 Weaker theories cannot prove the consistency of PA

Since we are now going to be talking about different theories we use subscripts to keep track: so Con_T is the canonical consistency statement for T, constructed on the same lines as $\mathsf{Con}_{\mathsf{PA}}$ (our original sentence Con for PA). Corollary (a) of our key Theorem 31.1 is then:

Theorem 32.1 *If T is a consistent sub-theory of* PA *then* $T \nvdash$ $\mathsf{Con}_{\mathsf{PA}}$.

Evidently, if the sub-theory T doesn't have enough of the language of basic arithmetic, then it won't even be able to frame the PA-wff we've abbreviated $\mathsf{Con}_{\mathsf{PA}}$; so it certainly won't be able to prove it. The more interesting case is where T is a theory which does share the language of PA but doesn't have all the induction axioms and/or uses a weaker deductive system than classical first-order logic. Such a theory T can't prove *more* than PA. So, by Theorem 31.1, assuming T is consistent, T can't prove $\mathsf{Con}_{\mathsf{PA}}$ either.

Recall our brief discussion in Section 13.4, where we first raised the issue of PA's consistency. We noted that arguing for consistency by appealing to the existence of a putative interpretation that makes all the axioms true might perhaps be thought risky (we might worry that the appeal is potentially vulnerable to the discovery that our intuitions are deceptive and that there is a lurking incoherence in the interpretation). So, if we are moved by such considerations, the question naturally arises whether we can give a demonstration of PA's consistency that depends on something supposedly more secure. And once we've got the idea of coding up facts about provability using Gödel numbering, we might wonder whether we could, so to speak, lever ourselves up to establishing PA's consistency (not its truth, but at least its consistency) by assuming the truth of some weaker arithmetic such as the minimally inductive theory $\mathsf{I}\Delta_0$ and trying to prove $\mathsf{Con}_{\mathsf{PA}}$ in that theory. Theorem 32.1 shows that this can't be done.

32.3   PA cannot prove the consistency of stronger theories

Corollary (b) of the Second Theorem for PA is:

Theorem 32.2 *If T extends* PA, *then* PA $\nvdash$ Con_T.

That's because, if PA could establish the consistency of the stronger theory T, it would thereby establish the consistency of PA as part of that theory, contrary to the Second Theorem for PA.[1]

We will be returning to consider the significance of this second corollary in the next Interlude. It matters crucially for the assessment of Hilbert's Programme, which we briefly mentioned in Section 1.6: for that Programme turns on the hope that we might show that a theory with a highly infinitary subject-matter

---

[1] If T is presented so its axioms explicitly include those of PA, then Theorem 32.2 follows immediately from Theorem 34.1. It can be left as a testing exercise to think how to prove our theorem here in the general case where PA's axioms are T-theorems but not necessarily given as T-axioms.

is consistent by using non-infinitary reasoning about the theory. But the headline point is already clear: we *can't* take some problematic rich theory which extends arithmetic (set theory, for example) and show that it is consistent by (i) talking about its proofs using arithmetical coding tricks and then (ii) using less contentious reasoning already available in some weaker, purely arithmetical, presumptively consistent theory like PA.

32.4 Introducing Gentzen

Back in Section 13.4 we stressed that, just because we *can* go wrong in judging a theory's consistency, we shouldn't be too quick to cast doubt on PA's consistency. What reason have we to suppose that it might really be in trouble?

Still, we might wonder whether there are illuminating and informative ways of proving PA to be consistent. Trying to do this by appealing to a *stronger* theory which already contains PA won't be a good strategy for quieting doubts (for doubts about PA will presumably carry over to become doubts about the stronger theory). And the Second Theorem shows that it is impossible to prove PA's consistency by appealing to a *weaker* theory which is contained inside PA. But another possibility remains open (this was corollary (c) that we mentioned at the beginning of the chapter). It isn't ruled out that we can prove PA's consistency by appeal to an attractive theory which is weaker than PA in some respects but stronger in others.

And this is what Gerhard Gentzen gives us in his consistency proof for arithmetic.[2] To explore his type of argument at any length would sadly take us far too far away from our main themes: but it is perhaps worth pausing to indicate the main idea – though nothing later depends on this, so by all means skip on to the next chapter.

Here then is a very sketchy outline, due to Gentzen himself. As with our sketched proof of Goodstein's Theorem in Section 30.2 and the Kirby-Paris Theorem in Section 30.3, if you don't know much about the ordinals, you probably won't carry much away from these few brief hints. But don't worry: it is only the overall *structure* of the argument that we are really interested in here.

We start with a 'sequent calculus' formulation of PA, a formulation which is easily seen to be equivalent to our official Hilbert-style version. Then, Gentzen writes,

> The 'correctness' of a proof [in particular, the lack of contradiction] depends on the correctness of certain other simpler proofs contained in it as special cases or constituent parts. This fact motivates the arrangement of proofs in linear *order* in such a way

---

[2] Gentzen's key papers (1936, 1938) are difficult: but the headline news is given in a wonderfully clear way in his wide-ranging lecture on 'The concept of infinity in mathematics' (Gentzen, 1937, pp. 230–233, from which we are quoting here). For later variants on Gentzen's proof, see Mendelson (1964, Appendix), Pohlers (1989, ch. 1), and Takeuti (1987, ch. 2).

that those proofs on whose correctness the correctness of another proof depends precede the latter proof in the sequence. This arrangement of the proofs is brought about by correlating with each proof a certain transfinite ordinal number.

The idea, then, is that the various proofs in this version of PA can be put into an ordering by a kind of 'dependency' relation, with more complex proofs coming after simpler proofs. But why is the relevant linear ordering of proofs said to be *transfinite* (in other words, why must it allow an item in the ordering to have an infinite number of predecessors)? Because

> [it] may happen that the correctness of a proof depends on the correctness of infinitely many simpler proofs. An example: Suppose that in the proof a proposition is proved for *all* natural numbers by complete [i.e. course-of-values] induction. In that case the correctness of the proof obviously depends on the correctness of every single one of the infinitely many individual proofs obtained by specializing to a particular natural number. Here a natural number is insufficient as an ordinal number for the proof, since each natural number is preceded by only finitely many other numbers in the natural ordering. We therefore need the transfinite ordinal numbers in order to represent the natural ordering of the proofs according to their complexity.

Think of it this way: a proof by course-of-values induction of $\forall x \varphi(x)$ leaps beyond all the proofs of $\varphi(0)$, $\varphi(1)$, $\varphi(2)$, ... And the result $\forall x \varphi(x)$ depends for its correctness on the correctness of the simpler results. So, in the sort of ordering of proofs which Gentzen has in mind, the proof by induction of $\forall x \varphi(x)$ must come infinitely far down the list, after all the proofs of the various $\varphi(\bar{n})$.

And now Gentzen's key step is to argue by an induction along this transfinite ordering of proofs. The very simplest proofs right at the beginning of the ordering transparently can't lead to contradiction. Then it is easy to see that

> once the correctness of all proofs preceding a particular proof in the sequence has been established, the proof in question is also correct precisely because the ordering was chosen in such a way that the correctness of a proof depends on the correctness of certain earlier proofs. From this we can now obviously infer the correctness of all proofs by means of a transfinite induction, and we have thus proved, in particular, the desired consistency.

Transfinite induction here is just the principle that, if we can show that a proof has a property P if all its predecessors in the ordering do, then *all* proofs in the ordering have property P.[3]

---

[3]See Section 30.3, fn. 4.

But what kind of transfinite ordering is involved here? Gentzen's ordering of possible proof-trees in his sequent calculus for PA turns out to have the order type of the ordinals less than ε_0 (i.e. of the ordinals which are sums of powers of ω). So, what Gentzen's proof needs is the assumption that a relatively modest amount of transfinite induction – induction up to ε_0 – is legitimate.

Now, the PA proofs which we are ordering are themselves all finite objects; we can code them up using Gödel numbers in the now familiar sort of way. So in ordering the proofs, we are in effect thinking about a whacky ordering of (ordinary, finite) code numbers. And whether one number precedes another in the whacky ordering is nothing mysterious; a computation without open-ended searches can settle the matter.

So what resources does a Gentzen-style argument use, if we want to code it up and formalize it? The assignment of a place in the ordering to a proof can be handled by p.r. functions, and facts about the dependency relations between proofs at different points in the ordering can be handled by p.r. functions too. A theory in which we can run a formalized version of Gentzen's proof will therefore be one in which we can (a) handle p.r. functions *and* (b) handle transfinite induction up to ε_0, maybe via coding tricks. It turns out to be enough to have all p.r. functions 'built in' (in the way that addition and multiplication are built into Q) together with transfinite induction just for simple quantifier-free wffs containing expressions for these p.r. functions. Such a theory is neither contained in PA (since it can prove $\mathsf{Con_{PA}}$ by Gentzen's method, which PA can't), nor does it contain PA (since it needn't be able to prove instances of the ordinary Induction Schema for arbitrarily complex wffs).

So, in this sense, we can indeed prove the consistency of PA by using a theory which is weaker in some respects and stronger in others.[4]

32.5 What do we learn from Gentzen's proof?

Of course, it is a very moot point whether – if you were *really* worried about the consistency of PA – a Gentzen-style proof when fully spelt out would help resolve your doubts. For example, if you are globally worried about the use of induction in general, then appealing to an argument which deploys an induction principle won't help! But global worries about induction are difficult to motivate, and perhaps your worry is more specifically that induction over arbitrarily complex wffs might engender trouble. You note that PA's induction principle applies, inter alia, to wffs φ that themselves quantify over all numbers; and you might worry that if you understand the numbers to be what induction applies to (cf. Frege's definition, mooted in Section 18.1) then there's a looming circularity here – numbers are what induction applies to, but understanding some cases of

---

[4]Later, in a rather opaque paper, Gödel (1958) offers another consistency proof, by appeal to a quite different theory which is neither weaker nor stronger than PA. For explanation, see the opening pages of the challenging Avigad and Feferman (1998).

induction involves understanding quantifying over numbers. If *that* is your worry, the fact that we can show that PA is consistent using an induction principle which is only applied to quantifier-free wffs (even though the induction runs over a novel ordering on the numbers) could soothe your worries.

Be that as it may: the Gentzen proof is a fascinating achievement, containing the seeds of wonderful modern work in proof theory. But the full story will have to be left for another occasion. We need to get back to the basics of the Second Theorem.

33 Generalizing the Second Theorem

In the Chapter 31, we gave some headline news about the Second Theorem for PA, and about how it rests on the Formalized First Theorem. In this chapter, we'll say something about how the Formalized First Theorem is proved. More exactly, we state the so-called Hilbert-Bernays-Löb derivability conditions on the provability predicate for theory T and show that these suffice for proving the Formalized First Theorem for T, and hence the Second Theorem.

33.1 More notation

To improve readability, we now introduce some neat notation:

We will henceforth abbreviate $\mathsf{Prov}_T(\ulcorner\varphi\urcorner)$ simply by $\Box_T\varphi$.

So the wff Con_T can alternatively be abbreviated as $\neg\Box_T\bot$.[1]

Two comments. First, note that our box symbol actually does a double job: it further abbreviates the long predicative expression already abbreviated by Prov_T, *and* it absorbs the corner quotes that turn a wff into the standard numeral for that wff's Gödel number. If you are logically pernickety, then you might be rather upset about introducing a notation which in this way rather disguises the complex logical character of what is going on.[2] But my line is that abbreviatory convenience here trumps notational perfectionism.

Second, we will very often drop the explicit subscript from the box symbol, and let context supply it. We'll also drop other subscripts in obvious ways. For example, here's a snappy version of the Formalized First Theorem for theory T:

$$T \vdash \mathsf{Con} \to \neg\Box\mathsf{G}.$$

We already take the undecorated turnstile '$\vdash$' to signify by default the existence of a proof in T's deductive system. Similarly we will now take the undecorated

---

[1] If you are familiar with modal logic, you will immediately recognize the conventional symbol for the necessity operator. And the parallels and differences between "'$1+1=2$' is provable (in T)' and 'It is necessarily true that $1 + 1 = 2$' are highly suggestive. They are investigated by 'provability logic', the subject of a contemporary classic (Boolos, 1993).

[2] One beef is this. The notation '$\Box\varphi$' *looks* as if it ought to be a complex wff embedding φ, so that as φ increases in logical complexity, so does $\Box\varphi$. But not so. However complex φ is, $\ulcorner\varphi\urcorner$ is just a numeral and $\Box\varphi$, i.e. $\mathsf{Prov}_T(\ulcorner\varphi\urcorner)$, stays resolutely a Σ_1 wff.

The logically kosher approach is not to regard the box as an abbreviation, but to introduce a *new* modal language, and then explore a mapping relation that links modal sentences to arithmetical 'realizations' via a $\Box/\mathsf{Prov}$ link. For a properly careful treatment of this, see (Boolos, 1993) again.

occurrences of 'Con', and '¬□G' to indicate by default the canonical consistency sentence Con_T and the unprovability-in-T of the Gödel sentence G_T for that same theory.

33.2   The Hilbert-Bernays-Löb derivability conditions

(a)   We haven't yet *proved* the Formalized First Theorem for PA. And, as we remarked before, Gödel himself didn't prove the corresponding result for his particular formal theory P. The hard work was first done by David Hilbert and Paul Bernays in their *Grundlagen der Mathematik* (1939): the details of their proof are in fact due to Bernays, who had discussed it with Gödel during a transatlantic voyage.

Hilbert and Bernays helpfully isolated three conditions on the predicate Prov_T, conditions whose satisfaction is enough for a nice theory T to prove the Formalized First Theorem. Later, Martin H. Löb (1955) gave a rather neater version of these conditions. Here they are in Löb's version, and in our snappy notation:

C1. If $T \vdash \varphi$, then $T \vdash \Box\varphi$,

C2. $T \vdash \Box(\varphi \to \psi) \to (\Box\varphi \to \Box\psi)$,

C3. $T \vdash \Box\varphi \to \Box\Box\varphi$,

where φ and ψ are, of course, any sentences. Henceforth, we'll call these the *(HBL) derivability conditions*. We'll see in Chapter 35 why they hold for PA.

We note a little warm-up lemma:

> **Theorem 33.1** *If φ and ψ are logically equivalent, and the derivability conditions hold, then $T \vdash \Box\varphi \leftrightarrow \Box\psi$.*

Proof   By hypothesis $T \vdash \varphi \to \psi$, hence by (C1), $T \vdash \Box(\varphi \to \psi)$, so by (C2) $T \vdash \Box\varphi \to \Box\psi$. Similarly for the other half of the biconditional.   ⊠

(b)   Before showing that these conditions indeed do the job, let's consider why we might expect such conditions to suffice.

Start from Theorem 24.3, the special case of the Diagonalization Lemma. That biconditional gives us, in our new notation, both

D1. $T \vdash \mathsf{G} \to \neg\Box\mathsf{G}$,

D2. $T \vdash \neg\Box\mathsf{G} \to \mathsf{G}$.

And here again is a proof of the first part of the First Theorem (cp. Section 24.5). The principle (C1) – already proved in Theorem 24.1 – tells us that if $T \vdash \mathsf{G}$ then $T \vdash \Box\mathsf{G}$. So, by (D1), if $T \vdash \mathsf{G}$, it is immediate that T is inconsistent. Therefore, assuming T is consistent, $T \nvdash \mathsf{G}$.

Now, given that (D1) and (C1) immediately imply half of the First Theorem, we might reasonably expect to be able to argue from the claims that T *'knows' that (D1) holds* and T *'knows' that (C1) holds* to the conclusion T *'knows' that half the First Theorem holds.*

Let's put that more carefully! The thought that T 'knows' that (D1) holds is tantamount to

K. $T \vdash \Box(\mathsf{G} \to \neg\Box\mathsf{G})$.

But that already follows from (D1) by (C1). The thought that T 'knows' that (C1) holds is captured by (C3). The hoped for inference is therefore from (K) and (C3) to the Formalized First Theorem. That should go through so long as T is able to cope with the idea that what follows from provable propositions is itself provable. Which is just what (C2) requires.

So, putting everything together, the three derivability conditions – together with (D1), which is provable for any nice theory – look as if they ought to give us the Formalized First Theorem.

(c)   And indeed they do, as we'll now prove.

Theorem 33.2 *If T is nice and the derivability conditions hold, then $T \vdash \mathsf{Con} \to \neg\Box\mathsf{G}$.*

Proof   Simple logic plus the definition of '$\bot$' ensures that, for any wff φ,

$$T \vdash \neg\varphi \to (\varphi \to \bot).$$

Given the latter and (C1), this means

$$T \vdash \Box(\neg\varphi \to (\varphi \to \bot)).$$

So given (C2) and using modus ponens, it follows that for any φ

A. $T \vdash \Box\neg\varphi \to \Box(\varphi \to \bot)$.

We now argue as follows:

1.	$T \vdash \mathsf{G} \to \neg\Box\mathsf{G}$	D1 again
2.	$T \vdash \Box(\mathsf{G} \to \neg\Box\mathsf{G})$	From 1, given C1
3.	$T \vdash \Box\mathsf{G} \to \Box\neg\Box\mathsf{G}$	From 2, given C2
4.	$T \vdash \Box\neg\Box\mathsf{G} \to \Box(\Box\mathsf{G} \to \bot)$	Instance of A
5.	$T \vdash \Box\mathsf{G} \to \Box(\Box\mathsf{G} \to \bot)$	From 3 and 4
6.	$T \vdash \Box\mathsf{G} \to (\Box\Box\mathsf{G} \to \Box\bot)$	From 5, given C2
7.	$T \vdash \Box\mathsf{G} \to \Box\Box\mathsf{G}$	Instance of C3
8.	$T \vdash \Box\mathsf{G} \to \Box\bot$	From 6 and 7
9.	$T \vdash \neg\Box\bot \to \neg\Box\mathsf{G}$	Contraposing
10.	$T \vdash \mathsf{Con} \to \neg\Box\mathsf{G}$	Definition of Con.

Which gives us our general version of the Formalized First Theorem.   ⊠

And that of course immediately yields a version of the *Second Incompleteness Theorem for T*:

Theorem 33.3 *If T is nice and the derivability conditions hold, then $T \nvdash$ Con.*

Proof The argument goes as before:

1.	$T \vdash$ Con $\to \neg\Box$G	Just proved
2.	$T \vdash \neg\Box$G $\to$ G	D2
3.	$T \vdash$ Con $\to$ G	From 1 and 2
4.	$T \nvdash$ G	The First Theorem!
5.	$T \nvdash$ Con	From 3 and 4. ⊠

33.3 T's ignorance about what it can't prove

A nice theory T is omniscient about what it *can* prove: if $T \vdash \varphi$, then $T \vdash \Box\varphi$. To repeat, that's (C1), already proved as part of Theorem 24.1.

But T is totally ignorant about what it *can't* prove. Even if $T \nvdash \varphi$, we never get $T \vdash \neg\Box\varphi$.

To show this, we first note:

Theorem 33.4 *If T is nice and the derivability conditions hold, then for any sentence φ, $T \vdash \neg\Box\varphi \to$ Con.*

Proof We argue as follows:

1.	$T \vdash \bot \to \varphi$	By definition of $\bot$
2.	$T \vdash \Box(\bot \to \varphi)$	From 1, given C1
3.	$T \vdash \Box\bot \to \Box\varphi$	From 2, given C2
4.	$T \vdash \neg\Box\varphi \to \neg\Box\bot$	Contraposing
5.	$T \vdash \neg\Box\varphi \to$ Con	Definition of Con. ⊠

So T knows that *if* it can't prove φ *then* it must be consistent. So if T could indeed demonstrate that it can't prove φ, it could prove its own consistency, contrary to the Second Theorem, i.e. Theorem 33.3. Which shows that

Theorem 33.5 *If T is nice and the derivability conditions hold, then for no sentence φ do we have $T \vdash \neg\Box\varphi$.*

33.4 The Formalized Second Theorem

(a) One half of the First Theorem says that, for suitable T, if T is consistent, then T doesn't prove G. We can express that claim inside T with the sentence Con $\to \neg\Box$G. And *that* sentence can in fact be proved by T itself: $T \vdash$ Con $\to \neg\Box$G. We've called that the Formalized First Theorem.

The Second Theorem says that, for suitable T, if T is consistent, then T doesn't prove Con. We can express that claim too inside T, with the sentence

Con → ¬□Con. We can now show that this again is provable by T itself, for suitable theories. In other words, we have

Theorem 33.6 *If T is nice and the derivability conditions hold, then $T \vdash$ Con → ¬□Con.*

We can naturally call this *the Formalized Second Theorem*. Its proof is almost immediate from previous results:

1.	$T \vdash$ Con → ¬□G	Cf. proof of Thm 33.2
2.	$T \vdash$ ¬□G → G	D2
3.	$T \vdash$ Con → G	From 1, 2
4.	$T \vdash$ □(Con → G)	From 3, by C1
5.	$T \vdash$ □Con → □G	From 4, by C2
6.	$T \vdash$ Con → ¬□Con	From 1 and 5. ⊠

(b)   Theorem 33.4 tells us that $T \vdash$ ¬□Con → Con. Which together with our last theorem and with the box unpacked establishes $T \vdash$ Con ↔ ¬Prov(⌜Con⌝). Hence

Theorem 33.7 *If T is nice and the derivability conditions hold, then Con_T is a fixed point of $\neg\mathrm{Prov}_T$.*

Suppose we do use the phrase 'Gödel sentence' in a wide sense to refer to any fixed point for $\neg\mathrm{Prov}_T$ (see Section 24.6). Then we've just shown that there are Gödel sentences, like Con, which are *not* constructed using self-referential trickery. But, as we noted in Section 24.5, *any* fixed point for $\neg\mathrm{Prov}_T$ will be formally undecidable (assuming T is ω-consistent). *So there are formally undecidable sentences which aren't self-referential.* That observation ought to scotch once and for all any lingering suspicion that the incompleteness phenomena are somehow inevitably tainted by self-referential paradox.

33.5   Jeroslow's Lemma and the Second Theorem

We end this chapter with an aside for enthusiasts: we introduce a slightly different route to the Second Theorem which is due to R. G. Jeroslow (1973).

One pay-off is that we no longer rely on the derivability condition (C2) – though given that that condition is not the difficult one to establish for sensibly constructed provability predicates, that's perhaps not really a big gain. Still, Jeroslow's Lemma is quite fun to know about and it is mentioned from time to time in the secondary literature. So here goes . . .

(a)   Let's remind ourselves of the classic version of the Diagonalization Lemma:

Theorem 24.4 *If T is a nice theory and $\varphi(\mathsf{x})$ is any wff of its language with one free variable, then there is a sentence γ of T's language such that $T \vdash \gamma \leftrightarrow \varphi(⌜\gamma⌝)$.*

And now here's the variant Lemma due to Jeroslow. Let's say T is a *nice$^+$* theory if it is nice and also has a built-in function expression for each p.r. function. (Recall: a nice theory can strongly capture every p.r. function by some relational expression – by Theorems 16.4 and 17.2. And as noted in Section 16.3, when the condition for strong capturing is satisfied, we can harmlessly add to the theory's language a functional expression defined in terms of the capturing relation. That means that any nice theory can readily be made nice$^+$ by definitional extension while essentially still saying the same.)

> **Theorem 33.8** *If T is a nice$^+$ theory, and $\varphi(\mathsf{x})$ is any wff of its language with one free variable, then there is a term τ of T's language such that $T \vdash \tau = \ulcorner\varphi(\tau)\urcorner$.*

So while the original Diagonalization Lemma was about provable *equivalences*, Jeroslow's Diagonalization Lemma is about provable *identities*. The argument is very quick.

Proof   Given an open wff $\varphi(\mathsf{x})$, we can define a corresponding diagonalizing function d which takes the g.n. of a function symbol f, and returns the g.n. of the wff $\varphi(\mathsf{f}(\ulcorner\mathsf{f}\urcorner))$ (for arguments which aren't codes for monadic function symbols, the value of d can default to zero). This function is evidently p.r. because its value can be computed without unbounded loopings.

Since d is p.r., it is captured in T by some function symbol d. Hence by its definition, $d(\ulcorner\mathsf{d}\urcorner) = \ulcorner\varphi(\mathsf{d}(\ulcorner\mathsf{d}\urcorner))\urcorner$. And therefore, by the definition of capturing, $T \vdash \mathsf{d}(\ulcorner\mathsf{d}\urcorner) = \ulcorner\varphi(\mathsf{d}(\ulcorner\mathsf{d}\urcorner))\urcorner$. Put $\tau = \mathsf{d}(\ulcorner\mathsf{d}\urcorner)$ and Jeroslow's Lemma is proved.   ⊠

(b)   For a preliminary application of this Lemma, take the instance of our new theorem where $\varphi(\mathsf{x}) =_{\text{def}} \neg\mathsf{Prov}_T(\mathsf{x})$. So, assuming T satisfies the condition in the Lemma, there's a term τ such that

1. $T \vdash \tau = \ulcorner\neg\mathsf{Prov}_T(\tau)\urcorner$.

Dropping subscripts for readability, put $\mathsf{G}' =_{\text{def}} \neg\mathsf{Prov}(\tau)$, and suppose

2. $T \vdash \mathsf{G}'$.

That means that for some number m, $Prf(m, \ulcorner\neg\mathsf{Prov}(\tau)\urcorner)$ is true (just by the definition of the *Prf* relation). Hence, since Prf captures *Prf* in T, we have

3. For some m, $T \vdash \mathsf{Prf}(\overline{m}, \ulcorner\neg\mathsf{Prov}(\tau)\urcorner)$.

Whence, using (1), we get

4. For some m, $T \vdash \mathsf{Prf}(\overline{m}, \tau)$.

But since G' is equivalent to $\forall\mathsf{v}\neg\mathsf{Prf}(\mathsf{v}, \tau)$, (4) contradicts (2). So, given that T is consistent, (2) must be false: T cannot prove G'. Similarly, we can show that T cannot prove $\neg\mathsf{G}'$, on pain of being ω-inconsistent (exercise: check that claim).

In sum, a version of the First Theorem quickly follows again from Jeroslow's version of the Diagonalization Lemma.

(c) That was mildly diverting, but doesn't tell us anything novel. However, we can now use Jeroslow's Lemma to derive a version of the Second Theorem, and this time there *is* a new twist.

First, let's write '$\dot{\neg}x$' as T's functional expression for the p.r. function which maps the g.n. of a wff φ to the g.n. of its negation: thus $\dot{\neg}\ulcorner\varphi\urcorner = \ulcorner\neg\varphi\urcorner$ (cf. Section 25.3, fn. 3). And, as we noted in Section 31.6, one natural way of expressing a theory's consistency is

$$\mathsf{Con}'_T =_{\mathrm{def}} \neg\exists x(\mathsf{Prov}_T(x) \wedge \mathsf{Prov}_T(\dot{\neg}x)).$$

Let's suppose for reductio that

   1. $T \vdash \mathsf{Con}'_T$.

And now take the wff $\varphi(x)$ in Jeroslow's Lemma to be $\mathsf{Prov}_T(\dot{\neg}x)$. So, dropping subscripts again, for some term τ,

   2. $T \vdash \tau = \ulcorner\mathsf{Prov}(\dot{\neg}\tau)\urcorner$.

We can then argue as follows, assuming that conditions (C1) and (C3) apply to Prov_T and using the fact that nice$^+$ theories have some basic logic.

3.	$T \vdash \mathsf{Prov}(\dot{\neg}\tau) \to \mathsf{Prov}(\dot{\neg}\tau)$	Logic!
4.	$T \vdash \mathsf{Prov}(\dot{\neg}\tau) \to \mathsf{Prov}(\dot{\neg}\ulcorner\mathsf{Prov}(\dot{\neg}\tau)\urcorner)$	From 2, 3
5.	$T \vdash \mathsf{Prov}(\dot{\neg}\tau) \to \mathsf{Prov}(\ulcorner\mathsf{Prov}(\dot{\neg}\tau)\urcorner)$	Instance of C3
6.	$T \vdash \mathsf{Prov}(\dot{\neg}\tau) \to$	
	$\quad (\mathsf{Prov}(\ulcorner\mathsf{Prov}(\dot{\neg}\tau)\urcorner) \wedge \mathsf{Prov}(\dot{\neg}\ulcorner\mathsf{Prov}(\dot{\neg}\tau)\urcorner))$	From 4 and 5
7.	$T \vdash \mathsf{Prov}(\dot{\neg}\tau) \to \neg\mathsf{Con}'$	From 6, defn. Con$'$
8.	$T \vdash \neg\mathsf{Prov}(\dot{\neg}\tau)$	From 1, 7
9.	$T \vdash \mathsf{Prov}(\ulcorner\neg\mathsf{Prov}(\dot{\neg}\tau)\urcorner)$	From 8, by C1
10.	$T \vdash \mathsf{Prov}(\dot{\neg}\ulcorner\mathsf{Prov}(\dot{\neg}\tau)\urcorner)$	Defn. of $\dot{\neg}$
11.	$T \vdash \mathsf{Prov}(\dot{\neg}\tau)$	From 2, 10
12.	Contradiction!	From 8, 11.

So we can conclude that (1) can't be true: $T \nvdash \mathsf{Con}'_T$.[3]

(d) Which all goes to establish the following theorem:

> **Theorem 33.9** *If T is nice$^+$ and the derivability conditions (C1) and (C3) hold, then $T \nvdash \mathsf{Con}'_T$.*

Hence (C2) *doesn't* always have to hold for a provability predicate Prov_T for the corresponding sentence Con'_T to be unprovable in T: it isn't an *essential* derivability condition for a version of the Second Theorem to apply. Neat!

---

[3]An exercise to check your understanding: why haven't we made our argument look prettier by using our abbreviatory $\Box$ symbol?

34 Löb's Theorem and other matters

We now draw an assortment of interesting consequences from the assumption
that a provability predicate satisfies the derivability conditions and from the
resulting Second Theorem (we delay until the next chapter the business of investigating which theories have provability predicates that *do* satisfy the derivability
conditions). The high point of this chapter is Löb's Theorem. But that comes
later: first, we say something about ...

34.1 Theories that 'prove' their own inconsistency

(a) If PA is ω-consistent, it can't prove $\neg\mathsf{Con}$. That was Theorem 31.2: and
inspecting the easy proof we see that the result generalizes: an ω-consistent T
can't prove $\neg\mathsf{Con}_T$. By contrast, a consistent but *ω-inconsistent* T might well
have $\neg\mathsf{Con}_T$ as a theorem.

The proof of this odd result is also easy, once we note a simple lemma. Assume
a Gödel-numbering scheme is fixed. Then

> **Theorem 34.1** *Suppose S and R are two p.r. axiomatized theo-
> ries, where R is defined as being theory S augmented with addi-
> tional axioms. Then $\vdash \mathsf{Con}_R \to \mathsf{Con}_S$.*

Evidently, if the richer R is consistent, then the simpler S must be consistent
too. The theorem says that the canonical arithmetical expression of this fact is
a logical theorem. Why so?

Proof The relation $Prfs_S$ is defined from $Prfseq_S$ which has a definition including
$Axiom_S$ as a disjunct (see Section 20.4). And $Axiom_S$ – the property which holds
of n if n numbers an S-axiom – is itself defined disjunctively, with a clause for
each type of S-axiom as given in the definition of theory S.

The definition of Prf_R will be exactly similar, except that it instead embeds
$Axiom_R$, which is again defined disjunctively, still with a clause for each type of
S-axiom but with additional clauses for the additional axioms presented in the
definition of theory R.[1]

Now Prf_S canonically captures $Prfs_S$ by tracking its p.r. definition: so inside Prf_S there will be a disjunct featuring Axiom_S (which canonically captures
$Axiom_S$). Likewise Prf_R will be built in the same way except for a disjunct featuring Axiom_R (which canonically captures $Axiom_R$). Since Axiom_R is just Axiom_S
with added disjuncts, $\mathsf{Axiom}_S(\mathsf{x}) \vdash \mathsf{Axiom}_R(\mathsf{x})$. Whence $\mathsf{Prf}_S(\mathsf{x},\mathsf{y}) \vdash \mathsf{Prf}_R(\mathsf{x},\mathsf{y})$.

---

[1]So here we do rely on R's being explicitly presented as an augmentation of S.

The contrapositive of our lemma now immediately follows. For suppose $\neg\mathsf{Con}_S$, i.e. suppose $\exists \mathsf{v}\, \mathsf{Prf}_S(\mathsf{v}, \bot)$. Hence for some a, $\mathsf{Prf}_S(\mathsf{a}, \bot)$. That implies $\mathsf{Prf}_R(\mathsf{a}, \bot)$. So it follows that $\exists \mathsf{v}\, \mathsf{Prf}_R(\mathsf{v}, \bot)$, i.e. $\neg\mathsf{Con}_R$. And the inferences here only use first-order logic. ⊠

(b)   Let's put our lemma to use. Take the simpler theory S to be PA and let the richer theory R be PA augmented by the extra axiom $\neg\mathsf{Con}_{\mathsf{PA}}$.

By definition, we trivially have $R \vdash \neg\mathsf{Con}_{\mathsf{PA}}$. So using our lemma we can conclude $R \vdash \neg\mathsf{Con}_R$. R is ω-inconsistent (why? cf. Theorem 31.2). But it *is* consistent if PA is (if R proved a contradiction, i.e. $\mathsf{PA} + \neg\mathsf{Con}_{\mathsf{PA}} \vdash \bot$, then PA $\vdash \mathsf{Con}_{\mathsf{PA}}$, which by the Second Theorem would make PA inconsistent).

Putting everything together, we have

> **Theorem 34.2** *Assuming* PA *is consistent, the theory* $R =_{\mathrm{def}}$ PA $+ \neg\mathsf{Con}_{\mathsf{PA}}$ *is (i) consistent, though (ii) R 'proves' its own inconsistency, i.e. $R \vdash \neg\mathsf{Con}_R$, so (iii) the theory $R + \mathsf{Con}_R$ is inconsistent.*

What are we to make of these apparent absurdities? Well, giving the language of R its standard arithmetical interpretation, the theory is just wrong in what it says about its inconsistency! But on reflection that shouldn't be much of a surprise. Believing, as we no doubt do, that PA is consistent, we already know that the theory R gets things wrong right at the outset, since it has the false axiom $\neg\mathsf{Con}_{\mathsf{PA}}$. So R doesn't really *prove* (establish-as-true) its own inconsistency, since we don't accept the theory as correct on the standard interpretation.

34.2   The equivalence of fixed points for ¬Prov

(a)   Theorem 33.4 tells us that, assuming T is nice and the derivability conditions hold, $T \vdash \neg\square\mathsf{G} \to \mathsf{Con}$. Since $T \vdash \mathsf{G} \to \neg\square\mathsf{G}$, we therefore have $T \vdash \mathsf{G} \to \mathsf{Con}$. While in proving Theorem 33.3 we showed $T \vdash \mathsf{Con} \to \mathsf{G}$. Whence $T \vdash \mathsf{G} \leftrightarrow \mathsf{Con}$.

We can now generalize.

> **Theorem 34.3** *Given that the derivability conditions hold, any fixed point for $\neg\mathsf{Prov}_T(\mathsf{x})$ will be provably equivalent in T to the corresponding* Con_T.

Proof   Writing it with boxes, we assume $T \vdash \gamma \leftrightarrow \neg\square\gamma$ and need to derive $T \vdash \gamma \leftrightarrow \neg\square\bot$.

1.	$T \vdash \gamma \leftrightarrow \neg\square\gamma$	Premiss
2.	$T \vdash \gamma \to (\square\gamma \to \bot)$	From 1, by logic
3.	$T \vdash \square(\gamma \to (\square\gamma \to \bot))$	From 2, by C1
4.	$T \vdash \square\gamma \to \square(\square\gamma \to \bot))$	From 3, by C2
5.	$T \vdash \square\gamma \to (\square\square\gamma \to \square\bot)$	From 4, by C2

6. $T \vdash \Box\gamma \to \Box\Box\gamma$	From C3
7. $T \vdash \Box\gamma \to \Box\bot$	From 5 and 6
8. $T \vdash \bot \to \gamma$	Logic
9. $T \vdash \Box(\bot \to \gamma)$	From 8, by C1
10. $T \vdash \Box\bot \to \Box\gamma$	From 9, by C2
11. $T \vdash \Box\gamma \leftrightarrow \Box\bot$	From 7 and 10
12. $T \vdash \gamma \leftrightarrow \neg\Box\bot$	From 1 and 11. ⊠

(b)   Two quick but important comments on this result.

First, you might momentarily be tempted to think: 'Assuming T is nice, its canonical consistency sentence Con_T is true, as is its canonical Gödel sentence G_T. But we've just seen that all of T's other Gödel sentences – in the sense of all the fixed points for $\neg\mathsf{Prov}_T$ – are provably equivalent to these truths. So these other Gödel sentences (fixed points of $\neg\mathsf{Prov}$) must be true too.' But not so: that forgets that a nice T may be a consistent-but-*unsound* theory. Such a T will have some false Gödel sentences mixed in with the true ones (and hence some false biconditional theorems relating the false ones to the true ones), as we noted before in Section 24.6.

Second, as we will see in the next chapter, the derivability conditions hold for theories that contain PA, so they will hold for the theory $R =_{\mathrm{def}} \mathsf{PA} + \neg\mathsf{Con}_{\mathsf{PA}}$. Hence by Theorem 34.3, $R \vdash \mathsf{G}_R \leftrightarrow \mathsf{Con}_R$. Since by Theorem 34.2 $R \vdash \neg\mathsf{Con}_R$, it follows that $R \vdash \neg\mathsf{G}_R$. Hence the ω-inconsistent R also 'disproves' its own true canonical Gödel sentence. *That's why the requirement of ω-consistency – or at least 1-consistency – has to be assumed in proving the First Theorem, if we are to do it by constructing an original-style Gödel sentence like G_R.*

34.3   Consistency extensions

Let's pause over the provable equivalence in T of G_T and Con_T.

Assume PA is sound. If we add its canonical Gödel sentence G as a new axiom, the resulting theory PA + G trivially proves PA's Gödel sentence G. But the new theory remains sound and p.r. axiomatized, has its own canonical Gödel sentence $\mathsf{G}_{\mathsf{PA+G}}$, and so must remain incomplete. We can then, of course, construct a yet richer theory PA + G + $\mathsf{G}_{\mathsf{PA+G}}$, which can prove the Gödel sentences of both the two preceding theories: but this sound theory too has its own Gödel sentence and is incomplete. And so it goes. We can continue in the same vein, forming an infinite sequence of theories, each theory trivially proving the canonical Gödel sentences of its predecessors but remaining incomplete and incompletable.

Now, we've just shown that Con and G are provably equivalent in PA. Similarly for any nice theory T when the derivability conditions hold, and we'll later be showing that the derivability conditions indeed hold for theories extending PA. So our sequence of theories got by adding Gödel sentences is equivalent to the sequence PA, PA + Con, PA + Con + $\mathsf{Con}_{\mathsf{PA+Con}}$, $\ldots$, where the $n + 1$-th theory on the list adds the canonical consistency sentence of the n-th theory as a new axiom. So, each theory in the sequence proves the canonical consistency sentences of its predecessors, but still remains incomplete and incompletable.

The interest in putting things in terms of 'consistency extensions' rather than in terms of the addition of Gödel sentences lies in the following thought. Suppose again we accept that PA is sound. Then, reflecting on this, we'll readily come to accept that PA + Con is sound too – and we don't need to follow a Gödelian proof to get to *that* realization. And reflecting on that, we'll readily come to accept that PA + Con + Con$_{PA+Con}$ is sound too. And so on. In sum, starting from our initial thought that PA is sound, reflection drives us along the sequence of consistency extensions to accept that they are sound too: so there now seems a new naturalness to the process of extending PA, if you think in terms of adding consistency sentences rather than those tricksy Gödel sentences.

But the theories we arrive at by pursuing this natural line of thought *still* remain incomplete and incompletable. What if we consider 'consistency extensions' of PA which add not just a finite number of such consistency statements, but all of them together? What if we now add the consistency statement for that infinitely augmented theory, and keep on going again after that? No matter. So long as our theory remains p.r. axiomatized and sound (or indeed just consistent), it must remain incomplete.[2]

34.4   Henkin's Problem and Löb's Theorem

(a)   Our last theorem tells us about the fixed points of $\neg\mathsf{Prov}_T$: they are all provably equivalent, and hence all undecidable assuming T is nice and ω-consistent.

But what about fixed points for the unnegated Prov_T? By the Diagonalization Lemma, there is a sentence H such that $T \vdash \mathsf{H} \leftrightarrow \mathsf{Prov}(\ulcorner\mathsf{H}\urcorner)$: in other words, we can use diagonalization again to construct a sentence H that 'says' that it is provable.[3] Leon Henkin (1952) asked: *is* H provable?

This question was answered positively by Martin Löb (1955), who proved a more general result:

> **Theorem 34.4** *If T is nice and the derivability conditions hold, then if $T \vdash \Box\varphi \to \varphi$ then $T \vdash \varphi$.*

Since by definition $T \vdash \mathsf{Prov}(\ulcorner\mathsf{H}\urcorner) \to \mathsf{H}$ for Henkin's sentence H, i.e. $T \vdash \Box\mathsf{H} \to \mathsf{H}$, Löb's theorem immediately tells us that $T \vdash \mathsf{H}$, answering Henkin's problem.

As Kreisel (1965) first noted, this result also gives us another route to the Second Theorem. Suppose $T \vdash \mathsf{Con}$. Then $T \vdash \neg\mathsf{Con}_T \to \bot$, i.e. $T \vdash \Box\bot \to \bot$.

---

[2]Which is not at all to say that the general theory of consistency extensions isn't interesting, for there are troublesome complications, classically explored in Feferman (1960). For a more recent discussion, see Franzén (2004).

[3]If the Gödel sentence G is reminiscent of the Liar sentence 'This sentence is false', then Henkin's sentence H is reminiscent of the Truth-teller sentence 'This sentence is true'. For discussion of the Truth-teller, see e.g. Simmons (1993) or Yaqub (1993).

Hence, given Löb's Theorem, we can conclude $T \vdash \bot$, so T is inconsistent. So, if T is consistent then $T \nvdash \mathsf{Con}_T$, which is the Second Theorem again.

Proof for Löb's Theorem   Assume that, for a given φ,

   1. $T \vdash \Box\varphi \to \varphi$.

Now consider the wff $\mathsf{Prov}(\mathsf{x}) \to \varphi$. By hypothesis, T is nice, so we can invoke the general Diagonalization Lemma and apply it to this wff. Hence for some γ, T proves $\gamma \leftrightarrow (\mathsf{Prov}(\ulcorner\gamma\urcorner) \to \varphi)$.[4] Or, in our new notation,

2.	$T \vdash \gamma \leftrightarrow (\Box\gamma \to \varphi)$	
3.	$T \vdash \gamma \to (\Box\gamma \to \varphi)$	From 2
4.	$T \vdash \Box(\gamma \to (\Box\gamma \to \varphi))$	From 3, by C1
5.	$T \vdash \Box\gamma \to \Box(\Box\gamma \to \varphi)$	From 4, by C2
6.	$T \vdash \Box\gamma \to (\Box\Box\gamma \to \Box\varphi)$	From 5, by C2
7.	$T \vdash \Box\gamma \to \Box\Box\gamma$	By C3
8.	$T \vdash \Box\gamma \to \Box\varphi$	From 6 and 7
9.	$T \vdash \Box\gamma \to \varphi$	From 1 and 8
10.	$T \vdash \gamma$	From 2 and 9
11.	$T \vdash \Box\gamma$	From 10, by C1
12.	$T \vdash \varphi$	From 9 and 11. $\boxtimes$

(b)   We noted at the very outset that the reasoning for Gödel's First Theorem has echoes of the Liar paradox. The proof of Löb's theorem echoes the reasoning which leads to another logical paradox.

Suppose we temporarily reinterpret the symbol '$\Box$' as expressing a truth-predicate, so we read $\Box S$ as 'S' is true. Now let φ express any proposition you like, e.g. *The moon is made of green cheese*. Then

   $1'$. $\Box\varphi \to \varphi$

is surely a truism about truth.

Now suppose that the sentence γ says: *if γ is true, then the moon is made of green cheese* – and why shouldn't there be such a sentence? Then by definition

   $2'$. $\gamma \leftrightarrow (\Box\gamma \to \varphi)$.

From here on, we can argue as before (but now informally, deleting the initial '$T \vdash$'), appealing to conditions (C1) to (C3) reinterpreted as uncontentious-seeming principles about truth:

C1$'$. if φ then φ is true;

C2$'$. if $\varphi \to \psi$ is true, then if φ is true, ψ is true too;

C3$'$. if φ is true, then it is true that φ is true.

---

[4]Informally, then, γ 'says': if there is a T-proof of me, then φ.

Using these principles and starting from (1′) and (2′), the same ten-step argument will get us eventually to:

12′. φ.

So, from truisms about truth and a definitional equivalence like (2′) we can, it seems, prove that the moon is made of green cheese. Or prove anything else you like, since the interpretation of φ was arbitrary. (Exercise: check through the details.)

This line of reasoning is a version of what is nowadays usually known as 'Curry's paradox', after Haskell B. Curry who presented it in his (1942); but very close relations of it were certainly known to medieval logicians such as Albert of Saxony.[5] It isn't obvious what the best way is to block this kind of paradox, any more than it is obvious what the best way is to block the Liar. There is no doubt *something* fishy about postulating a sentence γ such that (2′) holds: but *what* exactly?

Very fortunately, answering that last question is not our business. We merely remark that Löb's Theorem, like Gödel's, is not a semantic paradox but a limitative result, a result about a theory's inability to prove certain propositions about its own provability properties.

34.5 Löb's Theorem and the Second Theorem

It was fun and illuminating to prove Löb's Theorem by direct appeal to the derivability conditions, and to note that it entails the Second Theorem. But in fact, we could have proceeded the other way about and derived Löb's Theorem from the Second Incompleteness Theorem.[6]

> **Theorem 34.5** *Assuming the Second Incompleteness Theorem applies widely enough (including to single-axiom extensions of T), then Löb's Theorem's holds for T.*

Proof sketch Suppose $T \vdash \Box_T \varphi \to \varphi$. And now consider the theory T' you get adding $\neg\varphi$ as a new axiom to T.

T' is consistent iff $T \nvdash \varphi$. And this elementary logical observation can be proved in a sensible T', so we have (a) $T' \vdash \neg\Box_T\varphi \to \mathsf{Con}_{T'}$. From our initial assumptions, however, we have (b) $T' \vdash \neg\varphi$ and (c) $T' \vdash \Box_T\varphi \to \varphi$. And (a), (b) and (c) together imply $T' \vdash \mathsf{Con}_{T'}$. Assuming the Second Incompleteness Theorem applies to T', T' must be inconsistent. So $T \vdash \varphi$.

In short, given the Second Theorem, if $T \vdash \Box_T\varphi \to \varphi$ then $T \vdash \varphi$, which is Löb's Theorem. ⊠

---

[5]See his terrific *Insolubilia* of 1490, translated in Kretzmann and Stump (1988).

[6]The argument was presented in a talk by Saul Kripke in 1966.

35 Deriving the derivability conditions

We have seen that $T \nvdash \mathrm{Con}_T$, i.e. $T \nvdash \neg\Box\bot$, holds for any nice theory T so long as the box satisfies the Hilbert-Bernays-Löb (HBL) derivability conditions

C1. if $T \vdash \varphi$, then $T \vdash \Box\varphi$,

C2. $T \vdash \Box(\varphi \to \psi) \to (\Box\varphi \to \Box\psi)$,

C3. $T \vdash \Box\varphi \to \Box\Box\varphi$.

But we already know from Theorem 24.1 that (C1) holds for any nice theory: indeed, as the second proof of the theorem showed us, if $T \vdash \varphi$, even $Q \vdash \Box_T\varphi$. So what does it take for conditions (C2) and (C3) to obtain as well? We start by considering the case of PA, and then generalize.

35.1 The second derivability condition for PA

We want to show

> **Theorem 35.1** PA $\vdash \Box(\varphi \to \psi) \to (\Box\varphi \to \Box\psi)$, *for any sentences* φ, ψ.

Preliminary result We first establish a background fact:

    A. If $Prf(m, \ulcorner(\varphi \to \psi)\urcorner)$ and $Prf(n, \ulcorner\varphi\urcorner)$, then $Prf(o, \ulcorner\psi\urcorner)$.

where $o = m \star n \star 2^{\ulcorner\psi\urcorner}$ (the star indicates the concatenation function introduced in Section 20.1). That ought to hold because, if m numbers a PA-proof of the sentence $(\varphi \to \psi)$, and n numbers a proof of φ, then $m \star n \star 2^{\ulcorner\psi\urcorner}$ numbers the sequence of wffs you get by writing down the proof of $(\varphi \to \psi)$ followed by the proof of φ followed by the one-wff sequence ψ. But this longer sequence is, of course, a PA-proof of the sentence ψ by modus ponens.

Review our outline definition of Prf in Section 20.4; then we see that it indeed implies (A). For brevity, put $i = len(m) - 1$, $j = len(n) - 1$, $k = len(o) - 1$. Assume we have (i) $Prf(m, \ulcorner(\varphi \to \psi)\urcorner)$ and (ii) $Prf(n, \ulcorner\varphi\urcorner)$.

Then by the definition of Prf, (i) implies (iii) $exf(m, i) = \ulcorner(\varphi \to \psi)\urcorner$. And (ii) implies (iv) $exf(n, j) = \ulcorner\varphi\urcorner$.

It is then quite easy to see that (iii) implies (v) $exf(o, i) = \ulcorner(\varphi \to \psi)\urcorner$. Likewise (iv) implies (vi) $exf(o, i+j+1) = \ulcorner\varphi\urcorner$. Trivially, (vii) $exf(o, k) = \ulcorner\psi\urcorner$.

So putting things together we get (viii) $MP(exf(o, i), exf(o, i+j+1), exf(o, k))$. And that's the final crucial step towards establishing (ix) $Prf(o, \ulcorner\psi\urcorner)$.   ⊠

Proof sketch for the theorem Now look at our target result. This says in effect that within PA we can argue from the assumptions $\exists v\,\mathsf{Prf}(v, \ulcorner(\varphi \to \psi)\urcorner)$ and $\exists v\,\mathsf{Prf}(v, \ulcorner\varphi\urcorner)$ to the conclusion $\exists v\,\mathsf{Prf}(v, \ulcorner\psi\urcorner)$.

We are evidently going to need to instantiate the two existential assumptions with free variables (acting as parameters), and show within PA that we can get from the assumptions $\mathsf{Prf}(a, \ulcorner(\varphi \to \psi)\urcorner)$ and $\mathsf{Prf}(b, \ulcorner\varphi\urcorner)$ to something of the form $\mathsf{Prf}(c, \ulcorner\psi\urcorner)$, so we can then infer $\exists v\,\mathsf{Prf}(v, \ulcorner\psi\urcorner)$.

Inspired by (A) the obvious route forward is to try to establish its formal analogue:

B. We can argue inside PA from premises of the form $\mathsf{Prf}(a, \ulcorner(\varphi \to \psi)\urcorner)$ and $\mathsf{Prf}(b, \ulcorner\varphi\urcorner)$ to the corresponding conclusion $\mathsf{Prf}((a * b) * 2^{\ulcorner\psi\urcorner}, \ulcorner\psi\urcorner)$,

where the star sign '$*$' is a function expression which captures the p.r. concatenation function $\star$ (we can definitionally augment PA with such a function expression). If we can establish (B) then we are done.

But now reflect that Prf canonically captures Prf, i.e. is structured to track the p.r. definition of Prf. So Prf will be built from expressions exf and MP which capture *exf* and *MP*, in such a way that we can track inside PA the informal chain of steps from (i) and (ii) to (ix) sketched above. To check that, it is just a question of hacking through the (quite unilluminating) details![1] ⊠

One comment on this proof. As we will see, it does depend crucially on Prf *canonically* capturing Prf. Suppose Prf° is Prf plus some 'noise'. It could still be the case that Prf° captures Prf, but because of the extra noise, the formal version of (viii) no longer implies the formal version of (ix) using Prf°. We will in fact find an example of this at the beginning of the next chapter.

35.2 The third derivability condition for PA

Suppose we can establish

Theorem 35.2 *For any Σ_1 sentence ψ, PA $\vdash \psi \to \Box\psi$.*

Whatever kind of sentence φ is, $\Box\varphi$, i.e. $\mathsf{Prov}(\ulcorner\varphi\urcorner)$, is Σ_1 (as we noted in Section 33.1, fn. 2). So, if we put $\Box\varphi$ for ψ, we can show as a corollary that the derivability condition (C3) holds for PA, i.e.

Theorem 35.3 *For any φ, PA $\vdash \Box\varphi \to \Box\Box\varphi$.*

---

[1]There's no way of making the proof any prettier. Masochists who want to complete the story for themselves can start by looking at Grandy (1977, p. 75) for a few more details. See also Rautenberg (2006, Sec. 7.1).

We should emphasize that filling out the argument *does* require work, as e.g. Smoryński (1977, p. 839) also emphasizes and as Grandy's treatment reveals. The *very* brief gestures towards proofs in e.g. Boolos (1993, p. 44) and Takeuti (1987, p. 86) are therefore perhaps somewhat misleading.

So how do we prove our Theorem 35.2?

Sketch of a proof sketch Recall from Section 11.6 that we showed that even Q is Σ_1-complete. Hence if ψ is a true Σ_1 sentence, then PA $\vdash \psi$. So what our new theorem says is essentially that PA 'knows' that it is Σ_1-complete. In other words, when ψ is Σ_1, PA 'knows' that $\psi \to \Box\psi$.

And that gives us the key to proving our theorem. We just need to show that the argument for PA's Σ_1-completeness can be formalized inside PA. Which, needless to say, is a great deal easier said than done: as always, the devil is in the details.

The original argument for the Σ_1-completeness of any theory containing Q works by an informal induction (from outside Q of course) on the complexity of the relevant wffs. We first showed that Q can prove the simplest true Σ_1 wffs; and then we showed that Q can prove more complex Σ_1 truths built up using the connectives, bounded quantifiers and unbounded existentials, given that it can prove all simpler Σ_1 truths.

We now need to run the same kind of induction again (from outside PA). We first show that PA can prove $\psi \to \Box\psi$ for the simplest Σ_1 wffs ψ. Then we show that PA can prove $\psi \to \Box\psi$ for more complex Σ_1 wffs built by conjunction, disjunction, bounded quantification, and existential quantification so long as it can prove the instances for simpler Σ_1 wffs.

However, to make this argument work, there's a major wrinkle. Up to now, the box has been applied to *sentences*: but to get our proof to work, it turns out that we need to introduce a generalized version of the box, $\boxdot$, which can apply to wffs with free variables as well. We will need then to show the likes of PA $\vdash x + y = z \to \boxdot(x + y = z)$. And to derive such implicitly general wffs inside PA, we will have to appeal to instances of PA's Induction Schema.

We'll not get bogged down by following through any more of the details here: we just emphasize again that last point that completing the proof requires using a modicum of induction *inside* PA.[2] $\boxtimes$

Assuming that our argument-strategy really can be made good, that completes the proof that the derivability conditions hold for PA. So, assuming PA is consistent, we have indeed established PA $\nvdash$ Con.

35.3   Generalizing to nice* theories

The argument that the derivability conditions hold for PA generalizes. Take the most troublesome case, the third condition. So long as T again has a little induction for us to use inside the theory – inspection of details reveals that $\mathrm{I}\Sigma_1$ induction will suffice – the same argument can be run again, and we get $T \vdash \Box_T\varphi \to \Box_T\Box_T\varphi$. Indeed, $\mathrm{I}\Sigma_1 \vdash \Box_T\varphi \to \Box_T\Box_T\varphi$.

---

[2]Those whose mathematical masochism wasn't sated by following up the last footnote can get a fuller story in Boolos (1993, pp. 44–49). Alternatively, see Rautenberg (2006, Sec. 7.1).

And containing $I\Sigma_1$ is more than enough to ensure the other two derivability conditions hold too (we didn't need induction inside PA to establish that the conditions hold for PA).

So let's introduce a bit of jargon. If T is consistent, p.r. axiomatized, and contains Q, we said it is 'nice'. Let's now say:

> A *nice** theory T is one which is nice and also includes induction at least for Σ_1 wffs and so extends $I\Sigma_1$.

Given what we've just said,

> **Theorem 35.4** *The HBL derivability conditions hold for any nice* theory.*

But Theorem 33.3 tells us that if T is nice and satisfies the derivability conditions then $T \nvdash \mathsf{Con}_T$. Hence we can derive an attractively tidy general version of the Second Incompleteness Theorem:

> **Theorem 35.5** *If T is nice*, then $T \nvdash \mathsf{Con}_T$.*

35.4 The Second Theorem for weaker arithmetics

The last theorem is good enough. But what about theories that are too weak to be nice*?

Well, the derivability conditions still hold even for some weaker arithmetics. In particular, they hold for EA, the theory of so-called elementary arithmetic, which is what you get by taking Q and adding Δ_0-induction plus the axiom which says that 2^n is a total function (cf. Section 12.4). Hence the proof of the Second Theorem goes through in the same way here too.

What about unaugmented $I\Delta_0$ or our old friend Q itself? Given the weakness of these theories, it will be no surprise at all to learn that the Second Theorem applies again: i.e. $I\Delta_0 \nvdash \mathsf{Con}_{I\Delta_0}$, $Q \nvdash \mathsf{Con}_Q$. But the published proofs for these cases go quite differently, because $I\Delta_0$, and so Q, isn't strong enough for the usual proof via the derivability conditions. We need not go into details.[3]

---

[3] See Wilkie and Paris (1987) and Bezboruah and Shepherdson (1976) respectively for the proofs of the Second Theorem for $I\Delta_0$ and Q. See Willard (2001) for an extended exploration of some hobbled and perhaps not-very-natural arithmetics which *can* prove (some of) their own consistency sentences.

36 'The best and most general version'

After the glory days of the 1930s, Gödel's comments on the details of his incompleteness theorems were few and far between. However, he did add a brief footnote to the 1967 translation of a much earlier piece on 'Completeness and consistency'. And Gödel thought that his brisk remarks in that footnote were sufficiently important to repeat them in a short paper in 1972, in a section entitled 'The best and most general version of the unprovability of consistency in the same system'.[1]

Gödel makes two main points. The first alludes to variant consistency sentences which *are* provable: we discuss these first. We then go on to prove some results about so-called reflection principles which hopefully throw light on Gödel's second point. (We will return to develop that second point further in the following Interlude, when we discuss Hilbert's Programme.)

36.1 There are provable consistency sentences

(a) Back in Section 31.6, we left the question hanging whether there could be some non-canonical way of expressing PA's consistency using a sentence Con° such that PA *does* prove Con°.

The short answer is 'yes'. To take the simplest example, suppose we put

$$\mathsf{Prf}^\circ(\mathsf{x},\mathsf{y}) =_{\mathrm{def}} \mathsf{Prf}(\mathsf{x},\mathsf{y}) \wedge (\forall \mathsf{v} \leq \mathsf{x})\neg\mathsf{Prf}(\mathsf{v}, \ulcorner\bot\urcorner)$$

where Prf as usual canonically captures the relation Prf. Now,

1. Suppose $Prf(m, n)$. Then (i) $\mathsf{PA} \vdash \mathsf{Prf}(\overline{\mathsf{m}}, \overline{\mathsf{n}})$. Given that PA is consistent, no number numbers a proof of $\bot$, so we have each of $\mathsf{PA} \vdash \neg\mathsf{Prf}(0, \ulcorner\bot\urcorner)$, $\mathsf{PA} \vdash \neg\mathsf{Prf}(1, \ulcorner\bot\urcorner)$, $\mathsf{PA} \vdash \neg\mathsf{Prf}(2, \ulcorner\bot\urcorner)$, ..., $\mathsf{PA} \vdash \neg\mathsf{Prf}(\overline{\mathsf{m}}, \ulcorner\bot\urcorner)$, and therefore (ii) $\mathsf{PA} \vdash (\forall \mathsf{v} \leq \overline{\mathsf{m}})\neg\mathsf{Prf}(\overline{\mathsf{m}}, \ulcorner\bot\urcorner)$. So from (i) and (ii) we have $\mathsf{PA} \vdash \mathsf{Prf}^\circ(\overline{\mathsf{m}}, \overline{\mathsf{n}})$, assuming PA's consistency.

2. Suppose not-$Prf(m, n)$. Then (i) $\mathsf{PA} \vdash \neg\mathsf{Prf}(\overline{\mathsf{m}}, \overline{\mathsf{n}})$, so $\mathsf{PA} \vdash \neg\mathsf{Prf}^\circ(\overline{\mathsf{m}}, \overline{\mathsf{n}})$.

Hence, assuming PA is indeed consistent, our doctored wff Prf° is another wff which captures Prf, and it is easily checked that Prf° expresses Prf too. Put $\mathsf{Prov}^\circ(\mathsf{y}) =_{\mathrm{def}} \exists \mathsf{v}\,\mathsf{Prf}^\circ(\mathsf{v}, \mathsf{y})$: then, on the same assumption, Prov° expresses the property of being a theorem of PA.

Now, predictably, we define

---

[1]The original note 'Completeness and consistency' is Gödel (1932). The added material first appears in van Heijenoort (1967, pp. 616–617) and is then reused in Gödel (1972).

$$\mathsf{Con}^{\circ} =_{\mathrm{def}} \neg\mathsf{Prov}^{\circ}(\ulcorner\bot\urcorner),$$

and Con° (at first sight) 'says' that PA is consistent. But it is immediate that

Theorem 36.1 PA $\vdash \mathsf{Con}^{\circ}$.

Proof Arguing inside PA, logic gives us $\neg(\mathsf{Prf}(\mathsf{a}, \ulcorner\bot\urcorner) \wedge \neg\mathsf{Prf}(\mathsf{a}, \ulcorner\bot\urcorner))$. That easily entails $\neg(\mathsf{Prf}(\mathsf{a}, \ulcorner\bot\urcorner) \wedge (\forall v \leq \mathsf{a})\neg\mathsf{Prf}(\mathsf{v}, \ulcorner\bot\urcorner))$. Apply universal quantifier introduction to conclude Con°. $\boxtimes$

Obviously, this same trick can be turned for an arbitrary theory T. Define Prf°_{T} in the same way in terms of the canonical Prf_{T}, construct the corresponding Con°_{T} and we'll get $T \vdash \mathsf{Con}^{\circ}_{T}$. So, as Gödel puts it in his note:

> ... the consistency (in the sense of non-demonstrability of both a proposition and its negation), even of very strong systems S, *may* be provable in S. (Gödel, 1972, p. 305)

(b) Since the doctored consistency sentence for T is provable, the corresponding $\square^{\circ}_{T}$ defined in terms of Prf°_{T} can't satisfy all the derivability conditions. For a start (as it is the easy case), the second condition can fail. Just imagine that, ordered by their Gödel numbers, the 'smallest' T-proofs of φ and $\varphi \to \psi$ precede the first proof of absurdity which in turn precedes the first proof of ψ. Then we will have $T \vdash \square^{\circ}_{T}\varphi$ and $T \vdash \square^{\circ}_{T}(\varphi \to \psi)$ without having $T \vdash \square^{\circ}_{T}\psi$.

(c) Take T to be any theory of interest, e.g. Zermelo-Fraenkel set theory (extended to define arithmetical vocabulary). Inspecting the argument above, we see that not only do we have $T \vdash \mathsf{Con}^{\circ}_{T}$ but also e.g. PA $\vdash \mathsf{Con}^{\circ}_{T}$. What does that tell us about the consistency of T? *Nothing*, of course. For we have the same result whether or not T is inconsistent.

Now, if T is indeed consistent, Prf°_{T} expresses the proof relation Prf_{T}, and Con°_{T} 'says' that T is consistent. But if T isn't consistent, then Prf°_{T} doesn't express that proof relation and Con°_{T} can't be read as expressing normal consistency. *But that means that we need to know whether T is consistent before we can interpret what we've proved in deriving Con°_{T}.* Therefore the proof can't tell us whether T is consistent.

(d) It might seem, then, that the construction of provable consistency sentences like Con°_{T} is mere trickery. Yet there is an initially appealing line of thought in the background here.

When trying to establish an as-yet-unproved conjecture, mathematicians will use any tools to hand, bringing to bear whatever background assumptions that they are prepared to accept in the context.[2] The more improvisatory the approach, the less well-attested the assumptions, then the greater the risk of lurking inconsistencies emerging, requiring our working assumptions to be revised. We

---

[2]See the classic Lakatos (1976) for a wonderfully lively exploration of the process of the growth of mathematical knowledge.

should therefore ideally keep a running check on whether apparent new results cohere with secure background knowledge. Only a derivation which passes the coherence test has a chance of being accepted as a *proof*. So how might we best reflect something of the idea that a genuine proof should be (as we might put it) consistency-minded, i.e. should come with a certificate of consistency with what's gone before?

As a first shot, we might suggest: there is a consistency-minded proof of φ in the axiomatized formal system T iff (i) there is an ordinary T-derivation of φ with super g.n. m, while (ii) there isn't already a T-derivation of absurdity with a code number less than m. That's the idea which is formally reflected in the provability predicate Prov_T°.

As another shot, equally natural, we could say there is a consistency-minded proof of φ in T iff (i) there is an ordinary T-derivation of φ with super g.n. m, while (ii) there isn't already a T-derivation of $\neg\varphi$ with a code number less than m. That's the idea which is formally reflected in an old acquaintance from Section 25.3, the Rosser provability predicate RProv_T.

There are other ways of 'Rosserizing' provability predicates; and as with Prov_T°, we can again construct provable consistency sentences using them. But nothing is thereby gained. Even though the initial motivating idea of keeping our proofs consistency-minded is plausible, implementing that idea using Rosserized proof predicates and then proving the corresponding consistency sentences can't settle anything, for basically the same reason that our simpler first shot proved hopeless. We can't get any informative news about consistency from the provability of the Rosserized sentences, because whether the sentences genuinely assert consistency depends again on whether the theory in question is consistent. So we won't discuss this kind of trickery any further.[3]

36.2   The 'intensionality' of the Second Theorem

We should pause to highlight again the point illustrated in (b) in the last section.

In showing that the derivability conditions hold for $\Box_T\varphi$, i.e. $\mathsf{Prov}_T(\ulcorner\varphi\urcorner)$, we have to rely on the fact that Prov_T is defined in terms of a predicate expression Prf_T which *canonically* captures Prf_T. Putting it roughly, we need $\mathsf{Prf}_T(\overline{m}, \overline{n})$ to reflect the details of what it takes for m to code a proof of the wff with g.n. n, and no more. Putting it even more roughly, we need Prf_T to have the right intended *meaning*. For if we doctor Prf_T to get another wff Prf_T° which still captures Prf_T but in a 'noisy' way, then – as we saw – the corresponding Prov_T° need not satisfy the derivability conditions.

So we have the following important difference between the First and the Second Theorem:

---

[3]Solomon Feferman's introduction in Gödel (1990, pp. 282–287) notes some more ways in which theories can in one sense or another prove their own consistency. For further discussion, see e.g. Visser (1989), Willard (2001).

1. Take *any* old wff Prf°_T which captures Prf_T, in however 'noisy' a way. Form the corresponding proof predicate Prov°_T. Take a fixed point γ for $\neg\mathsf{Prov}^\circ_T(\mathsf{x})$. Then we can show, just as before, that $T \nvdash \gamma$, assuming T is nice, and $T \nvdash \neg\gamma$ assuming ω-consistency as well. (The proof depends just on the fact that Prf°_T captures Prf_T, not at all on *how* it does the job.)

2. By contrast, if we want to prove $T \nvdash \mathsf{Con}_T$ (assuming T is nice*), we have to be a lot more picky. We need to start with a wff Prf_T which does 'have the right meaning', i.e. which *canonically* captures Prf_T, and then form the canonical consistency sentence Con_T from *that*. It we capture Prf_T in the wrong way, using a doctored Prf°_T, then the corresponding Con°_T may indeed be provable after all.

Solomon Feferman (1960, p. 35) – perhaps just a bit dangerously – describes results like the Second Theorem which depend on Prf_T 'more fully express[ing] the notion involved' as 'intensional'.[4] The label doesn't matter, but the distinction it marks does.

36.3   Reflection

(a)   We now start working towards Gödel's second point in his note. First we fix terminology. We start with three generic definitions:

i. The *reflection schema* for a theory T is

$$\Box_T\varphi \to \varphi$$

where φ holds the place for any sentence of T's language.

ii. We'll also say that the Π_1 *reflection schema* for T is the reflection schema restricted to cases where φ is a Π_1 sentence.

iii. A theory S *proves* Π_1 *reflection for* T iff, for every Π_1 sentence φ, $S \vdash \Box_T\varphi \to \varphi$.

Note that, by our definition at the beginning of Section 11.6, a theory T is Π_1-sound when, for every Π_1 sentence φ, if $T \vdash \varphi$ then φ is true. So it is immediate that a theory T is Π_1-sound when every instance of its Π_1 reflection schema is true.

Now two more specific definitions, concerning ways to extend the theory PA:

iv. The theory $\mathsf{PA} + \mathsf{Con}$ is as before the theory you get by adding PA's canonical consistency sentence Con as a new axiom to PA.

---

[4]'Intension' contrasts with 'extension'. A predicate's intension is its sense or meaning, its extension the set of things to which it applies.

v. The theory PAΠ is the theory you get by adding to PA all instances of its Π₁ reflection schema as additional axioms.

(b)   Suppose T is a nice *sound* theory. Then if φ is a provable sentence, φ is true. So each instance of the reflection schema for T, i.e. each instance of $\Box_T\varphi \to \varphi$, will be true. And you might have wondered if a rich enough sound theory could, on reflection (so to speak), believe in its own soundness; that is to say, you might have wondered if a nice sound T could ever prove all instances of its reflection schema. But not so.

Theorem 36.2 *If T is nice, then T cannot even prove all instances of the Π₁ reflection schema for T.*

Proof   Consider G, a Π₁ Gödel sentence such that $T \vdash G \leftrightarrow \neg\Box G$. If we also had $T \vdash \Box G \to G$ then, by elementary logic, $T \vdash G$, contradicting the First Theorem. Therefore, if T is nice, $T \nvdash \Box G \to G$.   ⊠

In fact we know more. If the derivability conditions hold for T and so we can derive Löb's Theorem, then reflection fails as badly as possible. For then, $T \vdash \Box\varphi \to \varphi$ implies $T \vdash \varphi$. Hence, if $T \vdash \varphi$ then, by the logic of the material conditional, $T \vdash \Box\varphi \to \varphi$: but in *every* other case, $T \vdash \Box\varphi \to \varphi$ is false.

(c)   Theorem 36.2 is the really important result for us. But let's pursue the topic of reflection just a bit further.

Our theorem tells us that a nice theory T cannot prove all instances of its Π₁ reflection schema. This result applies to PA in particular. However, if we believe that PA is sound – as we surely do! – then we'll believe that whatever it proves is true. So we'll think that every instance of its Π₁ reflection schema is in fact *true*.

Let's consider then the theory PAΠ which we get by adding to PA all those instances of PA's Π₁ reflection schema as further axioms. That's still a p.r. axiomatized theory, since we can test whether a wff is one of the new axioms without an unbounded search. And if we think that PA is sound, we'll accept PAΠ as sound too.

Theorem 36.3 PAΠ *proves* Con, *the consistency sentence for* PA.

Proof   ⊥ is Π₁ (why?). Since PAΠ proves all instances of PA's Π₁ reflection schema, PAΠ $\vdash \Box\bot \to \bot$, so PAΠ $\vdash \neg\Box\bot$.[5] In other symbols, PAΠ $\vdash$ Con.   ⊠

We have already shown that PA $\vdash$ Con $\to$ G (see the proof of Theorem 33.3). Hence, a fortiori, PAΠ $\vdash$ Con $\to$ G. So PAΠ $\vdash$ G.

More interestingly, there's a converse theorem:

Theorem 36.4 PA + Con *proves* Π₁ *reflection for* PA.

---

[5] Here, and throughout the rest of this section, unsubscripted '$\Box$' indicates provability in PA.

266

Proof Suppose φ is some Π_1 sentence. Then $\neg\varphi$ is logically equivalent to some Σ_1 sentence ψ. So we can apply Theorem 35.2 to show $\mathsf{PA} \vdash \psi \to \Box\psi$, and so (by the little lemma Theorem 33.1 which says we can substitute logical equivalents inside the box), we can start the following proof:

1. $\mathsf{PA} \vdash \neg\varphi \to \Box\neg\varphi$
2. $\mathsf{PA} \vdash \neg\varphi \to (\varphi \to \bot)$ Logic
3. $\mathsf{PA} \vdash \Box(\neg\varphi \to (\varphi \to \bot))$ From 2, by C1
4. $\mathsf{PA} \vdash \Box\neg\varphi \to \Box(\varphi \to \bot)$ From 3, using C2
5. $\mathsf{PA} \vdash \Box(\varphi \to \bot) \to (\Box\varphi \to \Box\bot)$ By C2
6. $\mathsf{PA} \vdash \neg\varphi \to (\Box\varphi \to \Box\bot)$ From 1, 4, 5.

But everything provable in PA is provable in $\mathsf{PA} + \mathsf{Con}$, so

7. $\mathsf{PA} + \mathsf{Con} \vdash \neg\varphi \to (\Box\varphi \to \Box\bot)$
8. $\mathsf{PA} + \mathsf{Con} \vdash \neg\Box\bot$ Trivial!
9. $\mathsf{PA} + \mathsf{Con} \vdash \Box\varphi \to \varphi$ From 7, 8, by logic.

Since φ was an arbitrary Π_1 sentence, that shows that $\mathsf{PA} + \mathsf{Con}$ proves any instance of Π_1 reflection for PA.   ⊠

(d)   In sum, we've just shown that the result of adding to PA all the instances of Π_1 reflection entails Con; and the result of adding Con to PA entails each instance of Π_1 reflection for PA. Which evidently is enough to establish

Theorem 36.5 *The theories* $\mathsf{PA\Pi}$ *and* $\mathsf{PA} + \mathsf{Con}$ *are equivalent.*

Recall then our neat little Theorem 11.7 which told us that if T extends Q, T is consistent iff it is Π_1-sound.. Equivalently, for theories which extend Q, T is consistent iff all the instances of T's Π_1 reflection schema are true. Thus, in particular, PA is consistent iff all instances of its Π_1 reflection schema are true. Theorem 36.5 can be thought of as one formal reflection of this fact.

36.4   The best version?

(a)   Suppose we are interested in Hilbert's project of trying to use uncontentious reasoning to show that the wilder reaches of mathematics are still 'safe' (see Section 1.6). What does 'safety' consist in? Well, you might very reasonably suppose that one condition any acceptably safe theory T should satisfy is this: *it shouldn't actually get things wrong about the Π_1 truths of arithmetic.* After all, those are essentially just the true equations involving particular numbers, and universally quantified versions of such equations. So we will surely want it to be the case that T is consistent with true Π_1 arithmetic in the sense that, if $T \vdash \varphi$ and φ is Π_1, then φ is true.

Thus, being Π_1-sound looks to be a natural minimal condition for being a 'safe' theory. Gödel calls what is in effect this same condition 'outer consistency' (though the label certainly hasn't caught on). And he remarks that 'for the usual systems [outer consistency] is trivially equivalent with consistency', which is just Theorem 11.7 again (Gödel, 1972, p. 305).

To establish formally that a theory T is Π_1-sound and hence to that extent 'safe' is therefore a matter of proving Π_1 reflection for T. But Theorem 36.2 tells us that a nice T can't prove every instance of its own Π_1 reflection schema. It follows that an arithmetic like PA certainly can't prove Π_1 reflection for a stronger, more infinitary, theory T either, unless T is inconsistent and proves everything. *In sum, we can't use relatively modest arithmetical reasoning to prove even the Π_1-soundness of a (consistent) stronger theory T.*

(b) As we'll see in the following Interlude, if we are interested in the Hilbertian project of certifying the safety of some infinitary theory T, then this result that we can't establish T's Π_1-soundness using non-infinitary reasoning is arguably the really crucial one. This is already enough to fatally undermine the project.

Yet note that the unprovability of 'outer consistency', as Gödel calls it, is an easier result than the unprovability of 'inner consistency', i.e. the unprovability of Con_T. Look again at the proof of Theorem 36.2: all it requires is that T is nice so that the Diagonalization Lemma and hence half the *First* Theorem hold for T. So, to show the unprovability of *outer* consistency/Π_1-soundness for T, we don't have to do the hard work of showing that the derivability conditions hold. By contrast, as we've seen, a full-blown proof of the unprovability of *inner* consistency is tougher.

Which is why the result that a theory T can't prove Π_1 reflection for T (if the conditions for the First Theorem apply to T) might be said to be 'the best and most general version of the unprovability of consistency in the same system' (if we are concerned with consistency-as-it-matters-to-Hilbert's-project).

And that is almost what Gödel himself says in his 1967/1972 note. What he actually asserts to be the best result is this (with just a trivial change):

> [W]hat can be shown to be unprovable in T is the fact that the rules of the equational calculus applied to equations demonstrable in T between primitive recursive terms yield only correct numerical equations (provided only that T possesses the property which is asserted to be unprovable). (Gödel, 1972, p. 305)

Here Gödel's T is a properly axiomatized theory which includes enough arithmetic to be p.r. adequate. But 'equations between primitive recursive terms' are equations of the form $\forall x(fx = gx)$, where f and g are p.r. functions. Assuming T contains enough arithmetic to know about p.r. functions, such equations can be represented by the likes of $\forall x \forall y \forall z(\mathsf{F}(x,y) \wedge \mathsf{G}(x,z) \to y = z)$ where F and G are Σ_1 wffs capturing f and g. So the whole wff here is Π_1 equivalent. (And there's a converse: any Π_1 sentence is equivalent to one saying that two p.r. functions always take the same values.[6]) So what Gödel is saying is in effect that T can't

---

[6] Proof hint: Use Theorem 12.4 to show that, if T has some induction, any Π_1 sentence

prove Π_1 reflection for T. And he claims that the key condition under which this holds is that Π_1 reflection is actually true for T, i.e. T is Π_1-sound, which – given that T includes Q – is equivalent to T's being consistent. But T's being consistent is just what is essential to the relevant half of the First Theorem holding for T, so that we can prove Theorem 36.2, the unprovability of so-called outer consistency.

Hence Gödel's remark indeed seems to be making a version of our point above: so long as T is nice, T can't prove Π_1 reflection for T (and that is enough to damage Hilbertian ambitions).[7]

36.5 Another route to accepting a Gödel sentence?

We'll return to say quite a bit more about Hilbert's Programme in the following Interlude. But before moving on, let's pause to consider whether our recent discussions throw any further light on the question we raised in Section 23.2: what routes are there that can lead us to accept a canonical Gödel sentence G_T as true?

We know that *one* kind of route goes via an explicit judgement that T is consistent: and we stressed before that there is quite a variety in the mathematical reasons we might have for forming that judgement. But now let's ask: do we *have* to go via explicit reflections about consistency? Can we perhaps go instead via thoughts about Π_1 reflection and rely on results such as that PAΠ ⊢ G?

Let me spin a just-so story. Imagine someone (we'll call him Kurt) who is a devotee of mathematical rigour and who delights in regimenting his knowledge into systems of Bauhaus austerity. Kurt's explorations, let's suppose, lead him to work within PA as an elegantly neat framework in which he can deduce all the familiar facts about the basic arithmetic of addition and multiplication, and lots of the less familiar facts too. Moreover, Kurt discovers the β-function trick which allows him to introduce definitions for all the *other* p.r. functions he knows about; so he can deduce what he knows about those p.r. functions too. Thoroughly immersed in the theory, Kurt enthusiastically follows deductions in PA wherever they take him: whenever he can derive φ in PA, then he adds φ to his stock of arithmetical beliefs – that is how he likes to do arithmetic. Compare: Kurt also follows where his eyes take him – in the sense that, if he takes himself to see that ψ, he (generally) comes to believe that ψ.

---

is provably equivalent to a wff of the form $\forall x \varphi(x)$, where φ is Δ_0. Then this $\forall x \varphi(x)$ can be massaged into a wff saying that the p.r. characteristic function for the property expressed by φ always equals the zero function.

[7]Feferman's introduction in Gödel (1990, pp. 282–287) seems to attribute to Gödel a more complex line of argument, and he briefly suggests using Jeroslow's variant version of the Second Theorem to throw light on Gödel's thinking. Michael Potter follows Feferman and attempts to develop this interpretation at greater length in Potter (2000, Section 10.3); however Richard Zach has shown that Potter's treatment is technically flawed (Zach, 2005). Which makes me hesitate to go beyond the simple reading of Gödel's remarks that I've given. See also Kreisel's comments in his review (1990, pp. 620–622).

Kurt, let's suppose, now gets reflective about his fact-gathering. For example, he comes to realize that, for lots of instances, when he seems to see that ψ, then it *is* the case that ψ (at least, that's how he supposes the world to go – and what other vantage point can he take?). And he finds no reason not to continue generally trusting his eyes, meaning that he is prepared more generally to endorse instances of the schema: when he seems to see that ψ, then ψ.

Similarly, Kurt reflects that when he can derive φ in PA, then it is the case that φ (at least, that's how he takes the arithmetical world to go). When it isn't the case that φ, he can't ever derive φ in PA: and so he finds no reason not to endorse his own continuing confidence in PA. So he is disposed to accept the conditional: when PA entails φ, then φ. Suppose that, for whatever reason, Kurt is especially interested in Π_1 arithmetical claims. Then, in particular, Kurt is disposed to accept instances of that conditional when φ is Π_1.

Next, Kurt hits one day on the idea of systematically introducing a code-numbering scheme which associates wffs and sequences of wffs with numbers (he discovers how to arithmetize syntax). And he realizes that the relation $Prf(m, n)$ – which holds when m codes for a PA derivation of the wff with number n – can be captured in PA by a long and complicated wff $\mathsf{Prf}(\mathsf{x}, \mathsf{y})$; and hence Kurt comes to see that there is a wff $\mathsf{Prov}(\mathsf{x}) =_{\text{def}} \exists \mathsf{v}\,\mathsf{Prf}(\mathsf{v}, \mathsf{x})$ which expresses provability in PA. Since for any Π_1 sentence φ, Kurt will happily accept that *if* PA *entails* φ *then* φ, he will now equally happily accept its arithmetical correlate $\mathsf{Prov}(\ulcorner\varphi\urcorner) \to \varphi$. Kurt, however, although bold in his enthusiasm for PA, is fundamentally a cautious man, as befits someone with his concern for rigour: so he *doesn't* suppose that all these instances of the reflection schema are themselves already proved by PA (which is a good thing, since we know from Löb's Theorem that PA *doesn't* prove such an instance unless it also proves φ itself). In other words, Kurt cheerfully allows that what follows from his original theory PA alone might be less than what follows from PA plus an arithmetization of his new thought that PA is reliable for Π_1 sentences: he never supposed that PA had to be the last word about the truths of L_A.

So Kurt has now moved on to the position of accepting the axioms of PAΠ, i.e. the axioms of PA plus all instances of its reflection schema $\Box\varphi \to \varphi$ where φ is a Π_1 sentence. Hence, Kurt has come to accept a theory in which he can produce a derivation of PA's canonical Gödel sentence (if and when he gets round to spotting the construction). And such a derivation will, for him, count as a *proof* of G, as it is done within the framework of PAΠ which he now accepts.

We could even, if we like, give the story a fanciful dramatic twist if we imagine Kurt proceeding as follows. We could imagine him *first* proving PA's canonical Gödel sentence in PAΠ, before he slaps his forehead in surprise as he sees that he also has a simple argument that PA doesn't prove G – thus showing that his earlier caution in not assuming that PA was the last word about arithmetic was well placed. (Though perhaps, like Isaacson, he still thinks that PA is the last word on purely arithmetical reasoning about numbers, for he recognizes that his new assumption which takes him beyond PA comes from reflections not just on numbers but on a formal arithmetical theory.)

So we seem to have got to the following point: Kurt *could* come to accept G *without* going via explicit thoughts about models for PA, or indeed thoughts about PA's syntactic consistency. Rather, he just notes his own confidence in PA's arithmetical reliability, and endorses it.

You might protest: 'Hold on! Kurt wouldn't accept that $0 = 1$ should it turn out that PA is inconsistent and proves $0 = S0$, would he? So it is only reasonable for him to be confident in instances of PA's reflection schema if he has a reason for thinking he isn't in for a really nasty surprise, i.e. if he *already* has a reason for thinking that PA is at least consistent. In other words, to follow through the suggested line of reasoning to the conclusion that G is true, Kurt after all does have to engage with some sort of argument – e.g. the specification of a model for PA – that could justify a belief in PA's consistency.'

But while this protest at first sight looks compelling, on reflection it is based on what is – to say the least – a deeply problematic epistemological assumption. To be sure, were it to turn out that PA 'proves' $0 = S0$, Kurt would abandon his confidence in PA. But why should we assume that it follows from *that* that Kurt needs some guarantee that PA won't deliver a nasty surprise if he is to be reasonable in moving from accepting PA (as he does) to accepting instances of its reflection schema? Compare: should I suddenly start seeing a crazy world of flying pigs and talking donkeys, I'll stop believing my eyes. But why should we assume that it follows from that that I need some guarantee in advance that things won't go crazy and that my eyes are (generally) reliable when I endorse the thought that in fact what my eyes tell me is the case (generally) *is* the case? In the world as it actually is, it is reasonable for me to reflectively endorse the presumption that my eyes are reliable, in the absence of countervailing considerations ('reasonable' in the sense that it is quite appropriate default behaviour for a responsible cognitive agent): it is similarly reasonable for Kurt to put his continued trust in PA in the absence of nasty surprises.

Of course, the route we've described which ends up with Kurt believing G, starting from an acceptance of PA's reflection principle, isn't available to get Kurt to endorse a canonical Gödel sentence for a theory he *doesn't* accept (like the theory $Q^{\dagger}$ which we defined in Section 23.2): in *that* sort of case, Kurt has to have sophisticated ideas about truth-in-a-model, or some such. The point we are making here is that this isn't how it *has* to be.[8]

---

[8]See here the debate between Ketland (1999, 2005) and Tennant (2002, 2005).

37 Interlude: The Second Theorem, Hilbert, minds and machines

The title of Gödel's great paper is 'On formally undecidable propositions of *Principia Mathematica* and related systems I'. And as we noted in Section 23.4, his First Incompleteness Theorem does indeed undermine *Principia*'s logicist ambitions. But logicism wasn't really Gödel's main target. For, by 1931, much of the steam had already gone out of the logicist project. Instead, the dominant project for showing that classical infinitary mathematics is in good order was Hilbert's Programme, which we mentioned at the outset (Section 1.6). This provided the real impetus for Gödel's early work; it is time we filled out more of the story.

However, this book certainly isn't the place for a detailed treatment of the changing ideas of Hilbert and his followers as their ideas developed pre- and post-Gödel; nor is it the place for an extended discussion of the later fate of Hilbertian ideas.[1] So our necessarily brief remarks will do no more than sketch the logical geography of some broadly Hilbertian territory: those with more of a bent for the history of logic can be left to fight over the question of Hilbert's precise path through the landscape.

Another, quite different, topic which we will take up in this Interlude is the vexed one of the impact of the incompleteness theorems, and in particular the Second Theorem, on the issue of mechanism: do Gödelian results show that minds cannot be machines?

37.1 'Real' vs. 'ideal' mathematics

What does it take to grasp the truths of arithmetic – what does it take, for example, to grasp that every number has a successor?

Kant famously thought that it involves the exercise of a special cognitive resource, *'intuition'*, whatever exactly that is. That is to say, it requires more than can be given just by analytic reflection on the logical concepts which we deploy in thinking about any subject matter at all.

Frege equally famously disagreed. His fundamental claim is that

> Pure thought (irrespective of any content given by the senses or even by an intuition a priori) can, solely from the content that results from its own constitution, bring forth judgements that at

---

[1] For more, see Zach (2003), much expanded to Zach (2006).

first sight appear to be possible only on the basis of some intuition.
(Frege, 1972, §23)

But sadly, as we've already noted, his logicist attempt to derive all of arithmetic from logic-plus-definitions became tangled in contradiction (Section 13.4). And Russell and Whitehead's attempt to develop a paradox-free foundation for arithmetic is also highly problematic when regarded as an attempt to vindicate the logicist project: for example, how can *Principia*'s Axiom of Infinity genuinely be counted as a logical axiom?

Still, even if we don't regard it as a purely logical system, it is still highly interesting that *Principia* points us towards a unified type-theoretic framework in which we can regiment not just arithmetic but a great deal of mathematics. Another (almost) all-embracing foundational framework is ZFC, Zermelo-Fraenkel set theory ZF plus the Axiom of Choice – which is, indeed, in some ways much nicer to use.

But hold on! If neither *Principia*'s system nor ZF with or without the Axiom of Choice can be advertised as belonging to pure logic, how can we choose which to use? Which is '*true*'? Both of them? Or should we perhaps use neither, but adopt some variant theory on the market as the 'correct' one?

Faced with puzzling questions like that, it is rather tempting to suppose that they are pretty intractable because they are based on misguided assumptions. Perhaps we just shouldn't think of infinitary type theories or set theories and the like as really being in the business of truth or falsity (so we can't ask which is correct). Rather, to put it in Hilbertian terms, we should divide mathematics into a core of uncontentious *real* mathematics and a superstructure of *ideal* mathematics. 'Real' mathematics is to be taken at face value (it expresses contentful propositions, or is 'contentual' for short); and the propositions of real mathematics are straightforwardly true or false. So, for example, very elementary arithmetic is real – it really is the case that two plus three is five, and it really isn't the case that two plus three is six. Perhaps more generally, Π_1-statements of arithmetic are real.

By contrast, 'ideal' mathematics shouldn't be thought of as being strictly speaking true or false at all – some parts of ideal mathematics are instrumentally useful tools, helping us to establish real truths, while other parts are just intellectual *jeux d'esprit*. Of course, there will be straightforward truths about what follows from what in a particular ideal mathematical game: the claim, though, is that the statements made *within* the game are rather like statements within a fiction, not themselves straightforwardly either true or false. So, for example, perhaps even the more ambitious parts of arithmetic, and certainly the wilder reaches of infinitary set theory, are to be treated as ideal in this sense.

In pursuing this idea, Hilbert himself was inclined to take a *very* restricted view of real mathematics. In part, that seems to have been a strategic ploy: the plan is to count as real mathematics only some uncontroversial core of arithmetic which even the most stringent critic of infinitary mathematics is happy to accept, before going on to argue that ideal mathematics can still be instrumentally useful. But Hilbert was also swayed by Kantian ideas. He thinks of our grasp of core 'real' arithmetic as grounded in 'intuitive' knowledge, in our apprehension of finite sequences of signs and of the results of manipulating these. And he thinks that this 'intuitive' apprehension yields knowledge of simple arithmetic operations on particular numbers, and also of Π_1 generalizations about these operations. But we can't pursue here the vexed question of how far 'intuition' can take us,[2] and so we'll put on hold the question of where exactly we might want to place the dividing line between core real mathematics and ideal mathematics. We will soon see, however, that given his wider purposes Hilbert needs to be right on one thing: real mathematics needs to include at least the arithmetic of Π_1 statements.

37.2   A quick aside: Gödel's caution

Hilbertians, then, thought that the status of most mathematics is to be sharply distinguished from that of some small central core of real, true, elementary arithmetic. Most mathematics is merely ideal: we can talk about what demonstrably follows from what within the game, but shouldn't talk of the statements made in the game as being true. Partly for that reason (as he later put it in a letter), when Gödel published his incompleteness paper

> ... a concept of objective mathematical truth as opposed to demonstrability was viewed with greatest suspicion and widely rejected as meaningless. (Gödel, 2003a, p. 10, fn. c)

Which probably explains why – as we've remarked before – Gödel in 1931 very cautiously downplayed the version of his incompleteness theorems that depended on the assumption that the theories being proved incomplete are *sound* (i.e. have *true* theorems), and instead put all the weight on the purely syntactic assumptions of consistency or ω-consistency.[3]

37.3   Relating the real and the ideal

Let's now ask: what relations might hold between Hilbert's two domains of mathematics? For brevity, we'll use the following symbols:

i. I is a particular ideal theory ('ideal' need not mean 'purely ideal': an ideal theory could extend a contentual one).

ii. C is our best correct theory of contentual real mathematics. Being correct, C is consistent, and all its deductive consequences are true. (We can leave it open for now whether C is a tidily axiomatizable theory.)

---

[2] For discussion, see for example Parsons (1980, 1998) and Tait (1981, 2002).
[3] For more on this historical point, see Feferman (1984).

So here are four relations that an ideal theory I might have to C.

1. I is *real-conservative*: for any real φ, if $I \vdash \varphi$, then $C \vdash \varphi$.

2. I is *real-sound*: for any real φ, if $I \vdash \varphi$, then φ is true.

3. I is *real-consistent*: for any real φ, if $I \vdash \varphi$, then $C \nvdash \neg\varphi$.

4. I is *weakly-conservative*: for any real φ, if $I \vdash \varphi$ and C decides φ, then $C \vdash \varphi$.

In the first case, the ideal theory I can only prove real propositions that we can already prove in our contentual theory C. In the second case, I can only prove *true* contentual propositions. In the third case, I might prove more real propositions than C, but it can't prove anything inconsistent with theorems of core real mathematics. Finally, in the fourth case, I agrees with C at least on the contentual real propositions that C can decide one way or the other.

Trivially, all four conditions require I to be consistent. And the relations between the conditions are now easily seen to be as follows:

$$(1) \to (2) \to (3) \leftrightarrow (4)$$

Proof If I is real-conservative it must be real-sound (since all C's entailments are true). But not vice-versa: C might prove fewer real truths than I.

If I is real-sound it is evidently real-consistent. But again not necessarily vice-versa. For suppose that C isn't negation-complete, and doesn't decide the contentual proposition φ. So we might have real-consistent I and I' such that $I \vdash \varphi$ and $I' \vdash \neg\varphi$: and I and I' can't both be real-sound.

Finally, if I is real-consistent it is weakly-conservative. Assume that $C \nvdash \neg\varphi$ if $I \vdash \varphi$; then if C decides φ, that means $C \vdash \varphi$. While for the reverse implication, suppose that I is weakly-conservative. Suppose too that $I \vdash \varphi$ but also $C \vdash \neg\varphi$ for some contentual φ. Then C decides φ; so by weak conservativeness $C \vdash \varphi$ making C inconsistent, contrary to the assumption that C is correct. Hence, if $I \vdash \varphi$, then $C \nvdash \neg\varphi$, and so I is real-consistent. ⊠

37.4 Proving real-soundness?

(a) Suppose that you *are* attracted by this plan of distinguishing a core of real, straightforwardly true, mathematics from the great superstructure of ideal mathematics. Then you'll want to know which bits of ideal mathematics are safe to use, i.e. don't lead you to false real beliefs, i.e. are real-sound. And, to the extent that you can ratify theories as real-sound, you will then have vindicated the practice of infinitary mathematics. Even though, *sotto voce*, you'll say to yourself that only real mathematics is genuinely true, you can plunge in and play the ratified games of ideal mathematics with a clear conscience, knowing that they can't lead you astray.

Which suggests a Hilbertian programme, very different from the logicist programme of trying to derive everything, including standard infinitary mathematics, from logic-plus-definitions. The new programme is: *seek to defend those parts of ideal mathematics we take to be useful in extending our contentual knowledge by showing them to be real-sound.*[4]

Now, we have already seen that there are Gödelian limitations on the provability of real-soundness. But rather than jump straight to pressing that point, let's proceed more slowly, pretending for a moment that we are still in a state of pre-Gödelian innocence.

(b) Even prescinding from Gödelian considerations, it might seem that the Hilbertian project is doomed from the very outset. For how can we possibly show that an ideal theory I has true contentual consequences without assuming that all the relevant axioms used in deriving these consequences are true and hence are contentual and hence I (or at least the fragment of I with contentual implications) is not really ideal after all?

But in fact, there *are* a couple of routes by which we could in principle lever ourselves up from a belief that I is *consistent* and has *some* correct real content to the conclusion that I is overall *real-sound.* We rely on a couple of easy theorems:

Theorem 37.1 *If I is consistent and extends C, and C is negation-complete, then I is real-sound.*

Proof Suppose φ is a contentual proposition and $I \vdash \varphi$ but φ is false. Since C by hypothesis is correct, $C \nvdash \varphi$. Since C is negation-complete, that implies $C \vdash \neg\varphi$. But by the definition of 'extends', if $C \vdash \neg\varphi$, $I \vdash \neg\varphi$. So I is inconsistent contrary to hypothesis. So if $I \vdash \varphi$ then φ is true.   ⊠

Theorem 37.2 *If I is consistent and extends Q, and if contentual mathematics extends no further than Π_1 propositions of arithmetic, then I is real-sound.*

Proof This is just half of Theorem 11.7 in our new jargon, since Π_1-soundness implies real-soundness if contentual mathematics extends no further than the Π_1 propositions of arithmetic.   ⊠

Given these mini-theorems, we can immediately discern the shape of two possible lines of argument for defending the use of an ideal theory in establishing contentual truths. We won't worry too much about whether either is the historical Hilbert's own mature programme for defending the useful branches of ideal mathematics: but they are both Hilbertian arguments in a broad sense.

H1. First option. We start by characterizing real contentual mathematics (perhaps quite generously). We then establish (i) that there is in fact

---

[4]Those bits of ideal mathematics which are hermetically sealed games, with no contentual implications, can be left to look after themselves!

a negation-complete theory C for this real mathematics, and also establish (ii) that our favourite useful theories of ideal mathematics like ZFC set theory both extend C and are consistent. Then by Theorem 37.1 we'll be entitled to believe the contentual implications of ZFC (or whatever) because we'll have a warrant for the claim that they are already entailed by C and so are true.[5] (The ideal theories can't prove anything that wasn't already provable in C; but going via the ideal theories might offer much shorter and/or much clearer proofs.)

H2. Second option. We restrict real mathematics to Π_1 claims of arithmetic – as it seems did Hilbert himself. Establish that favourite theories in ideal mathematics like ZFC set theory both extend Q and are consistent. Then by Theorem 37.2 we'll be entitled to believe the contentual implications of ZFC (even though, in this case, we *won't* always already be able to deduce them in a restricted contentual theory).

(c)   Now, both these lines of argument require us to establish the consistency of axiomatized theories in ideal mathematics in order to prove that they are 'safe'.[6] But what does 'establish' mean here? Given that the overall project is to find a respectable place for ideal mathematics (in particular, infinitistic mathematics) as an instrumentally useful supplement to real mathematics, 'establishing' can't involve appeal to the very same infinitary ideas which we are trying to legitimate. So consistency will have to be established by appeal to nothing more exotic than the kosher 'safe' reasoning already involved in contentual mathematics.

But how can we get common-or-garden contentual mathematics to touch directly on questions of the consistency of formal axiomatized theories? By the arithmetization of syntax, of course. But recall, a consistency claim about an ideal theory I is canonically expressible by a Π_1-equivalent sentence Con_I. So *if consistency proofs are to be within reach of contentual mathematics, then contentual mathematics must – as we said – be able to cope at least with some Π_1 claims of arithmetic* (so presumably must include at least Q).

---

[5]Here is Hilbert, seemingly endorsing a general argument from consistency to real-soundness:

> For there is a condition, a single but absolutely necessary one, to which the use of the method of ideal elements is subject, and that is the proof of consistency; for, extension by the addition of ideals is legitimate only if no contradiction is thereby brought about in the old, narrower domain, that is, if the relations that result for the old objects whenever the ideal objects are eliminated are valid in the old domain. (Hilbert, 1926, p. 383)

Hilbert doesn't here fully explain his thought, nor does he explicitly assert the negation-completeness of real mathematics which is needed for the obvious argument to go through. However, his pupil and collaborator Paul Bernays does:

> In the case of a finitistic proposition ... the determination of its irrefutability is equivalent to determination of its truth. (Bernays, 1930, p. 259)

It seems quite a reasonable bet that Hilbert agreed. See Raatikainen (2003).

[6]In fact, Gödel discovered his incompleteness theorems while trying to prove the consistency of classical analysis.

37.5   The impact of Gödel

(a)   Now trouble! First, if our contentual mathematics is to be regimented sufficiently well for the question of establishing that it is negation-complete to arise, then it will need to be a properly axiomatized theory C. But no properly axiomatized contentual theory C including Q can be negation-complete, by Gödel's First Theorem. So the First Theorem is *already* enough to sabotage the first Hilbertian programme (H1).

That leaves the second programme (H2) still in the hunt, as *that* doesn't require any assumptions about completeness. However we now know from Gödel's Second Theorem that no modest formal arithmetic can establish the consistency of a fancy ideal theory. So the second programme (H2) must fail too as the desired 'contentual' consistency proofs for branches of ideal mathematics won't be forthcoming.

(b)   Those are the obvious claims about the impact of Gödel's Theorems on the general Hilbertian project of trying to establish the real-soundness of ideal theories by giving consistency proofs. Is there any wriggle room left?

Michael Detlefsen (1986) has mounted a rearguard defence of Hilbert that, in part, plays with the thought that we should ensure that proofs are consistency-minded – i.e. a sequence of wffs counts as a proof of φ only if there is no 'earlier' proof of $\neg\varphi$, etc. But while we can trivially prove such Rosserized theories to be consistent (see Section 36.1), this idea can't in fact be used to rescue a version of our second Hilbertian programme, for the following reason.

(H2) depends on Theorem 37.2, which tells us that if I is consistent and extends Q, then I is real-sound – assuming real mathematics goes no further than Π_1 truths. Now, for normal theories, it can of course be easy to show that I extends Q (just show that I proves Q's axioms). If, however, I_R is a Rosserized theory, then – while we can trivially see that it can't Rosser-prove contradictions – *we can't in general effectively decide whether it extends* Q. For suppose $Q \vdash \varphi$. Then even if I_R proves Q's axioms, it doesn't follow that there is a consistency-minded proof of φ, because for all we know there could be an I_R proof of $\neg\varphi$ which is shorter than the shortest Q proof of φ. So if we Rosserize our theories – or fiddle with proof predicates in similar ways – then we can't make use of (H2) to show that our ideal theories are real-sound.

(c)   So, is there any *other* route to establishing real-soundness for ideal theories, using only relatively modest arithmetic reasoning? Well, if we continue to suppose that real mathematics must be able to do Π_1 arithmetic, so real-soundness embraces Π_1-soundness, then we know that there can't be. That is, of course, what is shown by Gödel's 'best and most general version of the unprovability of consistency in the same system' (see Section 36.4). *Modest arithmetic reasoning can't even prove the Π_1-soundness of modest arithmetics, let alone the Π_1-soundness of more fancy theories.* So Gödel's 'best' version would really seem to mark the end of the story.

But perhaps not entirely. For remember, in Section 32.4 we briefly outlined Gentzen's consistency proof for PA. To be sure, that uses reasoning which goes beyond the most elementary; but it might perhaps still be defended as belonging to 'safe' real mathematics in an extended sense. Exploring such prospects for consistency proofs using safe-but-not-strictly-finitary methods would, however, take us far too far afield. And let's not complicate matters. Whatever the options for descendants of Hilbert's Programme, the headline news remains this: the Hilbertian project in anything very close to its original form is sunk by Gödel.[7]

37.6   Minds and computers

(a)   We now turn to a very different matter on which Gödel's Theorems have been thought to impact. Consider the following first-shot argument:

> Call the set of mathematical sentences which I accept, or at least could derive from what I accept, my *mathematical output O*. And consider the hypothesis that there is some kind of computing machine which can in principle list off my mathematical output – i.e., it can effectively enumerate (the Gödel numbers for the sentences in) O. Then O is effectively enumerable, and by Theorem 26.1 it follows that there is a p.r. axiomatized theory M whose theorems are exactly my mathematical output. Since I accept the axioms of Q plus, indeed, some induction, M is at least as strong as $I\Sigma_1$, and so M is p.r. adequate. So I can now go on to prove that M can't prove its canonical Gödel sentence G_M. But in going through that proof, I will come to establish by mathematical reasoning that G_M is true. Hence M does not, after all, entail *all* my mathematical output. Contradiction. So no computer can effectively generate my mathematical output. Even if we just concentrate on my mathematical abilities and potential mathematical output, I can't be emulated by a mere computer!

This style of argument is often presented as leading to the conclusion that I can't be emulated by a 'Turing machine' in particular (see Section 3.1(c)). But note that if there is any force to the sketched argument, it will apply to computing devices more generally – which is why we can discuss it here, before we get round to explaining the special notion of a Turing machine.

One immediate problem with this kind of argument, of course, is the unclarity of the idea of my 'mathematical output'. Is it to contain just what I could derive, given my limited cognitive abilities, my limited life, etc? In that case, my output

---

[7]From the start, Gödel himself left open the possibility that there could be a Hilbertian project which relied on a richer kind of consistency proof: see Gödel (1931, p. 195). For some relevant later investigations, see Gödel (1958). And for general discussion see also Giaquinto (2002, Part V, ch. 2), as well as Zach (2003, 2006).

is finite; and then quite trivially we know the argument goes wrong – because for any finite set of Gödel numbers, there will trivially be a computer that can enumerate it, given enough memory storage (just build the finite list into its data store). So to get an argument going here, we'll have to radically idealize my mathematical abilities, for a start by allowing me to follow a proof of arbitrary length and complexity. But what other idealizations are allowed? That's very unclear.

However, even setting aside that point, the argument is in bad shape, ultimately for a very simple reason already prefigured in Section 23.3(a). Grant that I can in general establish that *if* the axiomatized theory T is consistent and contains enough arithmetic, then G_T is true. But of course, assuming T has enough induction, T itself can also prove $Con_T \rightarrow G_T$. Apply that to the particular case where T is the theory M that generates my mathematical output. Then even if I can establish that *if* M is consistent, then G_M is true, M can prove that too (since it will contain $I\Sigma_1$): there is no difference yet between my output and M's.

Now, if I could now go on to establish that M *is* consistent, then that would indeed distinguish me from M, because I can then establish that M's canonical Gödel sentence is true, and M can't. But we've so far been given no reason to suppose that I *can* show that M is consistent, even if idealized.[8]

(b)   Can we improve the first-shot argument? To make progress, we need to be entitled to the thought that the relevant M is consistent. Of course, we *hope* that our mathematical output O is consistent, and so correspondingly we will *hope* that M is consistent. But wishful thinking isn't an argument.[9] However, perhaps we can get somewhere if we think of our mathematical output not as defined in terms of what we *accept* and can derive from what we accept, but in

---

[8]The locus classicus for this point is Putnam (1960).

[9]John Lucas has urged that the hypothesis that there is a machine that emulates me is only worth considering given that the mechanist makes a consistency assumption:

> Putnam's objection fails on account of the dialectical nature of the Gödelian argument. . . . [T]here is a claim being seriously maintained by the mechanist that the mind can be represented by some machine. Before wasting time on the mechanist's claim, it is reasonable to ask him some questions about his machine to see whether his seriously maintained claim has serious backing. It is reasonable to ask him not only what the specification of the machine is, but whether it is consistent. Unless it is consistent, the claim will not get off the ground. If it is warranted to be consistent, then that gives the mind the premiss it needs. The consistency of the machine is established not by the mathematical ability of the mind but on the word of the mechanist. (Lucas, 1996, p. 113)

But the issue isn't whether the *machine* which is supposed to be emulating my mathematical output is consistent – it churns away, effectively enumerating a set of Gödel numbers, and it can be in as good order as any other computing machine. The question is whether the sentences which the Gödel numbers encode form a consistent collection. Even if they don't, those sentences could still be my idealized mathematical output: for example, perhaps the set theory I accept is inconsistent, but the shortest proof of inconsistency is far too long for anyone actually to grasp – which is why, as real-world unidealized mathematicians, we haven't noticed the contradictions which lurk over the horizon, far down the road.

terms of what we *know* because we can prove it true (since every sentence we can prove true must be consistent with every other sentence we can prove true).

So let's consider the following argument, which is essentially due to Paul Benacerraf (1967). Let's now use K for the set of mathematical truths that are knowable by me (idealizing, take it to be the deductive closure of what I can prove true by mathematical argument). And again suppose that there is a computer which emulates me in the sense that it effectively enumerates K. By the same argument as before, this entails the assumption

1. There is a p.r. axiomatized theory N such that, for all φ, $\varphi \in K \leftrightarrow N \vdash \overline{\varphi}$.

where $\overline{\varphi}$ is N's formal counterpart for the informal claim φ. And now let's *also* assume that one of the broadly mathematical things that I know is that the theory N indeed generates my output of mathematical knowledge. Equivalently,

2. 'for all φ, $\varphi \in K \leftrightarrow N \vdash \overline{\varphi}$' $\in K$.

We can then continue as follows:

3. By hypothesis, everything in K is true: so K is consistent – or Con_K for short. And since that's a mini-proof of Con_K, then 'Con_K' $\in K$.

4. Since Con_K, and a sentence is in K if and only if its formal counterpart is provable in N, then N is consistent too, i.e. Con_N for short. So, since we've just proved *that*, '$Con_K \wedge$ (for all φ, $\varphi \in K \leftrightarrow N \vdash \overline{\varphi}$) $\rightarrow Con_N$' $\in K$.

5. Since K by hypothesis is deductively closed, (2), (3) and (4) imply 'Con_N' $\in K$.

6. So by (1) again, $N \vdash \mathsf{Con}_N$, where Con_N formally expresses the consistency of N.

7. But since I know quite a bit of arithmetic to be true (more than $\mathsf{I\Sigma_1}$), enough arithmetic must be built into N for the Second Theorem to apply. Hence $N \nvdash \mathsf{Con}_N$. Contradiction!

8. So either assumption (1) or assumption (2) has to be false.

In other words, either my (idealized) mathematical knowledge isn't capturable in a theory describing the potential output of a computing machine or, if it is, I don't know which theory, and hence which machine, does the trick. Which is quite neat,[10] but also perhaps fairly unexciting. After all, it isn't exactly easy to tell which bits of my putative mathematical knowledge really *are* knowledge (perhaps ZFC is inconsistent after all!): so why on earth should we suppose that

---

[10] If it is indeed legitimate to assume that it is the same formal consistency statement that is involved at steps (6) and (7): but let's grant that it is.

I'd have the god-like ability to correctly spot the computer program that actually gets things right in selecting out (the deductive closure of) what I truly know?

So we can and should cheerfully embrace the second limb of the disjunctive conclusion.

(c) In his Gibbs lecture, Gödel himself considered the impact of the Second Theorem on issues about minds and machines (Gödel, 1951). Like Benacerraf, he reaches a disjunctive conclusion.[11] Gödel starts by remarking that the Second Theorem

> ... makes it impossible that someone should set up a certain well-defined system of axioms and rules and consistently make the following assertion: All of these axioms and rules I perceive (with mathematical certitude) to be correct, and moreover I believe that they contain all of mathematics. If someone makes such a statement he contradicts himself. For if he perceives the axioms under consideration to be correct, he also perceives (with the same certainty) that they are consistent. Hence he has a mathematical insight not derivable from his axioms. (Gödel, 1951, p. 309)

So this is the now familiar thought that we can keep on extending sound theories to get new ones by adding their consistency sentences as new axioms. But how far can we follow through this process? Either we say 'infinitely far': at least in principle, idealizing away from limitations on memory and time and so forth, we can keep on going for ever, grasping ever more extensive systems of arithmetic as evidently correct, but never completing the task. Or we say 'the human mind (even if we bolt on more memory and abstract from time constraints etc.) loses its grip at some point': and then there are further truths that remain forever beyond the reach of proof. Gödel puts the alternatives this way:

> Either mathematics is incompletable in this sense, that its evident axioms can never be comprised in a finite rule, that is to say, the human mind (even within the realm of pure mathematics) infinitely surpasses the powers of any finite machine, or else there exist absolutely [unprovable Π_1 sentences] ... where the epithet 'absolutely' means that they would be [unprovable], not just within some particular axiomatic system, but by *any* mathematical proof the human mind can conceive.[12]

Now, there are questions which can be raised about this argument:[13] but perhaps the principal point to make is that, even if the argument works, its

---

[11] An outline of Gödel's position was reported by Wang (1974, pp. 324–326); but the lecture wasn't published until 1995.

[12] From Gödel (1951, p. 310). I've reversed the order of the passages either side of the lacuna. Also, Gödel talks of 'unsolvable diophantine equations' rather than, equivalently, about unprovable Π_1 sentences: see (Gödel, 1995, p. 157) for an explanation of the connection.

[13] See, for example, the discussion in Feferman (2006).

disjunctive conclusion is again anodyne. Anyone of naturalist inclinations will be happy enough to agree that there are limits on the possibilities of human mathematical cognition, even if we abstract from constraints of memory and time. Gödel himself was famously not naturalistically inclined and, according to Hao Wang, he was inclined to reject the second disjunct (Wang, 1974, pp. 324–326): but for once there seems no evident good reason to follow Gödel here.

(d) So are there *other* arguments that lead from thoughts about Gödelian incompleteness to more substantial, and non-disjunctive, conclusions?

Well, there is indeed a battery of attempts to find such arguments, starting with a much-cited paper by John Lucas (1961) and latterly continuing in books by Roger Penrose (1989, 1994). We certainly haven't space to follow all the twists and turns in the debates here. But it is fair to say that these more intricate anti-mechanist arguments based on the incompleteness theorems have so far produced *very* little conviction indeed. If you want to explore further, Stewart Shapiro's rich (1998) makes an excellent place to start.

37.7 The rest of this book: another road-map

We have now proved Gödel's First Incompleteness Theorem and outlined a proof of his Second Theorem.

And it is worth stressing that the ingredients used in our discussions so far have really been *extremely* modest. We introduced the ideas of expressing and capturing properties and functions in a formal theory of arithmetic, the idea of a primitive recursive function, and the idea of coding up claims about relations between wffs into claims about relations between their code numbers. We showed that some key numerical relations coding proof relations for sensible theories are p.r., and hence can be expressed and indeed captured in any theory that includes Q. Then, in the last fifteen or so chapters, we have worked Gödelian wonders with these very limited ingredients. We haven't needed to deploy any of the more sophisticated tools from the logician's bag of tricks. Note, in particular, that in proving our formal theorems, we *haven't* yet had to call on a general theory of computable functions or (equivalently) on a general theory of effectively decidable properties and relations.

Compare our incompleteness theorems in Chapters 6 and 7. Theorem 6.3 says: If T is a sound effectively axiomatized theory whose language is sufficiently expressive, then T is not negation-complete. Theorem 7.2 says: if T is a consistent, sufficiently strong, effectively axiomatized theory of arithmetic, then T is not negation-complete. A 'sufficiently expressive language', remember, is one which could express at least every effectively computable one-place function, and a 'sufficiently strong theory' is one which can capture at least all effectively decidable numerical properties. So those informal theorems *do* deploy the notions of effective computability and effective decidability. And to get the introductory part of the book and its informal completeness theorem to fit together nicely with our later official Gödelian proofs, we'll therefore need to give a formal treatment of these notions.

So that's our main task in the remaining chapters. Of course, we are not aiming here for a very extensive coverage of the general theory of computability (that would require a book in itself); we'll just be concentrating on a handful of central topics which are most immediately relevant to developing our understanding of incompleteness theorems. In more detail, here's what lies ahead:

1. We first extend the idea of a primitive recursive function in a natural way, and define a wider class of intuitively computable functions, the μ-recursive functions. We give an initial argument for *Church's Thesis* that these μ-recursive functions comprise *all* total numerical functions which are effectively computable. (Chapter 38)

2. We already know that Q, and hence PA, can capture all the p.r. functions: we next show that they can capture all the μ-recursive functions. (Chapter 39)

3. The fact that Q and PA are recursively adequate immediately entails that neither theory is decidable – and it isn't mechanically decidable either what's a theorem of first-order logic. We quickly derive the formal counterpart of the informal syntactic incompleteness theorem of Chapter 7. (Chapter 40)

4. We then turn to introduce another way of defining a class of intuitively computable functions, the *Turing-computable* functions: *Turing's Thesis* is that these are exactly the effectively computable functions. We go on to outline a proof that the Turing-computable (total) functions are in fact just the μ-recursive functions again. (Chapters 41, 42)

5. Next we prove another key limitative result (i.e. a result, like Gödel's, about what *can't* be done). There can't be a Turing machine which solves the *halting problem*: there is no general effective way of telling in advance whether an arbitrary machine with program Π ever halts when it is run from input n. We show that the unsolvability of the halting problem gives us another proof that it isn't mechanically decidable what's a theorem of first-order logic, and it also entails Gödelian incompleteness again. (Chapter 43)

6. The fact that two independent ways of trying to characterize the class of computable functions coincide supports what we can now call the *Church-Turing Thesis*, which underlies the links we need to make e.g. between formal results about what a Turing machine can decide and results about what is effectively decidable in the intuitive sense. We finish the book by discussing the Church–Turing Thesis further, and consider its status. (Chapters 44, 45)

38 μ-Recursive functions

This chapter introduces the notion of a μ-recursive function – which is a natural extension of the idea of a primitive recursive function. Plausibly, the effectively computable functions are exactly the μ-recursive functions (and likewise, the effectively decidable properties are exactly those with μ-recursive characteristic functions).

38.1 Minimization and μ-recursive functions

The primitive recursive functions are the functions which can be defined using *composition* and *primitive recursion*, starting from the successor, zero, and identity functions. These functions are computable. But they are not the only computable functions defined over the natural numbers (see Section 14.5 for the neat diagonal argument which proves this). So the natural question to ask is: what other ways of defining new functions from old can we throw into the mix in order to get a broader class of computable numerical functions (hopefully, to get *all* of them)?

As explained in Section 14.4, p.r. functions can be calculated using *bounded* loops (as we enter each 'for' loop, we state in advance how many iterations are required). But as Section 4.6 illustrates, we also count *unbounded search* procedures – implemented by 'do until' loops – as computational. So, the obvious first way of extending the class of p.r. functions is to allow functions to be defined by means of some sort of 'do until' procedure. We'll explain how to do this in four steps.

(a)   Here's a simple example of a 'do until' loop in action. Suppose that G is a decidable numerical relation. And suppose that for every x there is a y such that Gxy. Then, given a number x, we can find a G-related number y by the brute-force algorithmic method of running through the numbers y from zero up and deciding in each case whether Gxy, until we get a positive result.

Suppose that G's characteristic function is the function g (so Gxy holds just when $g(x, y) = 0$). The algorithm can be presented like this:

1. $y := 0$
2.   do until $g(x, y) = 0$
3. $y := y + 1$
4.   end do
5. $f(x) := y$

Here, we set y initially to take the value 0. We then enter a loop. At each iteration, we do a computation to decide whether the Gxy holds, i.e. whether $g(x, y) = 0$.[1] If it does, we exit the loop and put $f(x)$ equal to the current value of y; otherwise we increment y by one and do the next test. By hypothesis, we do eventually hit a value of y such that $g(x, y) = 0$: the program is bound to terminate. So this 'do until' routine calculates the number $f(x)$ which is *the least y such that $g(x, y) = 0$*, i.e. the least number to which x is G-related. This algorithm, then, gives us a way of effectively calculating the values of a new total function f, given the function g.

(b)   Now we generalize. Let $\vec{x}$ stand in for n variables. Then we'll say that

> The $(n+1)$-place function $g(\vec{x}, y)$ is *regular* iff it is a total function and for all values of $\vec{x}$, there is a y such that $g(\vec{x}, y) = 0$.

Suppose g is a regular computable function. Then the following routine will effectively compute another function $f(\vec{x})$:

1. $y := 0$
2.   do until $g(\vec{x}, y) = 0$
3. $y := y + 1$
4.   end do
5. $f(\vec{x}) := y$

By hypothesis, g is a total computable function; so for each k, checking whether $g(\vec{x}, k) = 0$ is a mechanical business which must deliver a verdict. By hypothesis again, g is regular, so the looping procedure eventually terminates for each $\vec{x}$. Hence f is a total computable function, defined for all arguments $\vec{x}$.

   This motivates another definition:

> Suppose $g(\vec{x}, y)$ is an $(n + 1)$-place regular function. Let $f(\vec{x})$ be the n-place function which, for each $\vec{x}$, takes as its value the least y such that $g(\vec{x}, y) = 0$. Then we say that f is defined by *regular minimization* from g.

Then what we've just shown is that *if f is defined from the regular computable function g by regular minimization, then f is a total computable function too*, with values of f effectively computable using a 'do until' routine.[2]

(c)   Now some notation. Recall, in Section 14.7(c) we introduced the symbolism 'μy' to abbreviate 'the least y such that ...'. So now, when f is defined from g by regular minimization, we can write:

---

[1] Recall, for us 0 serves as the truth-value *true*: see Section 2.2 again.

[2] If we drop the requirement that g is regular, the 'do until' procedure may sometimes, or even always, fail to produce an output: it may compute a *partial* function $f(\vec{x})$ which is defined for only some or even for no values. The theory of partial computable functions is a very important rounding out of the general theory of computable functions. But we don't need to tangle with it in this book. Almost all the functions we'll be talking about are *total* functions, as we will keep emphasizing from time to time.

$$f(\vec{x}) = \mu y[g(\vec{x}, y) = 0].$$

(The square brackets here are strictly speaking unnecessary, but are not unusual and greatly aid readability.) When $g(\vec{x}, y)$ is the characteristic function of the relation $G(\vec{x}, y)$, we will occasionally write, equivalently,

$$f(\vec{x}) = \mu y[G(\vec{x}, y)].$$

Compare, then, the operation of bounded minimization which we met in Section 14.7: we are now concerned with a species of *unbounded* minimization.

(d)   Summarizing so far: we said that we can expect to expand the class of computable functions beyond the p.r. ones by considering functions that are computed using a 'do until' search procedure. We've just seen that when we define a function by regular minimization, this in effect specifies that its value is to be computed by just such a search procedure. Which suggests that a third mode of definition to throw into the mix for defining computable functions, alongside composition and primitive recursion, is definition by regular minimization.

With that motivation, let's say:

> The *μ-recursive* functions are those that can be defined from the initial functions by a chain of definitions by composition, primitive recursion and/or regular minimization.[3]

Or putting it more carefully, we can say

1. The initial functions S, Z, and I_i^k are μ-recursive;
2. if f can be defined from the μ-recursive functions g and h by composition, then f is μ-recursive;
3. if f can be defined from the μ-recursive functions g and h by recursion, then f is μ-recursive;
4. if g is a *regular* μ-recursive function, and f can be defined from g by regular minimization, then f is μ-recursive;
5. nothing else is a μ-recursive function.

Since regular minimization yields total functions, the μ-recursive functions are always total computable functions. Trivially, all p.r. functions also count as μ-recursive functions.

38.2   Another definition of μ-recursiveness

This little section is just to forestall a query which might already have occurred to you!

---

[3] Many, perhaps most, writers nowadays use plain 'recursive' instead of 'μ-recursive'. But the terminology hereabouts can be confusingly variable. It will do no harm, then, to stick to our explicit label.

A 'for' loop – i.e. a programming structure which instructs us to iterate some process as a counter is incremented from 0 to n – can of course be recast as a 'do until' loop which tells us to iterate the same process while incrementing the counter *until its value equals* n. So it looks as if definitions by primitive recursion, which call 'for' loops, could be subsumed under definitions by minimization, which call 'do until' loops. Hence you might well suspect that clause (3) in our definition is redundant. And you'd be *almost*, though not quite, right. By a theorem of Kleene's (1936c), you can indeed drop (3) *if* you add addition, multiplication and the characteristic function of the less-than relation to the list of initial functions. Some books do define μ-recursive functions this way; see e.g. Shoenfield (1967, p. 109).

Still, I for one don't find this approach as natural or illuminating, so let's stick to the more conventional mode of presentation given in the previous section.

38.3   The Ackermann–Péter function

Since μ-recursive functions can be defined using unbounded searches and p.r. functions can't, we'd expect there to be μ-recursive functions which aren't primitive recursive. But can we give some examples?

Well, the computable-but-not-p.r. function $d(n)$ that we constructed by the diagonalization trick in Section 14.5 is in fact an example. But it isn't immediately obvious *why* the diagonal function is μ-recursive. So in this section and the next we'll look at another example, which is both more tractable and also mathematically more natural. The basic idea is due to Wilhelm Ackermann (1928).

Let's begin with a simple observation. Recall that any p.r. function f can be specified by a chain of definitions in terms of primitive recursion and composition leading back to initial functions. This definition won't be unique: there will always be various ways of defining f (for a start, by throwing in unnecessary detours). But take the shortest definitional chain – or, if there are ties for first place, take one of the shortest. Now, the length of this shortest definitional chain for f will evidently put a limit on how fast $f(n)$ can grow as n grows. That's because it puts a limit on how complicated the computation can be – in particular, it restricts the number of loops-within-loops-within-loops that we have to play with. So it limits the number of times we ultimately get to apply the successor function, depending on the initial input argument n. A similar point applies to two-place functions, etc.

That's a bit abstract, but the point is easily seen if we consider the two-place functions f_1, i.e. *sum* (repeated applications of the successor function), f_2, i.e. *product* (repeated sums), f_3, i.e. *exponentiation* (repeated products). These functions have increasingly long full definitional chains; and the full programs for computing them involve 'for' loops nested with increasing depth. And as their respective arguments grow, the value of f_1 of course grows comparatively slowly, f_2 grows faster, f_3 faster still.

This sequence of functions can obviously be continued. Next comes f_4, the *super-exponential*, defined by repeated exponentiation:

$$x \Uparrow 0 = x$$
$$x \Uparrow Sy = x^{x \Uparrow y}.$$

Thus, for example, $3 \Uparrow 4$ is $3^{3^{3^{3^3}}}$ with a 'tower' of four exponents. Similarly, we can define f_5 (super-duper-exponentiation, i.e. repeated super-exponentiation), f_6 (repeated super-duper-exponentiation), and so on. The full chain of definitions for each f_k gets longer and longer as k increases – and the values of the respective functions grow faster and faster as their arguments are increased.[4]

But now consider the function $a(x) = f_x(x, x)$. The value of $a(x)$ grows *explosively*, running away ever faster as x increases. Indeed, take any given one-place p.r. function; this has a maximum rate of growth determined by the length of its definition, i.e. a rate of growth comparable to some $f_n(x, x)$ in our hierarchy. But $f_x(x, x)$ eventually grows faster than any particular $f_n(x, x)$, when $x > n$. Hence $a(x)$ isn't primitive recursive. Yet it is evidently computable.

This idea of Ackermann's is *very* neat, and is worth pausing over and developing a bit. So consider again the recursive definitions of our functions f_1 to f_4 (look again at Section 14.1, and at our definition of '$\Uparrow$' above). We can rewrite the second, recursion, clauses in each of those definitions as follows:

$$f_1(y, Sz) = Sf_1(y, z)$$
$$= f_0(y, f_1(y, z)), \text{ if we cunningly define } f_0(y, z) = Sz$$
$$f_2(y, Sz) = f_1(y, f_2(y, z))$$
$$f_3(y, Sz) = f_2(y, f_3(y, z))$$
$$f_4(y, Sz) = f_3(y, f_4(y, z)).$$

There's a pattern here! So now suppose we put

$$f(x, y, z) =_{\text{def}} f_x(y, z).$$

Then the value of f gets fixed via a *double* recursion:

$$f(Sx, y, Sz) = f(x, y, f(Sx, y, z)).$$

However, nothing very exciting happens to the second variable, 'y'. So we'll now let it just drop out of the picture, and relabel the remaining variable to get a variant on Ackermann's construction due to Rósza Péter (1935). Consider, then, the function p governed by the clause

$$p(Sx, Sy) = p(x, p(Sx, y)).$$

Of course, this single clause doesn't yet fully define p – it doesn't tell us what to do when either argument is zero. So we need somehow to round out the definition. Let's adopt the following three equations:

---

[4]The claim, of course, isn't that longer definitions always *entail* faster growth, only that our examples show how longer definitions *permit* faster growth.

$$p(0, y) = Sy$$
$$p(Sx, 0) = p(x, S0)$$
$$p(Sx, Sy) = p(x, p(Sx, y)).$$

To see how these equations work together to determine the value of p for given arguments, work through the following calculation:

$$
\begin{aligned}
p(2, 1) &= p(1, p(2, 0)) \\
&= p(1, p(1, 1)) \\
&= p(1, p(0, p(1, 0))) \\
&= p(1, p(0, p(0, 1))) \\
&= p(1, p(0, 2)) \\
&= p(1, 3) \\
&= p(0, p(1, 2)) \\
&= p(0, p(0, p(1, 1))) \\
&= p(0, p(0, p(0, p(1, 0)))) \\
&= p(0, p(0, p(0, p(0, 1)))) \\
&= p(0, p(0, p(0, 2))) \\
&= p(0, p(0, 3)) \\
&= p(0, 4) \\
&= 5.
\end{aligned}
$$

To evaluate the function, the recipe is as follows. At each step look at the innermost occurrence of p, and apply whichever of the definitional clauses pertains – it's trivial that only one can. Keep on going until at last you reach something of the form $p(0, m)$ and then apply the first clause one last time and halt.

Two comments on this. First, inspection of the patterns in our sample computation should convince you that the computation of $p(m, n)$ always terminates.[5] Hence the function is well-defined. Second, our informal recipe for computing it evidently involves a *do until* procedure.

So – given everything we've said – it shouldn't be a surprise to learn that we indeed have the following theorem, for which we'll give an outline proof in the next section:

Theorem 38.1 *The Ackermann–Péter function is μ-recursive but not primitive recursive.*

---

[5]Hint: remove the 'p's and the brackets, so you can see the patterns in the numbers more easily.

But here, for enthusiasts, is a quick proof by reductio that the computation always terminates. Suppose that for some values of m, n, $p(m, n)$ is *not* defined. Let a be the smallest number such that, for some n, $p(a, n)$ is not defined. Then let b be the smallest number such that $p(a, b)$ is not defined. From the first clause defining p, $a > 0$; and hence from the second clause, $b > 0$. So we have, for some a' and b', $a = Sa'$, $b = Sb'$. Then the hypothesis is that $p(Sa', Sb')$ is undefined. But since b, i.e. Sb', is the smallest number n such that $p(Sa', n)$ is undefined, that means $p(Sa', b')$ *is* defined. And since a, i.e. Sa', is the smallest number m for which $p(m, n)$ is not always defined, $p(a', n)$ is always defined, and hence in particular $p(a', p(Sa', b'))$ is defined. But by the third clause, $p(Sa', Sb') = p(a', p(Sa', b'))$ so is defined after all. Contradiction.

38.4 Ackermann–Péter is μ-recursive but not p.r.

Proof sketch: $p(x, y)$ is μ-recursive The general strategy here is neat, and we'll later make use of it a number of times, so it's well worth getting the hang of this argument-by-coding. The outline proof has three stages.

(i) *Introducing more coding* Consider the successive terms in our calculation of the value of $p(2, 1)$. We can introduce code numbers representing these terms by a simple, two-step, procedure:

1. Transform each term like $p(1, p(0, p(1, 0)))$ into a corresponding sequence of numbers like $\langle 1, 0, 1, 0 \rangle$ by the simple expedient of deleting the brackets and occurrences of the function-symbol 'p'. (We can uniquely recover terms from such sequences in the obvious way.)

2. Code the resulting sequence $\langle 1, 0, 1, 0 \rangle$ by Gödel-style numbering, e.g. by using powers of primes. So we put e.g.

$$\langle l, m, n, o \rangle \Rightarrow 2^{l+1} \cdot 3^{m+1} \cdot 5^{n+1} \cdot 7^{o+1}$$

(where we need the '+1' in the exponents to handle the zeros).

Hence we can think of our computation of $p(2, 1)$ as generating in turn what we'll call the 'p-sequences'

$$\langle 2, 1 \rangle, \langle 1, 2, 0 \rangle, \langle 1, 1, 1 \rangle, \langle 1, 0, 1, 0 \rangle, \ldots, \langle 0, 4 \rangle, \langle 5 \rangle.$$

Then we code up each such p-sequence; so the successive steps in the calculation of $p(2, 1)$ will respectively receive what we'll call the π-code numbers

$$72, 540, 900, 2100, \ldots$$

(ii) *Coding/decoding functions* So far, that's just routine coding. Now we put it to work by defining a couple of coding/decoding functions as follows:

   i. $c(x, y, z)$ is the π-code of the p-sequence corresponding to the output of the z-th step (counting from zero) in the calculation of $p(x, y)$ if the calculation hasn't yet halted by step z; and otherwise $c(x, y, z) = 0$. So, for example, $c(2, 1, 0) = 72$, $c(2, 1, 3) = 2100$, and $c(2, 1, 20) = 0$.

   ii. $fr(x) =$ one less than the exponent of 2 in the prime factorization of x. Hence, if n is a π-code for a sequence of numbers, $fr(n)$ recovers the first member of the sequence.

(iii) *Facts about our coding functions*

1. The coding function $c(x, y, z)$ is primitive recursive. That's because the evaluation of c for arguments l, m, n evidently involves a step-by-step numerical computation tracking the first n steps in the calculation of $p(l, m)$,

using the very simple rules that take us from one step to the next. No step-to-step move involves flying off on an open-ended search. So 'for' loops will suffice to construct an algorithm for the computation of c. And such an algorithm will always determine a p.r. function.

That's quick-and-dirty: making the argument watertight is pretty tedious though not difficult. There's nothing to be learnt from spelling out the details here: so we won't.

2. The calculation of the function p for given arguments x, y eventually halts at the z-th step for some z, and then $c(x, y, Sz) = 0$. Hence c is regular.

3. $\mu z[c(x, y, Sz) = 0]$ is therefore the step-number of the final step in the calculation which delivers the value of $p(x, y)$. Since c is regular, it follows that $\mu z[c(x, y, Sz) = 0]$ defines a μ-recursive function.

4. Hence $c(x, y, \mu z[c(x, y, Sz) = 0])$ gives the π-code number of the final value of $p(x, y)$. Since this compounds a p.r. (hence μ-recursive) function with a μ-recursive one, it is also μ-recursive.

5. Hence, decoding, $p(x, y) = fr(c(x, y, \mu z[c(x, y, Sz) = 0]))$.

6. But the function fr is primitive recursive (in fact $fr(x) =_{\text{def}} exf(x, 0) \mathbin{\dot{-}} 1$, where exf is as introduced in Section 14.8).

7. Thus $p(x, y)$ is the composition of a p.r. function and a μ-recursive function. Hence it is μ-recursive.　　　　　　⊠

Proof sketch: $p(x, y)$ is not p.r. The quick and dirty argument for this notes that the functions $p(0, y)$, $p(1, y)$, $p(2, y)$, ... are faster and faster growing functions of y. Take any p.r. function $f(y)$. How fast this can grow as y increases will depend on the length of f's definition as a p.r. function. And there will always be some $p(k, y)$ which grows faster, if we take k large enough. So, a fortiori, for $y > k$, $p(y, y)$ grows faster than $f(y)$. Hence $p(y, y)$ is distinct from $f(y)$. But f was an arbitrary one-place p.r. function, so $p(y, y)$ is distinct from any p.r. function. Therefore $p(x, y)$ isn't p.r. either (for if it were, $p(y, y)$ would be p.r. after all).

To tidy up this argument, let's say a monadic function $a(y)$ dominates a function $g(\vec{x})$ if $g(\vec{m}) < a(n)$ whenever each of the numbers in $\vec{m}$ is no greater than n. Then, with a bit of effort, we can show:

1. Each of the initial functions S, Z, and I_i^k is dominated by $p(2, y)$.

2. If the p.r. functions $g(y)$ and $h(y)$ are dominated by $p(\hat{g}, y)$ and $p(\hat{h}, y)$ respectively (for corresponding numbers $\hat{g}$ and $\hat{h}$), then their composition $f(y) = h(g(y))$ is dominated by $p(j, y)$, where j depends on $\hat{g}$ and $\hat{h}$. Similarly for many-placed functions.

3. If the p.r. function $h(y)$ is dominated by $p(\hat{h}, y)$, then the function f defined from h by primitive recursion – so $f(0) = g, f(Sy) = h(f(y))$ – is dominated by $p(k, y)$, where k depends on g and $\hat{h}$. Similarly for more complex definitions by primitive recursion.

Hence, since every p.r. function is built out of the initial functions by a finite number of applications of composition and primitive recursion, it follows that for any p.r. $f(\vec{y})$ there is some $p(n, y)$ that dominates it. As before, it follows that p isn't p.r.[6] ⊠

38.5 Introducing Church's Thesis

The coding argument that shows that p is μ-recursive illustrates a very powerful general strategy. For example – although we won't give the details here – the computable-but-not-p.r. diagonal function $d(n)$ from Section 14.5 can similarly be shown to be μ-recursive by a broadly similar proof.[7]

And now generalizing, we might reasonably expect to be able to code up the step-by-little-step moves in *any* well-defined calculation using a primitive recursive coding function like c (primitive recursive because, when broken down to minimal steps, there again won't be any flying off on open-ended searches). If the output of the calculation is defined for every input, then – using the new c-function – exactly the same argument will be available to show that the mapping from input to output must be a μ-recursive function.

---

[6]To fill in the details, particularly about how the bounds j and k are calculated, see Hedman (2004, pp. 308–309).

[7]See Péter's classic (1951), with a revised edition translated as her (1967). The key idea is to use a double recursion again to define a function $\varphi(m, n)$ such that, for a given m, $\varphi(m, n) = f_m(n)$, where running through the f_i gives us our effective enumeration of the p.r. functions. And since φ is definable by a double recursion it can be shown to be μ-recursive by the same kind of argument which showed that the Ackermann–Péter function is μ-recursive. Hence $d(n) = \varphi(n, n) + 1$ is μ-recursive too.

Just for enthusiasts: it is perhaps worth footnoting that it would be quite wrong to take away from our discussion so far the impression that μ-recursive-but-not-p.r. functions must all suffer from explosive growth. Péter gives a beautiful counter-example. Take our enumeration f_i of p.r. functions, and now consider the functions $g_i(n) =_{\text{def}} sg(f_i(n))$, where sg is as defined in Section 14.8 (i.e. $sg(k) = 0$ for $k = 0$, and $sg(k) = 1$ otherwise). Evidently, running through the g_i gives us an effective enumeration – with many repetitions – of all the p.r. functions which only take the values 0 and 1. Now consider the μ-recursive function $\psi(n) = \overline{sg}(\varphi(n, n)) = |1 - sg(\varphi(n, n))|$ (where φ is as above). This function too only takes the values 0 and 1; but it can't be primitive recursive. For suppose otherwise. Then for some k, $\psi(n) = g_k(n) = sg(\varphi(k, n))$. So we'd have

$$sg(\varphi(k, n)) = \psi(n) = |1 - sg(\varphi(n, n))|$$

and hence

$$sg(\varphi(k, k)) = \psi(k) = |1 - sg(\varphi(k, k))|.$$

Which is impossible. Therefore there are μ-recursive-but-not-p.r. functions which only ever take the values 0 and 1, and hence do not suffer value explosion. However, while *values* of such functions can remain tame, *lengths of computations* don't, as we'll see in Section 43.6, fn. 5. There remains a sense, then, in which μ-recursive-but-not-p.r. functions are *wild*.

This line of thought – which has its roots in Church (1936b) – very strongly encourages the conjecture that in fact *all* effectively computable total functions will turn out to be μ-recursive.

Reflections on modern programming languages point in the same direction. For when things are reduced to basics, we see that the main programming structures available in such languages are (in effect) 'for' loops and 'do until' loops, which correspond to definitions by primitive recursion and minimization. Hence – given that our modern general-purpose programming languages have so far proved sufficient for specifying algorithms to generate any computable function we care to construct – it doesn't seem a very big leap to conjecture that every algorithmically computable total function should be definable in terms of composition (corresponding to the chaining of program modules), primitive recursion, and minimization.

In sum, such considerations certainly give a very high initial plausibility to what's called

> *Church's Thesis*   The total numerical functions that are effectively computable by some algorithmic routine are just the μ-recursive functions.[8]

And certainly all the evidence supports Church's Thesis. For a start, no one has ever been able to define an intuitively computable numerical total function which *isn't* μ-recursive.

We'll be saying a lot more about all this later. Pending further discussion, however, we'll for the moment just *assume* that Church's Thesis is true. Given this assumption, the class of μ-recursive functions is indeed of very special interest as it just *is* the class of effectively computable (total) numerical functions.

38.6   Why can't we diagonalize out?

You might find that last claim very puzzling (in fact, perhaps you *ought* to find it puzzling!). For don't we already have the materials to hand for a knock-down argument against Church's Thesis? Back in Section 14.5, we proved that not every computable function is primitive recursive by the trick of 'diagonalizing out'. That is to say, we used a diagonal construction which took us from a list of all the p.r. functions to a further computable function which *isn't* on the list. Why shouldn't we now use the same trick again to diagonalize out of the class of μ-recursive functions?

Well, the argument would have to go:

---

[8]The reason for the label will emerge in Chapter 44. Compare *Turing's Thesis* which we very briefly introduced in Section 3.1: that says that the (total) numerical functions that are effectively computable by some algorithmic routine are just those functions that are computable by a Turing machine (which is a computer following a very simple-minded type of program: for more explanation, see Chapter 41). It turns out that our two Theses are equivalent, because the μ-recursive functions are exactly the Turing-computable ones, as we'll show in Chapter 42.

Take an effective enumeration of the μ-recursive functions, f_0, f_1, f_2, ..., and define the diagonal function $d(n) = f_n(n) + 1$. Then d differs from each f_j (at least for the argument j). But d is computable (since to evaluate it for argument n, you just set a computer to enumerate the f_j until it reaches the n-th one, and then by hypothesis the value of $f_n(n) + 1$ is computable). So d is computable but not μ-recursive.

But this argument fails, and it is very important to see why. The crucial point is that we are not entitled to its initial assumption. While the p.r. functions are effectively enumerable, *we can't assume that there is an effective enumeration of the μ-recursive functions.*

What makes the difference? Well, remind yourself of the informal argument (in Section 14.5) that shows that we can mechanically list off the recipes for the p.r. functions. If we now try to run a parallel argument for the claim that the μ-recursive functions are effectively enumerable, things go just fine at the outset:

Every μ-recursive function has a 'recipe' in which it is defined by primitive recursion or composition or regular minimization from other functions which are defined by recursion or composition or regular minimization from other functions which are defined ultimately in terms of some primitive starter functions. So choose some standard formal specification language for representing these recipes. Then we can effectively generate 'in alphabetical order' all possible strings of symbols from this language ...

But at this point the parallel argument breaks down, since we *can't* continue

... and as we go along, we can mechanically select the strings that obey the rules for being a recipe for a μ-recursive function.

In order to determine mechanically whether a series of definitions obeys the rules for being the recipe for a μ-recursive function, we'd need an effective way of determining whether each application of the minimization operator is an application to a *regular* function. So we'd need a way of effectively determining whether a p.r. function $g(\vec{x}, y)$ is such that for each x there is a y such that $g(\vec{x}, y) = 0$. And there is in general no effective way of doing that.

It's worth adding another observation. Note that we *know* that there can't be an effective enumeration of the *effectively* computable total functions f_0, f_1, f_2, ... For if there were one, we could define $d(n) = f_n(n) + 1$, which would then evidently be an effectively computable function, but which would be distinct from all the f_i. Contradiction.

Since we know the effectively computable total functions are not effectively enumerable, we certainly can't just *assume* that there is an effective enumeration of the μ-recursive functions. To use the 'diagonalizing out' argument against

Church's Thesis, we'd *already* need some independent reason for thinking the enumeration can be done. There isn't any.

In sum: Church's Thesis that the μ-recursive functions are *all* the (total, numerical) computable functions lives to fight another day.[9]

38.7   Using Church's Thesis

Church's Thesis is a biconditional: a total numerical function is μ-recursive if and only if it is effectively computable in the intuitive sense. Half the Thesis is quite unexciting – if a function is μ-recursive, then it is certainly effectively computable. It is the other half which is the interesting claim, the half which says that if a total numerical function is *not* μ-recursive then it is *not* computable in the intuitive sense.

Over the coming chapters, we'll repeatedly be appealing to Church's Thesis, but in two quite different ways which we need to distinguish very clearly. Let's call these the *interpretive* and the *labour-saving* uses respectively.

The interpretive use relies on the Thesis to pass from technical claims about what is or isn't μ-recursive to claims about what is or isn't effectively computable. Here, then, the Thesis is being used to justify an informal gloss on our technical results. And if we are in general to interpret formal results about μ-recursiveness as telling us about effective computability in the intuitive sense, then necessarily we have to appeal to one or other half of the Thesis.

The labour-saving use relies on the Thesis to pass in the opposite direction, from the informal to the technical. In particular, it allows us to jump from a quick-and-dirty informal proof that something is effectively computable to conclude that a corresponding function is μ-recursive. This kind of fast-track argument for some technical claim is fine, given that Church's Thesis is entirely secure. However, any claim about μ-recursiveness which can be established using this informal fast-track method *must* also be provable the hard way, without appeal to the Thesis (otherwise we would have located a disconnect between the informal notion of computability and μ-recursiveness, contradicting the Thesis). Hence this labour-saving use of the Thesis must always be inessential.

For clarity's sake, we'll adopt the following convention in the rest of this book. When we simply say 'by Church's Thesis ...', or 'given Church's Thesis ...', etc. we'll always be appealing to Church's Thesis in the first way, to make a connection between a formal claim and a claim about computability in the intuitive sense. When we occasionally make use of Church's Thesis in the second way, to support a technical claim that some function is μ-recursive, then we'll explicitly signal what we are doing: we'll say 'by a labour-saving appeal to Church's Thesis' or some such.

---

[9]Let's stress again: it is important that we have been talking throughout about *total* computable functions.

39 Q is recursively adequate

Theorem 17.1 tells us that Q can capture all p.r. functions. This result was a key load-bearing element in the proof of the Gödel-Rosser Theorem that nice theories – consistent p.r. axiomatized theories which extend Q – must be incomplete.

In this chapter we prove that Q can in fact capture all μ-recursive functions too. This new result will be the corresponding load-bearing part of proofs of various central theorems in the next chapter.

39.1 Capturing a function defined by minimization

We prove a preliminary theorem about one-place functions (which has an obvious generalization to many-place functions):

> **Theorem 39.1** *Suppose the one-place function f is defined by regular minimization from the two-place function g, so $f(x) = \mu y[g(x, y) = 0]$. And suppose that $g(x, y)$ is captured in Q by the Σ_1 wff $G(x, y, z)$. Then f is captured by the Σ_1 wff*
>
> $$F(x, y) =_{\text{def}} G(x, y, 0) \land (\forall u \leq y)(u = y \lor \exists z(G(x, u, z) \land z \neq 0)).$$

Intuitively, F expresses f; i.e. $f(m) = n$ iff $F(\overline{m}, \overline{n})$ is true (think about it!). We see by inspection that, if G is Σ_1, then so is F.[1] To prove F also does the capturing job, we need to show that for any m, n,

    i. if $f(m) = n$, then $Q \vdash F(\overline{m}, \overline{n})$,
    ii. $Q \vdash \exists! w\, F(\overline{m}, w)$.

It is very tempting to say 'Exercise!'. But, if you *really* insist ...

Proof of (i) There are two cases to consider. Suppose $f(m) = 0$, i.e. $g(m, 0) = 0$. Then, because G captures g, we have $Q \vdash G(\overline{m}, 0, 0)$. It is then trivial that $Q \vdash (\forall u \leq 0)\, u = 0$ and hence $Q \vdash (\forall u \leq 0)(u = 0 \lor \exists z(G(x, u, z) \land z \neq 0))$. That shows $Q \vdash F(\overline{m}, 0)$.

Now suppose $f(m) = n$, where $n > 0$. Then we have $g(m, n) = 0$, but for each $k < n$, $g(m, k) = p$ for some $p \neq 0$. Because G captures g, we have $Q \vdash G(\overline{m}, \overline{n}, 0)$. We also have, for any $k < n$, for some p, $Q \vdash G(\overline{m}, \overline{k}, \overline{p}) \land \overline{p} \neq 0$, whence $Q \vdash \exists z(G(\overline{m}, \overline{k}, z) \land z \neq 0)$. So for any $k \leq n$ $Q \vdash \overline{k} = \overline{n} \lor \exists z(G(\overline{m}, \overline{k}, z) \land z \neq 0)$. We can then infer $Q \vdash (\forall u \leq \overline{n})(u = \overline{n} \lor \exists z(G(\overline{m}, u, z) \land z \neq 0))$ by appeal to (O4) of Section 11.3. And that shows $Q \vdash F(\overline{m}, \overline{n})$. $\boxtimes$

---

[1] Indeed, that's exactly why we have chosen our particular construction for F, rather than use the perhaps initially more appealing $G(x, y, 0) \land (\forall u \leq y)(u \neq y \to \neg G(x, u, 0))$.

Proof of (ii)   Assume $f(m) = n$, and it is enough to show that inside Q we can argue from $\mathsf{F}(\overline{m}, \overline{n})$ and $\mathsf{F}(\overline{m}, \mathsf{w})$ to $\mathsf{w} = \overline{n}$. By (O8) of Section 11.3 we have $\mathsf{w} \le \overline{n} \vee \overline{n} \le \mathsf{w}$. Argue by cases.

Suppose $\mathsf{w} \le \overline{n}$. Given $\mathsf{F}(\overline{m}, \overline{n})$ we can infer $\mathsf{w} = \overline{n} \vee \exists \mathsf{z}(\mathsf{G}(\overline{m}, \mathsf{w}, \mathsf{z}) \wedge \mathsf{z} \ne 0)$. But $\mathsf{F}(\overline{m}, \mathsf{w})$ implies $\mathsf{G}(\overline{m}, \mathsf{w}, 0)$ which rules out the second disjunct (why? because $\mathsf{w} \le \overline{n}$ we have $\mathsf{w} = \overline{k}$ for some $k \le n$, and so $\mathsf{G}(\overline{m}, \overline{k}, 0)$; and because G captures g we have $\exists ! \mathsf{z} \mathsf{G}(\overline{m}, \overline{k}, \mathsf{z})$). So $\mathsf{w} = \overline{n}$.

Suppose $\overline{n} \le \mathsf{w}$. Given $\mathsf{F}(\overline{m}, \mathsf{w})$ we can infer $\overline{n} = \mathsf{w} \vee \exists \mathsf{z}(\mathsf{G}(\overline{m}, \overline{n}, \mathsf{z}) \wedge \mathsf{z} \ne 0)$. But $\mathsf{F}(\overline{m}, \overline{n})$ rules out the second disjunct. So we can again infer $\mathsf{w} = \overline{n}$.   ⊠

39.2   The recursive adequacy theorem

(a)   Recall that we said that a theory is p.r. adequate if it captures each p.r. function (Section 17.1). Let's likewise say:

> A theory is *recursively adequate* iff it captures each μ-recursive function.

Theorems 17.1 and 17.2 tell us that Q is p.r. adequate, and indeed can capture any p.r. function using a Σ_1 wff. Similarly:

> **Theorem 39.2** Q *can capture any μ-recursive function using a* Σ_1 *wff.*

Proof   It is evidently enough to show that:

1. Q can capture the initial functions by Σ_1 wffs.

2. If Q can capture the functions g and h by Σ_1 wffs, then it can also capture by a Σ_1 wff a function f defined by composition from g and h.

3. If Q can capture the functions g and h by Σ_1 wffs, then it can also capture by a Σ_1 wff a function f defined by primitive recursion from g and h.

4. If Q can capture the function g by a Σ_1 wff, then Q can capture by a Σ_1 wff a function f defined by regular minimization from g.

Since any μ-recursive function f has a definition via composition, primitive recursion, and regular minimization, leading back to initial functions, it will follow that f can be captured by a Σ_1 wff.

Clauses (1), (2) and (3) were established in Chapter 17. In the previous section, we proved clause (4) for the case where f is a one-place function defined from a two-place function g: but the generalization to the case of an n-place function defined by regular minimization of an $n + 1$-place function is immediate.[2]   ⊠

---

[2]We noted in Section 38.2 that we can also define the μ-recursive functions using a somewhat richer class of initial functions, together with composition and minimization. If we show this alternative definition to be equivalent, we can then prove that Q captures the μ-recursive functions using just a version of (1) plus (2) and (4), avoiding the hard work of establishing (3). Among others, Epstein and Carnielli (2000) prove Q's recursive adequacy this way.

Given that Q captures all μ-recursive functions, so does *any* nice theory, including PA of course. And, by Theorem 16.1, such theories will also capture all properties with μ-recursive characteristic functions.

39.3   Sufficiently strong theories again

Recall from Section 7.1 the idea of a 'sufficiently strong' theory, i.e. the idea of a theory that captures all intuitively decidable one-place properties of numbers, i.e. the idea of a theory that captures all effectively computable two-place characteristic functions.

We now know – by Church's Thesis (used in interpretative mode) – that a theory which is recursively adequate can capture *all* effectively computable functions, and hence will be sufficiently strong.

So Theorem 39.2 (at very long last) redeems our earlier promise to vindicate the intuitive notion of a sufficiently strong theory: such a theory need be no richer than the decidedly tame theory Q. We will return to this point.

39.4   Nice theories can *only* capture μ-recursive functions

The result that Q, and hence PA, can capture all μ-recursive functions is the one which we will be repeatedly using. But there's also a converse result:

> **Theorem 39.3** *If T is nice, any total function which can be captured in T is μ-recursive.*

A proof using Church's Thesis   Take the monadic case (it will be obvious how to generalize). Suppose the total function $f(m)$ can be captured as a function in T. Then, by definition, there is a two-place open wff $\varphi(\mathsf{x}, \mathsf{y})$ which is such that if $f(m) = n$, $T \vdash \varphi(\overline{\mathsf{m}}, \overline{\mathsf{n}})$; and if $f(m) \neq n$, $T \vdash \neg\varphi(\overline{\mathsf{m}}, \overline{\mathsf{n}})$, so – since T is consistent by assumption – $T \nvdash \varphi(\overline{\mathsf{m}}, \overline{\mathsf{n}})$. Trivially, then, the value of $f(m)$ is the least number n (indeed, the only number n) such that $T \vdash \varphi(\overline{\mathsf{m}}, \overline{\mathsf{n}})$.

So, for given m, we can effectively compute the value of $f(m)$ using an open-ended search as follows. Start effectively enumerating the T-theorems, and keep on going until we output a theorem of the form $\varphi(\overline{\mathsf{m}}, \overline{\mathsf{n}})$ for some number n. The value of $f(m)$ will then be this resulting value n. (We know we can effectively enumerate the T-theorems by Theorem 4.1; and because f is total, the search for a theorem of the form $\varphi(\overline{\mathsf{m}}, \overline{\mathsf{n}})$ always terminates.)

Since there is therefore an algorithm for computing the value of f, it follows by Church's Thesis (used in labour-saving mode) that f is μ-recursive.   ☒

Note a slightly surprising corollary. Although PA can prove a lot more than Q, the additional power doesn't enable it to capture more (total) functions. Q and PA alike can capture all the μ-recursive functions but no more, so it follows that Q and PA capture exactly the same functions.

40 Undecidability and incompleteness

With a bit of help from Church's Thesis, our Theorem 39.2 – Q is recursively adequate – very quickly yields two new Big Results: first, any nice theory is undecidable; and second, theoremhood in first-order logic is undecidable too.

The old theorem that Q is p.r. adequate is, of course, the key result which underlies our previous incompleteness theorems for theories that are p.r. axiomatized and extend Q. Our new theorem correspondingly underlies some easy (but unexciting) generalizations to recursively axiomatized theories that extend Q. More interestingly, we can now prove a formal counterpart to the informal incompleteness theorem of Chapter 7.

40.1 Some more definitions

We pause for some reminders, interlaced with definitions for some fairly self-explanatory bits of new jargon.

(a) First, recall from Section 3.2 the informal idea of a decidable property, i.e. a property P whose characteristic function c_P is computable. In Section 14.6 we introduced a first formal counterpart to this idea, the notion of a p.r. property, i.e. one with a p.r. characteristic function. However, there can be decidable properties which aren't p.r. (as not all effective computations deliver primitive recursive functions). We now add:

> A numerical property P is *recursively decidable* iff its characteristic function c_P is μ-recursive.

That it is to say, P is recursively decidable iff there is a μ-recursive function which, given input n, delivers a 0/1, yes/no, verdict on whether n is P. The definition obviously extends in a natural way to cover recursively decidable numerical relations. And note that by the remark after the proof of Theorem 39.2, a recursively adequate theory can capture all recursively decidable properties of numbers.

Predicatably, we will say that a non-numerical property is recursively decidable iff some acceptable Gödel-style coding associates the property with a recursively decidable numerical property (and again similarly for relations).[1] We'll also say that a *set* is recursively decidable iff the property of being a member of the set is recursively decidable.

---

[1] By a variant of the argument of Section 19.2, whether a non-numerical property is recursively decidable does not depend on our choice of Gödel coding. (Exercise: check this!)

Finally, Church's Thesis (in interpretative mode) implies that the effectively decidable properties/relations/sets are just the recursively decidable ones.

(b) Here's some standard alternative jargon you should know. A (numerical) *decision problem* takes the following form: to tell of any arbitrary number n whether it is P (or to tell of an arbitrary pair of numbers m, n, whether m has relation R to n, etc.). Then:

> The decision problem for numerical property P is *recursively solvable* iff there is a μ-recursive function which, given any input n, delivers a $0/1$, yes/no, verdict on whether n is P. (Similarly for relations).

To say that the decision problem for P is recursively solvable is therefore just to say that the property P is recursively decidable.

(c) Now some definitions relating to theories.

When we very first characterized the idea of a formal theory, we said that for a properly constructed theory it must be effectively decidable what's a wff, what's an axiom, and what's a well-constructed logical derivation. If we use the Gödel-numbering trick, then the requirement becomes that the properties of numbering a wff, axiom or correct derivation must be decidable. Previously, we introduced the notion of a p.r. axiomatized theory, which is one for which those three numerical properties are primitive recursive. So now let's correspondingly say:

> A theory is *recursively axiomatized* when the properties of numbering a wff, axiom or correct derivation are recursively decidable.

By Church's Thesis again, the effectively axiomatized theories in the intuitive sense are exactly the recursively axiomatized ones.

A nice theory, recall, is one that is consistent, p.r. axiomatized, and extends Q. Correspondingly,

> A *nice'* theory is one that is consistent, recursively axiomatized, and extends Q.

We said in Section 4.4 that a theory is effectively decidable iff the property of being a theorem of that theory is effectively decidable. So correspondingly,

> A theory is *recursively decidable* iff it is recursively decidable whether a given sentence is a theorem.

(d) Finally, recall from Section 3.3 that we said that a set Σ is effectively enumerable (e.e.) iff it is either empty or there is an effectively computable (total) function which 'lists off' its members. Putting that more carefully, Σ is effectively enumerable iff it is either empty or there is a surjective effectively computable (total) function $f \colon \mathbb{N} \to \Sigma$. In other words, Σ is either empty or the range of an effectively computable function. Similarly, then, we'll now say:

A numerical set $\Sigma \subseteq \mathbb{N}$ is *recursively enumerable* iff it is either empty or there is a μ-recursive function which enumerates it – i.e. iff Σ is either empty or it is the range of a μ-recursive function.

And by extension, a non-numerical set Σ is recursively enumerable iff, under an acceptable system of Gödel coding, the set Σ' which contains the code numbers of members of Σ is recursively enumerable.

It is standard to abbreviate 'recursively enumerable' as 'r.e.'. Church's Thesis tells us, of course, that a set is r.e. if and only if it is e.e. (i.e. is effectively enumerable in the intuitive sense).

40.2   Q and PA are undecidable

Now down to new business! We start with a Big Result which is immediate, given what we've already shown:

Theorem 40.1 *If T is a nice theory, it isn't recursively decidable.*

Proof   Fix on an acceptable system of Gödel-numbering. To say that T is recursively decidable is then equivalent to saying that the numerical property $Prov_T$ of numbering a T-theorem is recursively decidable (i.e. it has a μ-recursive characteristic function).

Suppose, then, that $Prov_T$ *is* recursively decidable. Then $Prov_T$ would be capturable in T since T contains Q and so is recursively adequate by Theorem 39.2 and hence captures all recursively decidable properties. However, our Theorem 24.8 long ago told us that no open wff in a nice theory T can capture $Prov_T$. So the supposition is false. $\boxtimes$

If Q is consistent, it is nice. Hence if Q is consistent, it isn't recursively decidable. Likewise, if PA is consistent, it isn't recursively decidable.

An appeal to Church's Thesis in interpretative mode now links these results involving the formal notion of recursive decidability to results framed in terms of the informal notion of effective decidability:

Theorem 40.2 *If T is a nice theory, it isn't effectively decidable. In particular, assuming their consistency, neither Q nor PA is effectively decidable.*

40.3   The *Entscheidungsproblem*

(a)   Q is such a *very* simple theory we might quite reasonably have hoped that there *would* be some mechanical way of telling which wffs are and which aren't its theorems. But we now know that there isn't; and, in a sense, you can blame the underlying first-order logic. For, assuming Q's consistency, we get *Church's Theorem* as an immediate corollary of Q's recursive undecidability:

Theorem 40.3 *The property of being a theorem of first-order logic is not recursively decidable.*[2]

Proof Suppose first-order theoremhood is recursively decidable. Let $\hat{Q}$ be the conjunction of the seven non-logical axioms of Q. By our supposition, there is a μ-recursive function which, given any L_A sentence φ, decides whether $(\hat{Q} \to \varphi)$ is a logical theorem. But, trivially, $(\hat{Q} \to \varphi)$ is a logical theorem just if φ is a Q-theorem. So our supposition implies that there is a μ-recursive function which decides what's a theorem of Q. But we've just shown there can be no such function, assuming Q's consistency – which refutes the supposition. ⊠

We now see the particular interest in finding a recursively adequate (and hence undecidable) arithmetic like Q which has only a *finite* number of axioms. We couldn't have similarly shown the undecidability of first-order logic by invoking the undecidability of PA, for example, because we couldn't start 'Let $\hat{P}$ be the conjunction of the axioms of PA ...' since PA has an infinite number of axioms. And indeed, PA *essentially* has an infinite number of axioms: there is no equivalent theory with the same theorems which only has a finite number of axioms (by Theorem 31.3).

(b)   Hilbert and Ackermann's *Grundzüge der theoretischen Logik* (1928) is the first recognizably modern logic textbook, still very worth reading. In §11, they famously pose the so-called *Entscheidungsproblem*, the problem of coming up with an effective method for deciding of an arbitrary sentence of first-order logic whether it is valid or not.

Theorem 40.3 tells us that the *Entscheidungsproblem* is recursively unsolvable: there is no recursive function that takes (the g.n. of) a sentence and returns a 0/1, yes/no, verdict on theoremhood. But this doesn't *quite* answer Hilbert and Ackerman's problem as posed. But Church's Thesis can again be invoked to bridge the gap between our last theorem and this:

Theorem 40.4 *The property of being a theorem of first-order logic is not effectively decidable.*

40.4   Incompleteness theorems for nice' theories

In the rest of this chapter, we return to incompleteness theorems again. Our main task is to revisit the informal incompleteness argument of Chapter 7 which, recall, inferred incompleteness from an undecidability result. Since we now can prove the recursive undecidability of nice theories, we should be able to give a formal version of that earlier informal incompleteness proof.

But before doing this, let's very quickly revisit our earlier Gödelian incompleteness results (which we established without any reference to a general theory

---

[2]This was first shown, independently, by Church (1936a, b) and Turing (1936), by different routes: see also Section 43.3.

of computable functions) in order to link them to the versions which are found in many modern treatments (where computability and the theory of μ-recursive functions are often discussed first).

Given Church's Thesis, the formal analogue of the intuitive notion of a formalized theory is the idea of a *recursively* axiomatized theory. However, when we proved Gödel's results about the limitations of formal arithmetics, we didn't discuss recursively axiomatized theories of arithmetic but – at first sight, rather more narrowly – *primitive recursively* axiomatized theories. Does that matter?

Well, as we've already explained in Section 26.1, focusing on p.r. axiomatized theories isn't really a restriction either in practice or in principle. That's because recursively but not p.r. axiomatized theories will in practice be peculiar beasts (for example: to set things up so we have to check that a putative axiom *is* an axiom by an open-ended 'do until' search is just not a very natural way of constructing an axiomatized theory). And a version of Craig's Theorem tells us that, in any case, if T is an effectively axiomatized theory, then we can construct a p.r. axiomatized theory T' which will have the same theorems as T, and hence such that T' will be negation-complete iff T is. So the First Theorem which proves the incompleteness of consistent, rich enough, p.r. axiomatized theories thereby proves the incompleteness of *any* consistent, rich enough, effectively axiomatized theory. We could therefore very well leave matters as they were, happily concentrating on p.r. axiomatized theories, without missing out on anything very important.

But for the record, let's now note that the key theorems we have proved for *nice* theories (i.e. consistent p.r. axiomatized theories which include Q) can be extended – now without appealing to Craig's Theorem – to apply directly to *nice'* theories, where

> A theory T is nice' iff T is consistent, recursively axiomatized, and extends Q.

For just one example, corresponding to Theorem 22.2 we have

> **Theorem 40.5** *If T is a nice' theory then there is an L_A-sentence φ of Goldbach type such that $T \nvdash \varphi$ and (if T is also ω-consistent) $T \nvdash \neg\varphi$.*

Proof sketch   We essentially just replicate the arguments for the original theorem. The crucial point of difference will be that, if we are only told that T is recursively axiomatized (as opposed to p.r. axiomatized), then we can only show the corresponding relation Gdl_T to be μ-recursive (as opposed to being primitive recursive). That's because Gdl_T is defined in terms of Prf_T which is in turn defined in terms of the likes of $Axiom_T$, and we are now only given that such basic properties of the theory are recursively decidable. But no matter. T includes the axioms of Q, so T is recursively adequate by Theorem 39.2, so T can *still* capture Gdl_T by a Σ_1 wff, and *that's* the crucial fact from which the rest of the argument runs on exactly as before.   ☒

In the same way, we can go on to show that the Diagonalization Lemma and the Gödel-Rosser Theorem which we proved for nice theories apply equally to nice' ones. And, of course, since the requirement that a nice' theory includes Q is just there to ensure we are dealing with a recursively adequate theory, our extended theorems can be equally well stated as applying to all consistent, recursively axiomatized, recursively adequate theories.

But we won't pause to spell out any further all the obvious variant ways of reworking our old formal theorems by replacing talk of p.r. axiomatization by recursive axiomatization, and talk of p.r. adequacy by recursive adequacy. It just isn't a very illuminating task.

40.5 Negation-complete theories are recursively decidable

We now return, as promised, to the informal incompleteness argument of Chapter 7 which inferred incompleteness from undecidability. We start by giving formal analogues of two preliminary results from Chapter 4.

(a)   Here again is (part) of

> **Theorem 4.1** *If T is an effectively axiomatized theory, then the set of theorems of T is effectively enumerable.*

And from this we inferred

> **Theorem 4.2** *Any consistent, effectively axiomatized, negation-complete theory T is decidable.*

These results involve the informal ideas of an effectively axiomatized theory, of effective enumerability and of decidability. We can now prove counterpart theorems involving the ideas of a recursively axiomatized theory, and of recursive enumerability and recursive decidability. The formal counterpart of Theorem 4.1 is evidently

> **Theorem 40.6** *If T is a recursively axiomatized theory then the set of theorems of T is recursively enumerable.*

So we'll now establish this, and then go on to deduce the formal counterpart of Theorem 4.2, i.e.

> **Theorem 40.7** *Any consistent, recursively axiomatized, negation-complete formal theory T is recursively decidable.*

Proof for Theorem 40.6   We could just take Theorem 4.1 and then use Church's Thesis in labour-saving mode to pass from the informal theorem to the formal one! There is, however, some interest in showing how to do things the hard way (though you are certainly allowed to skip).

305

As we said, since T is recursively axiomatized (so the likes of $Axiom_T$ are μ-recursive), there's a μ-recursive relation Prf_T such that $Prf_T(m, n)$ holds when m is the super Gödel number of a T-proof of the wff with Gödel number n. And we can now give the following definition:[3]

$$code(0) = \mu z[(\exists p < z)(\exists w < z)(z = 2^p \cdot 3^w \wedge Prf_T(p, w)]$$
$$code(Sn) = \mu z[(\exists p < z)(\exists w < z)(z > code(n) \wedge$$
$$z = 2^p \cdot 3^w \wedge Prf_T(p, w)].$$

The idea here is that, as we run through successive values of n, $code(n)$ runs through all the values of z ($= 2^p \cdot 3^w$) which code up pairs of numbers p, w such that $Prf(p, w)$. Assuming we never run out of T-theorems, the minimization operator in the clause defining $code(Sn)$ always returns a value. So $code$ is properly defined by recursion from μ-recursive functions and is μ-recursive.

Now recall the familiar exf function from Section 14.8, where $exf(m, i)$ returns the exponent of the prime π_i in the factorization of m. Since 3 is π_1 (2 is the zero-th prime π_0 on our way of counting!), $exf(m, 1)$ in particular returns the power of 3 in the factorization. Hence, to extract the values of w from the pairing code numbers, i.e. to extract the Gödel numbers of theorems, we can put

$$enum(n) = exf(code(n), 1)$$

and $enum(n)$, as we want, is μ-recursive, and it enumerates the Gödel numbers of theorems.   ⊠

Proof for Theorem 40.7   Again, we could use Church's Thesis in labour-saving mode to pass from our old informal Theorem 4.2 to its formal counterpart. But just for fun we'll do things the harder way (by all means skip again).

Assume T is consistent, recursively axiomatized, and negation-complete. Recall that we can define a p.r. function

$$neg(n) = \ulcorner \neg \urcorner \star n.$$

If n numbers a wff φ, $neg(n)$ numbers its negation $\neg \varphi$. Consider, then, the relation defined disjunctively as follows:

$$R(j, n) \text{ iff not-}Sent(n) \vee enum(j) = n \vee enum(j) = neg(n),$$

where $Sent(n)$ is true iff n numbers a sentence of T's language. This is a recursively decidable relation (why?). And, given T is negation-complete, for every n there is some j for which $R(j, n)$ holds; for either n doesn't number a sentence, in which case $R(0, n)$ holds, or n does number a sentence φ, and either φ or $\neg \varphi$ is a theorem, so for some j either $enum(j) = \ulcorner \varphi \urcorner$ or $enum(j) = \ulcorner \neg \varphi \urcorner = neg(\ulcorner \varphi \urcorner)$.

So $\mu j[R(j, n)]$ is always defined. That means we can put

---

[3]For brevity, we will apply the minimization operator to a property rather than to its characteristic function – see Section 38.1(c). We'll also assume without significant loss of generality that T has an unlimited number of theorems: it's left as a boring exercise to cover the exceptional case where T is a hobbled theory with a non-standard logic and only a finite number of consequences.

$thm(n) = 0$ if (i) $0 < n$ and (ii) $enum(\mu j[Rjn]) = n$
$thm(n) = 1$ otherwise,

and thm will be μ-recursive (compare Section 14.7 (E)).

It is now easy to see that thm is the characteristic function of the property of numbering a T-theorem. For $thm(n) = 0$ can't hold when n is not a sentence, for then $\mu j[Rjn] = 0$ and that's ruled out by (i). And when n does number a sentence, $\mu j[Rjn]$ returns either (1) the position j in the enumeration of theorems where the sentence numbered n is proved or (2) the position where the negation of that sentence is proved; clause (ii) ensures that case (1) obtains. So we are done.   ⊠

(b)   Before moving on, let's recall another informal old theorem:

> **Theorem 3.9** *There is an effectively enumerable set of numbers which is not decidable.*

The obvious analogue of this, given Church's Thesis again, is

> **Theorem 40.8** *There is a recursively enumerable set of numbers which is not recursively decidable.*

We could prove this the hard way too, by giving a formal version of the proof of the old theorem. But a much snappier proof is available:

Proof By Theorem 40.6, the Gödel numbers of the theorems of any recursively axiomatized theory are r.e.; so the Gödel numbers of Q-theorems in particular are r.e. But Theorem 40.1 tells us that Q is recursively undecidable. So the set of Gödel numbers of Q theorems is r.e. but not recursively decidable.   ⊠

40.6   Recursively adequate theories are not recursively decidable

Back in Chapter 7 we showed:

> **Theorem 7.1** *No consistent, sufficiently strong, effectively axiomatized theory of arithmetic is decidable.*

Its formal counterpart, as you would expect from the remarks in Section 39.3, is the now easily proved generalization of Theorem 40.1:

> **Theorem 40.9** *No consistent, recursively adequate, recursively axiomatized theory is recursively decidable.*

Proof sketch   The argument is exactly parallel to the argument for the slightly less general Theorem 40.1 which said that *nice* theories are recursively undecidable. The proof of *that* theorem invoked Theorem 24.8: no open wff in a nice theory T can capture the corresponding numerical property $Prov_T$. So to prove our new theorem, we simply need to generalize the argument in order to prove that no open wff in a consistent, recursively axiomatized, recursively adequate theory T can capture $Prov_T$. (It is a useful reality check to make sure you understand how to do this, and hence how to complete the proof.)   ⊠

40.7   Incompleteness again

So now we just put everything together. In Chapter 7 we took Theorem 4.2 (any consistent, axiomatized, negation-complete formal theory T is decidable) and Theorem 7.1 (no consistent, sufficiently strong, axiomatized formal theory of arithmetic is decidable), and inferred

> **Theorem 7.2** *If T is a consistent, sufficiently strong, effectively axiomatized theory of arithmetic, then T is not negation-complete.*

Now we similarly can put together their formal counterparts Theorem 40.7 (any consistent, recursively axiomatized, negation-complete formal theory T is recursively decidable) and Theorem 40.9 (if T is a consistent, recursively axiomatized, recursively adequate theory, then it isn't recursively decidable), and we deduce

> **Theorem 40.10** *A consistent, recursively adequate, recursively axiomatized theory of arithmetic cannot be negation-complete.*

This is, of course, just a version of the Gödel-Rosser Theorem minus any information about what the undecidable sentences look like.

So we have at last fulfilled our promise to show that the informal argument of Chapter 7 is indeed in good order.

40.8   True Basic Arithmetic is not r.e.

Finding a formal counterpart to our 'syntactic' informal incompleteness argument of Chapter 7 was pretty straightforward, given what had gone before. So what about giving a formal counterpart of the informal 'semantic' argument for incompleteness in Chapter 6?

That argument depended on establishing

> **Theorem 6.1** *The set of truths of a sufficiently expressive arithmetical language L is not effectively enumerable.*

We now know that even Q can *capture* every recursive function. We can quickly add that Q's language L_A can *express* any recursive one-place function. (Why? If we assume Q is sound, we can appeal to the final remark of Section 5.6. Or more directly, we can just check that each step in our recipe for constructing a Σ_1 wff for capturing a μ-recursive function f gives us a wff which expresses f.) Hence, by Church's Thesis, L_A is 'sufficiently expressive'. We also know by Church's Thesis again that the formal counterpart of the notion of effective enumerability is that of recursive enumerability. As before, let $\mathcal{T}_A$ be True Basic Arithmetic, i.e. the set of truths of L_A. Then putting everything together gives us an argument for one natural formal counterpart to Theorem 6.1:

Theorem 40.11 $\mathcal{T}_A$ *is not recursively enumerable.*

But can we prove this theorem without relying on the informal argumentation which underlies Theorem 6.1 and without invoking Church's Thesis? Well, the following proof is now available to us:

Proof L_A suffices to express any recursive function. Now suppose for reductio there *is* a recursive function f that enumerates $\mathcal{T}_A$, i.e. enumerates the Gödel-numbers for the true L_A sentences. There will be a formal L_A wff $\mathsf{F}(\mathsf{x},\mathsf{y})$ which expresses that enumerating function f.

And then the formal wff $\exists\mathsf{x}\,\mathsf{F}(\mathsf{x},\mathsf{y})$ will be satisfied by a number if and only if it numbers a truth of L_A. But that contradicts Theorem 27.2, Tarski's Theorem which tells us that there cannot be such a wff. Hence there can be no enumerating recursive function f.   ⊠

That's neat. However, this proof depends on our already having established Tarski's Theorem, a close cousin of the incompleteness theorem. If we want to rely on Theorem 40.11 for an independent argument for incompleteness, then, we might prefer an alternative formal proof. Can we find a proof which doesn't rely on Church's Thesis but which still stays closer to the informal argument in Chapter 6? That earlier argument appealed to various intuitive claims about algorithms and computer programs. However, if we are going to sharpen up some such line of argument and make it rigorous, we'll have to give some theoretical treatment of a general-purpose programming framework.

So we now turn to make a start on doing this. In the last chapters of this book, we'll aim to say *just* enough about Alan Turing's classic analysis of algorithmic computation to throw some more light on incompleteness phenomena (and also to give crucial added support to Church's Thesis which connects the formal idea of μ-recursiveness with the informal ideas of computability/effective enumerability/decidability which we care about).

41 Turing machines

In this chapter, we introduce Turing's classic analysis of effective computability.[1] And then – in the next chapter – we will establish the crucial result that the Turing-computable total functions are exactly the μ-recursive functions. This result is fascinating in its own right; it is hugely important historically; and it enables us later to establish some further results about recursiveness and incompleteness in a particularly neat way. So let's dive in without more ado.

41.1 The basic conception

Think of executing an algorithmic computation 'by hand', using pen and paper. We follow strict rules for writing down symbols in various patterns. To keep things tidy, let's write the symbols neatly one-by-one in the squares of some suitable square-ruled paper. Eventually – assuming that we don't find ourselves carrying on generating output forever – the computation process stops and the result of the computation is left written down in some block of squares on the paper.

Now, Turing suggests, using a two-dimensional grid for writing down the computation is not of the essence. Imagine cutting up the paper into horizontal strips a square deep, and pasting these together into one long tape. We could use that as an equivalent workspace.

Using a rich repertoire of symbols is not of the essence either. Suppose some computational system uses 27 symbols. Number these off using a five-binary-digit code (so the 14th symbol, for example, gets the code '01110'). Then divide each of the original squares on our workspace tape into a row of five small cells. Instead of writing one of the original symbols into one of the original big squares, we could just as well write its binary code digit-by-digit into a block of five cells.

So – admittedly at some cost in user-friendliness – we can think of our original hand computation as essentially equivalent to following an algorithm (a set of instructions) for doing a computation by writing down or erasing binary digits one-at-a-time in the cells on a linear tape, as in our diagram on the next page. The arrow-head indicates the position of the *scanned cell*, i.e. the one we are examining as we are about to apply the next computational instruction. (We'll assume, by the way, that we can paste on more blank workspace as we need it – so, in effect, the tape is unlimited in length in both directions).

---

[1]See Turing (1936), which is handily reprinted with a very useful long introduction in Copeland (2004).

0	1	0	0	1		1	1	1			1	0	

▲

So much for our workspace. Let's now consider the *instructions* which govern our computation when broken down into minimal steps. We will list these instructions as labelled or numbered lines, where each line tells us what to do depending on the contents of the scanned cell. So we can think of a single line from our algorithm (our set of instructions) as having the form

q: if the scanned cell contains '0', do action A_0, then go to q_0;
   if the scanned cell contains '1', do action A_1, then go to q_1;
   if the scanned cell is blank, do action A_B, then go to q_B.

where the q are line-numbers or labels. (That's the general pattern: for a particular line-number q, some of the conditional clauses could be absent.)

It is convenient to distinguish two special line-numbers or labels, 1 and 0. The first will be used to mark the starting instruction of an algorithm, while the second is really a pseudo-label and 'go to 0' will actually mean 'stop the execution of the algorithm'.

What are the possible actions A here? There are two basic types. We can *write* in the scanned cell – i.e. over-write any contents in the scanned cell with '0' or '1', or 'write a blank' (i.e. erase the current contents). Or else we can *move* along the tape so that a new cell becomes the scanned cell. We'll take it that any moving is to be done step-wise, one cell at a time. So more complex actions – like e.g. copying the contents of the scanned cell into the next four cells to the right – are to be performed as a sequence of basic actions.

There are therefore five possible minimal actions A, which we can indicate as follows:

0: write a '0' in the scanned cell (overwriting any scanned content);
1: write a '1';
B: write a blank;
L: make the next cell to the left the scanned cell;
R: make the next cell to the right the scanned cell.

So let's now say, as a first shot, that a *Turing program* is just a collection of instructions of the very simple three-clause form we illustrated, which tell us which action to perform and which line to jump to next.

In executing such a program, we typically start with some number or numbers written as *input* on the work-tape (written in binary digits, of course). And we begin by following the relevant instruction given at the program line labelled 1. We then follow the further instructions we encounter as we are told to jump from line to line. The execution of a set of instructions can then stop for two reasons: we can reach the explicit 'go to 0' instruction to stop; or we can run out of lines to tell us what to do next. In the first case, we'll say that the execution *halts gracefully* or simply *halts*; in the second case, we'll say that it *freezes*. If and when the execution halts gracefully, we look at what's left on the tape – and in particular at the block of binary digits containing the final scanned cell – and those digits give the numerical *output* of our calculation.

Hence we can say, as a first shot, that a one-place total function f is computed by a Turing program Π if, given n in binary symbols on the tape as input, executing the program eventually yields $f(n)$ as output.

The idea of a Turing computation, then, is quite extraordinarily simple – it is basically a matter of running a program whose sole programming structure is the 'go to line q' instruction. In Section 41.3, we'll give some mini-examples of Turing programs in operation. But first we really need to refine the ideas we've just sketched. By the end of the next section, then, we will have introduced some sharper terminology and also cleaned up some details. Along the way, there's a number of fairly arbitrary and entirely non-significant choices to be made. We won't comment on these, but don't be surprised to find other choices made in other treatments: the basic conception, however, is always essentially the same. (Suggestion: it might help considerably to read the next two sections in parallel, using theory and practice to illuminate each other.)

41.2   Turing computation defined more carefully

(i) *I-quadruples*   We define an *instruction-quadruple* (or 'i-quadruple' for short) to be an ordered quadruple of the form

$$\langle q_1, S, A, q_2 \rangle$$

whose elements are as follows:

1. q_1 is a numeral other than '0'; we'll refer to this first element of an i-quadruple as its *label* – to emphasize the point that the numerals here aren't doing arithmetical work. An i-quadruple labelled '1' is an *initial* quadruple.

2. S – representing the contents of the scanned cell – is one of the symbols '0', '1', 'B'. ('B', of course, represents a blank cell.)

3. A is one of the symbols '0', '1', 'B', 'L', 'R': these represent the five possible minimal actions.

4. q_2 is a numeral, pointing to the next instructions to execute.

An i-quadruple is to be read as giving a labelled, conditional, two-part instruction as follows:

q_1: *if* the scanned cell contains S, *do* the action indicated by A, then *go to* the instructions with label q_2 – unless q_2 is '0', in which case halt the execution of the program.

So we can now compress our verbose tri-partite instruction line

q: if the scanned cell contains '0', do action A_0, then go to q_0;
     if the scanned cell contains '1', do action A_1, then go to q_1;
     if the scanned cell is blank, do action A_B, then go to q_B

into three i-quadruples which share the same initial label, thus:

$$\langle q, 0, A_0, q_0 \rangle, \ \langle q, 1, A_1, q_1 \rangle, \ \langle q, B, A_B, q_B \rangle.$$

(ii) *Turing programs* Our first shot at characterizing a Turing program therefore comes to this: it is a set of i-quadruples. But we plainly don't want *inconsistent* sets which contain i-quadruples with the same label which issue inconsistent instructions. So let's say more formally:

A set Π of i-quadruples is consistent iff there's no pair of i-quadruples $\langle q_1, S, A, q_2 \rangle$, $\langle q_1, S, A', q_2' \rangle$ in Π such that $A \neq A'$ or $q_2 \neq q_2'$.

Which leads to the following sharpened official definition:

A *Turing program* is a finite consistent set of i-quadruples.[2]

(iii) *Executing Turing programs* We execute a Turing program Π as follows:

1. We start with the work-tape in some *initial configuration* – i.e. with digits occupying a *finite* number of cells, and with some particular cell being scanned. Suppose the content of that initial scanned cell is S.

2. We then look for some appropriate initial quadruple to execute. That is to say, we look for an i-quadruple in Π of the form $\langle 1, S, A, q_2 \rangle$: by consistency there is at most one distinct such i-quadruple. We perform action A and jump to the instructions with label q_2.

3. We next look for an i-quadruple in Π of the form $\langle q_2, S', A', q_3 \rangle$, where S' is the content of the currently scanned cell. We perform action A' and jump to the instructions with label q_3. And we keep on going ...

4. ... unless and until the execution stops because either (a) we are explicitly told to halt – i.e. we encounter a 'jump to 0' instruction – or (b) the program freezes because there is no relevant i-quadruple for us to apply next.

We will often be particularly concerned with cases where we run a program starting and finishing in what we'll call *standard* modes. To explain:

---

[2]Each i-quadruple includes just *one* action instruction A, either a symbol-writing instruction from $\{0, 1, B\}$, or a head-moving instruction from $\{L, R\}$. An obviously equivalent and perhaps neater alternative is to define a program as a set of *i-quintuples*, where each i-quintuple includes *both* a symbol-writing instruction *and* a head-moving instruction.

5. We start with the work-tape in a *standard initial configuration*. In other words, the tape is blank except for containing as input one or more blocks of binary digits, with the blocks separated by single blank cells: and the initial scanned cell is the left-most cell of the left-most block.

6. A run of the program is said to *halt standardly* if (a) it reaches a 'jump to 0' instruction (so the program halts gracefully, rather than just freezes), and (b) it leaves the work-tape blank apart from a single block of binary digits, with (c) the scanned cell being the left-most cell of this block.

Note that if a program halts standardly, it finishes in a state which can serve as a standard initial configuration for other suitable programs. So we can always unproblematically chain together program-modules that start and stop in standard modes to form longer programs.

(iv) *Turing-computable functions*   Suppose that $f \colon \mathbb{N} \to \mathbb{N}$, i.e. f is a one-place total numerical function. Then we will say that

> The Turing program Π computes the function f just if, for all n, when Π is executed starting with the tape in a standard initial configuration with n in binary digits on the tape (and nothing else), then the execution halts standardly with the output $f(n)$ in binary digits on the tape.

We can generalize to cover many-place functions. For example, suppose that $g \colon \mathbb{N}, \mathbb{N} \to \mathbb{N}$, i.e. g is a two-place function, which maps two numbers to another number. Then,

> The Turing program Π computes the function g just if, for all m, n, when Π is executed starting with the tape in a standard initial configuration with m in binary digits on the tape, followed by a blank cell, followed by n in binary digits (and nothing else), then the execution halts standardly with the output $g(m, n)$ in binary digits on the tape.

Finally, we say:

> A total function is *Turing-computable* iff there is a Turing program that computes it.[3]

Every Turing-computable function is effectively computable (just follow the instructions!). The claim that the converse is also true, i.e. that every effectively computable total function is Turing-computable, is *Turing's Thesis*, which we first met in Section 3.1.

---

[3]There's a very natural generalization of this definition, where we say that a *partial* function f is Turing-computable if there is a Turing program which computes the right value for input n whenever $f(n)$ is defined and which doesn't halt otherwise. But as we said, we're not going to discuss partial computable functions here.

41.3   Some simple examples

In this section we'll work through three simple example Turing programs, which ought to be enough to illustrate at least the basic ideas. But do feel free to skim through very lightly if you have no taste for this kind of thing.

(a) *The successor function*   We'll start with a program that computes the successor function. So suppose the initial configuration of the tape is as follows

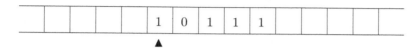

Then the program needs to deliver the result '11000'. The program task can be broken down into three stages:

Stage A   We need to make the scanned cell the right-most cell in the initial block of digits. Executing the following i-quadruples does the trick:

$$\langle 1, 0, R, 1 \rangle$$
$$\langle 1, 1, R, 1 \rangle$$
$$\langle 1, B, L, 2 \rangle$$

These initial instructions move the scanned cell to the right until we overshoot and hit the blank at the end of the initial block of digits; then we shift back one cell, and look for the instructions with label '2':

Stage B   Now for the core computation of the successor function. Adding one involves putting a '1' in the scanned final cell if it currently contains '0'; or else putting a '0' in the scanned cell if it currently contains '1', and 'carrying one' – i.e. moving the scanned cell one left and adding one again.

The following i-quadruples program for that little routine, which finishes by pointing us to instructions labelled '4':

$$\langle 2, 0, 1, 4 \rangle$$
$$\langle 2, 1, 0, 3 \rangle$$
$$\langle 3, 0, L, 2 \rangle$$
$$\langle 2, B, 1, 4 \rangle$$

Note, the fourth quadruple is to deal with the case where we keep on 'carrying 1' until we hit the blank at the front of the initial block of digits. Executing these instructions gets the tape in our example into the following state:

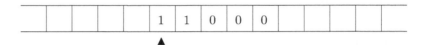

| | | | | 1 | 1 | 0 | 0 | 0 | | | | |

▲

And we now jump to execute the appropriate instruction with label '4'.

Stage C   Finishing up. We need to ensure that the scanned cell returns to be at the front of the block, so that the computation halts in standard configuration. Analogously to Stage A, we can write:

$$\langle 4, 0, L, 4\rangle$$
$$\langle 4, 1, L, 4\rangle$$
$$\langle 4, B, R, 0\rangle$$

Following these instructions, the scanned cell moves leftwards until it overshoots the block and then moves it right one cell, and finally the execution halts gracefully:

| | | | | 1 | 1 | 0 | 0 | 0 | | | | |

▲

The scanned cell is now the first in the block of digits. There is nothing else on the tape. So we have halted standardly. Our i-quadruples together give us a program which computes the successor function.

(b) *Another program for the successor function*   Note that our successor program, applied to '111', changes those digits to '000', then prefixes a '1' to give the correct output '1000'. So in this case, the output block of digits starts *one cell to the left* of the position of the original input block. We'll now – for future use (in Section 42.2) – describe a variant of the successor program, which this time always neatly yields an output block of digits starting in exactly the *same* cell as the original input block.

What we need to do, clearly, is to add a routine at the beginning of the program which, if but only if it detects an unbroken block of '1's, shifts that block one cell to the right (by adding a '1' at the end, and deleting a '1' at the beginning). The following will do the trick, and also – like Stage A of our previous program – it moves the current cell to the end of the block:

$$\langle 1, 0, R, 12\rangle$$
$$\langle 1, 1, R, 1\rangle$$
$$\langle 1, B, 1, 10\rangle$$
$$\langle 10, 1, L, 10\rangle$$
$$\langle 10, B, R, 11\rangle$$
$$\langle 11, 1, B, 11\rangle$$
$$\langle 11, B, R, 12\rangle$$
$$\langle 12, 0, R, 12\rangle$$

$$\langle 12, 1, R, 12 \rangle$$
$$\langle 12, B, L, 2 \rangle$$

The initial instructions start us off scanning through the input block to the right. If we encounter a '0' then we continue by following the instructions labelled '12' which take the scanned cell to the end of the block. If all we meet are '1's then we put another '1' at the end of the block, and then follow the instructions labelled in the tens, which first take us back to the beginning of the block, then get us to delete the initial '1' and move one cell right, and *then* we follow the instructions labelled '12' to get us to the end of the block

You can check that running through the instructions labelled '1', '10' and '11' changes the state of the tape from the first to the second state illustrated next:

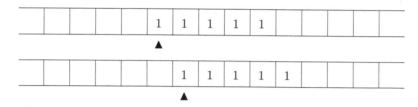

We then just follow the instructions labelled '12', so we end up scanning the cell at the end of the block (and start looking for instructions labelled '2').

We can now add on the same Stage B i-quadruples from our previous successor program to perform the task of adding one, and the same Stage C i-quadruples to get the scanned cell back to the beginning of the resulting block. Putting all those together gives us the desired program.

That's a lot of effort! – but then, programming at the level of 'machine code' (which is in effect what we are doing) *is* hard work. That's why, of course, we ordinarily use high-level programming languages and rely on compilers that work behind the scenes to translate our perspicuous and manageable programs into the necessary instructions for manipulating binary bits.

(c) *A copying program*   Our remaining example is a simple program which takes a block of input digits, and produces as output the same block (in the same place), followed by a blank, followed by a duplicate of the original block.

We obviously need to keep track of where we are in the copying process. We can do this by successively deleting a digit in the original block, going to the new block, writing a copy of the deleted digit, returning to the 'hole' we made to mark our place, replacing the deleted digit, and then moving on to copy the next digit. A program for doing all this can be broken down into the following four sub-programs:

1. *Choosing what to do*   Examine the scanned cell. If it contains a '0', delete it, and go to sub-program (2). If it contains a '1', delete it, and go to sub-program (3). If it is blank, then we've got to the end of the digits in the

317

original block, so we just need to call sub-program (4) to ensure that we halt gracefully.

2. *Copying a '0'*   This routine 'remembers' that we've just deleted a '0'. We scan on to the right until we find the *second* blank – which marks the end of our duplicate block (if and when it exists) – and write a '0'. Then we scan back leftwards until we again find the second blank (the blank created when we deleted the '0'). Rewrite a '0' there, which finishes copying that digit. So now move on to scan the next cell to the right, and return to sub-program (1).

3. *Copying a '1'*   Just like sub-program (2), except that this routine 'remembers' that we've just deleted a '1'.

4. *Finish up*   Move the scanned cell back to the beginning of the original block. (Note, though, this program won't halt standardly.)

To illustrate, suppose the current state of the tape is like this:

Sub-program (1) instructs us to delete the currently scanned digit and start executing sub-program (2):

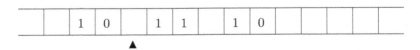

Sub-program (2) then takes us through the following stages: (a) we scan to the right to find the end of the second block, (b) we write '0' there, (c) we scan back to the left to find the 'hole' in the first block that we just created, (d) we rewrite '0' there, (e) we move one cell right. So these five stages produce these successive states of the tape:

| | 1 | 0 | 0 | 1 | 1 | | 1 | 0 | 0 | | | |

▲

| | 1 | 0 | 0 | 1 | 1 | | 1 | 0 | 0 | | | |

▲

And we've thereby copied another digit. We keep on going, until we run out of digits in the first block to copy – and then run a 'tidying up' module (4).

Here – just for the fun of it, if that's quite the right word – is how to code up our outlined program strategy into i-quadruples (we've marked the beginnings of the four sub-programs):

$\langle 1, 0, B, 10 \rangle$ (1)

$\langle 1, 1, B, 20 \rangle$

$\langle 1, B, L, 30 \rangle$

$\langle 10, B, R, 11 \rangle$ (2)

$\langle 11, 0, R, 11 \rangle$

$\langle 11, 1, R, 11 \rangle$

$\langle 11, B, R, 12 \rangle$

$\langle 12, 0, R, 12 \rangle$

$\langle 12, 1, R, 12 \rangle$

$\langle 12, B, 0, 13 \rangle$

$\langle 13, 0, L, 13 \rangle$

$\langle 13, 1, L, 13 \rangle$

$\langle 13, B, L, 14 \rangle$

$\langle 14, 0, L, 14 \rangle$

$\langle 14, 1, L, 14 \rangle$

$\langle 14, B, 0, 15 \rangle$

$\langle 15, 0, R, 1 \rangle$

$\langle 20, B, R, 21 \rangle$ (3)

$\langle 21, 0, R, 21 \rangle$

$\langle 21, 1, R, 21 \rangle$

$\langle 21, B, R, 22 \rangle$

$\langle 22, 0, R, 22 \rangle$

$\langle 22, 1, R, 22 \rangle$

$\langle 22, B, 1, 23 \rangle$

$\langle 23, 0, L, 23 \rangle$

$\langle 23, 1, L, 23 \rangle$

$\langle 23, B, L, 24 \rangle$

$\langle 24, 0, L, 24 \rangle$

$\langle 24, 1, L, 24 \rangle$

$\langle 24, B, 1, 25 \rangle$

$\langle 25, 1, R, 1 \rangle$

$\langle 30, 0, L, 30 \rangle$ (4)

$\langle 30, 1, L, 30 \rangle$

$\langle 30, B, R, 0 \rangle$

As we remarked, although this routine halts gracefully, it doesn't halt 'standardly' (in the special sense we defined) because it leaves more than one block of binary digits on the tape. But that's fine. And as we will see in the next chapter, it can be a useful subroutine in a longer program that *does* halt standardly and thus is apt for counting as computing a numerical function.

This last example, of course, just reinforces the point that writing Turing programs for performing even very simple tasks very quickly becomes *very* painful. So we won't give any more detailed examples.[4]

But our concern here isn't really with practical computing but rather with Turing's analysis of what a computation consists in when broken down to its

---

[4]Masochists can try their hands at programming one of the many on-line Turing machine simulators which are available (though be careful to read the fine details of how programs are to be specified). However, this is a game that quickly palls!

very smallest steps. If he is right that we can treat computations as ultimately symbol manipulation on a 'tape', looking at one cell at a time, etc., then *any genuinely algorithmic step-by-step computation can be replicated using a Turing program*. More on this anon.

41.4 'Turing machines' and their 'states'

We have so far imagined a human 'computer' executing a Turing program 'by hand', writing and erasing symbols from a paper tape, mechanically following the program's instructions. Evidently, a machine could do the same job. And any mechanism for running a Turing program might naturally be referred to as a 'Turing machine' (at least if we pretend that its 'tape' is inexhaustible).

But the theory of Turing computations just doesn't care about the hardware implementation. What matters about a Turing machine is always its program. Hence one standard practice is to think of a Turing machine as a type of idealized machine individuated by the Turing program it is running (i.e. same program, same Turing machine: different program, different Turing machine). Another, equally standard, practice is simply to *identify* a Turing machine with its program (it is common enough to read 'A Turing Machine is a set of quadruples such that . . . '). Nothing at all hangs on this. When we occasionally talk of Turing machines we'll in fact be thinking of them in the first way. But mostly – for clarity's sake – we'll carry on talking about programs rather than machines.

A final remark. Suppose that a Turing machine (in the first sense) is in the middle of executing a program. It is about to execute some i-quadruple in its program, while scanning a particular cell on a tape which has some configuration of cells filled with digits. We can think of this overall state-of-play as characterized by the 'internal' state of the machine (it is about to execute a quadruple with label q) combined with the 'external' state of the tape (the configuration of the tape, with one cell picked out as the 'current' cell). That's why q-labels are standardly said to identify (internal) *states* of the Turing machine.

42 Turing machines and recursiveness

We are not going to write any more programs to show, case by case, that this or that particular function is Turing-computable, not just because it gets painfully tedious, but because we can now fairly easily establish that *every* μ-recursive function is Turing-computable and, conversely, *every* Turing-computable function is μ-recursive.[1] This equivalence between our two different characterizations of computable functions is of key importance, and we'll be seeing its significance in the remaining chapters.

42.1 μ-Recursiveness entails Turing computability

Every μ-recursive function can be evaluated 'by hand', using pen and paper, prescinding from issues about the *size* of the computation. But we have tried to build into the idea of a Turing computation the essentials of any hand-computation. So we should certainly hope and expect to be able to prove:

Theorem 42.1 *Every μ-recursive function is Turing-computable.*

Proof sketch We'll say that a Turing program is *dextral* (i.e. 'right-handed') if

i. in executing the program – starting by scanning the leftmost of some block(s) of digits – we never have to write in any cell to the *left* of the initial scanned cell (or scan any cell more than one to the left of that initial cell); and

ii. if and when the program halts, the final scanned cell is the *same* cell as the initial scanned cell.

The key point about a dextral program, then, is that we can run it while storing other data safely on a leftwards portion of the tape, because the program doesn't touch that portion. So complicated computations can proceed by running a series of dextral sub-programs using leftwards portions of the tape to preserve data between sub-programs.

If a function is computable by a dextral Turing program, we'll say it is *dTuring-computable*. Suppose now that the following are all true:

1. The initial functions are dTuring-computable.

---

[1] If you happen to be browsing through, not having read the preceding few chapters, or if your attention flickered earlier, I'd better repeat: wherever you see bare talk of a 'function' it means a *total* function. We are not going to be talking about *partial* computable functions.

2. If the functions g and h are dTuring-computable, then so is a function f defined by composition from g and h.

3. If the functions g and h are dTuring-computable, then so is a function f defined by primitive recursion from g and h.

4. If the regular function g is dTuring-computable, so is the function f defined from g by regular minimization.

Then take any μ-recursive function f. This can be defined by some chain of definitions by composition and/or primitive recursion and/or regular minimization, beginning with initial functions. So as we follow through f's chain of definitions, we start with initial functions which are dTuring-computable – by (1) – and each successive definitional move takes us from dTuring-computable to dTuring-computable functions – by (2), (3), and (4). So f must be dTuring-computable. So a fortiori, the μ-recursive function f must be plain Turing-computable.

Hence to establish Theorem 42.1, it is enough to establish (1) to (4). But each of those is in fact more or less easy to prove.   ⊠

If this overall proof strategy seems familiar, that's because we used the same proof idea e.g. in Chapter 17 when showing that Q is p.r. adequate. The details needed to fill in the proof outline are given in the next section: however, although those details are rather pretty, they are technicalities without much conceptual interest, so by all means skip them.

42.2 μ-Recursiveness entails Turing computability: the details

For enthusiasts, we'll now explain how to fill in the details of our proof-sketch for Theorem 42.1 by establishing points (1) to (4).

Proof sketch for (1)   We proved in Section 41.3(b) that the successor function is not just Turing-computable but is dTuring-computable. It's trivial that the zero function $Z(x)$ is computable by a dextral program – just write a program that takes any block of digits, erases it from the right, and leaves a single '0' on the tape. It's also easily seen that the identity functions are dTuring-computable by erasing and moving blocks of digits.   ⊠

Proof sketch for (2)   For simplicity, we'll just consider the case of monadic functions. More complex cases can be dealt with using the same basic idea plus a few tricks.

Suppose the program Π_g dTuring computes g, and the program Π_h dTuring computes h. Then to dTuring compute the composite function $f(n) = h(g(n))$, we can just run Π_g on the input n to give $g(n)$, and then run Π_h on that output to calculate $h(g(n))$.

How exactly do we chain together two programs Π_g and Π_h into one composite program? We need to do two things. We first ensure that there is no clash of labels by changing the q-numbers in the i-quadruples in Π_h (doing it systematically, of course, to preserve the all-important cross-references between quadruples). And then we must just ensure that – rather than using the 'halt' label – Π_g ends by telling us to process the first instruction in our re-labelled Π_h. ⊠

Proof sketch for (3) We'll suppose f is defined by the recursion clauses

$$f(x, 0) = g(x)$$
$$f(x, Sy) = h(x, y, f(x, y))$$

where both g and h are dTuring computable total functions. We need to show that f is dTuring computable. (Simplifying the discussion for the case where the x-variable drops out of the picture, or generalizing it to cover the case where we have an array of variables $\vec{x}$, is straightforward.)

It is convenient to introduce an abbreviated way of representing the contents of a tape. We'll use $\boxed{n}$ to indicate a block of cells containing the binary digits for n, we'll use 'B' to indicate a single blank cell, and then e.g. $\boxed{m}\,B\,\boxed{n}$ represents a tape which is blank except for containing m in binary followed by a blank followed by n. So, what we need to describe is a dextral program that takes $\boxed{m}\,B\,\boxed{n}$ as input and delivers $\boxed{f(m,n)}$ as output. Here's a sketch:

1. Given the input $\boxed{m}\,B\,\boxed{n}$, use a combination of a copying program and a program for subtracting one to get the tape eventually to read as follows:

$$\boxed{m}\,B\,\boxed{n-1}\,B\,\boxed{m}\,B\,\boxed{n-2}\,B\,\boxed{m}\,B\,\boxed{n-3}\,B\dots$$
$$\dots B\,\boxed{m}\,B\,\boxed{2}\,B\,\boxed{m}\,B\,\boxed{1}\,B\,\boxed{m}\,B\,\boxed{0}\,B\,\boxed{m}$$

2. Now move to scan the first cell of the *right-most* block of digits $\boxed{m}$. Run our program for evaluating g starting in that position. Since this program is by hypothesis dextral, it doesn't visit the portion of the tape any further left than the blank before that last block. *So it is just as if the part of the tape to the left of the last block is completely empty.* Hence the program for evaluating g will run normally on the input m, and it will calculate $g(m)$, i.e. $f(m, 0)$. So after running g the tape will end up reading

$$\boxed{m}\,B\,\boxed{n-1}\,B\,\boxed{m}\,B\,\boxed{n-2}\,B\dots B\,\boxed{m}\,B\,\boxed{1}\,B\,\boxed{m}\,B\,\boxed{0}\,B\,\boxed{f(m,0)}$$

3. Scan the first cell of the concluding three blocks $\boxed{m}\,B\,\boxed{0}\,B\,\boxed{f(m,0)}$. Run our program for evaluating h starting in that position. Since this program too is by hypothesis dextral, it ignores the leftwards contents of the tape and the program will run normally on the three inputs m, 0, $f(m, 0)$ to calculate $h(m, 0, f(m, 0))$ – i.e. calculate $f(m, 1)$. So when we run h this first time, the tape will end up reading

$$\boxed{m}\,B\,\boxed{n-1}\,B\,\boxed{m}\,B\,\boxed{n-2}\,B\dots B\,\boxed{m}\,B\,\boxed{1}\,B\,\boxed{f(m,1)}$$

Now repeat the same operation. So next, scan the first cell of the last three blocks $\boxed{m}\,B\,\boxed{1}\,B\,\boxed{f(m,1)}$. Run the program for evaluating h again, and the tape will end up containing the shorter row of blocks

$$\boxed{m}\,B\,\boxed{n-1}\,B\,\boxed{m}\,B\,\boxed{n-2}\,B\ldots B\,\boxed{m}\,B\,\boxed{2}\,B\,\boxed{f(m,2)}$$

Keep on going, each time running h using the last three blocks of digits as input, and eventually we will arrive at just

$$\boxed{m}\,B\,\boxed{n-1}\,\boxed{f(m,n-1)}$$

One last run of h, with this as input, finally leaves us with

$$\boxed{f(m,n)}$$

on the tape (with the program halting standardly, with the scanned cell at the beginning of then block).

So we've outlined the shape of a program that gives a dTuring computation of the recursively defined f. ⊠

Proof sketch for (4) It remains to show that if the function $g(x,y)$ is dTuring-computable, then so is the function defined by $f(x) = \mu y[g(x,y) = 0]$, assuming g is regular. Hence we want to specify a program that takes $\boxed{m}$ as input and delivers the output $\boxed{\mu y[g(m,y)=0]}$.

Our task is to run the program g with successive inputs of the form $\boxed{m}\,B\,\boxed{n}$, starting with $n = 0$ and incrementing n by 1 on each cycle; and we keep on going until $g(m,n)$ gives the output '0', when we return the value of n. But note, we need to do this while 'remembering' at each stage the current values of m and n. So here's a four-stage strategy for doing the job. (Again, simplifying to cover the case where the x-variable drops out of the picture, or generalizing to cover the case where we have an array of variables $\vec{x}$, is straightforward.)

1. We are given $\boxed{m}$ on the tape. Use a modified copier to produce

$$\boxed{0}\,B\,\boxed{m}\,B\,\boxed{m}\,B\,\boxed{0}$$

   starting at the same point on the tape as the original block. This tape input is now to be fed into Stage (2).

2. Given any input of the kind

$$\boxed{n}\,B\,\boxed{m}\,B\,\boxed{m}\,B\,\boxed{n}$$

   move to scan the first occupied cell of the second block $\boxed{m}$. Now run the dextral program for g from that starting point, i.e. on the input $\boxed{m}\,B\,\boxed{n}$. By our characterization of Turing programs for functions, g halts standardly. See whether the output result is 0. If it isn't, go to Stage (3). If it is – and eventually it will be, since g is regular – we finish up with Stage (4).

3. The state of the tape is now

$$\boxed{n}\,B\,\boxed{m}\,B\,\boxed{g(m,n)}$$

Delete the final block, so we are left with just $\boxed{n}\,B\,\boxed{m}$ on the tape. Next increment the $\boxed{n}$ block by one, and then use a modified copier to yield

$$\boxed{Sn}\,B\,\boxed{m}\,B\,\boxed{m}\,B\,\boxed{Sn}$$

And repeat Stage (2) on this input.

4. The state of the tape is now

$$\boxed{n}\,B\,\boxed{m}\,B\,\boxed{0}$$

We just need to delete the blocks of digits after the first one, then ensure we end up scanning the first cell of the remaining block, and we are done!

These stages can clearly be combined into a composite program, which will halt standardly with $\mu y[g(m,y) = 0]$ on the tape. The program is dextral. So the composite program will indeed be a dTuring program for $f(x) = \mu y[g(x,y) = 0]$, and hence, once more, we are done. $\boxtimes$

Which is, all in all, a really rather elegant proof – which is why I just couldn't resist giving it here!

42.3   Turing computability entails μ-recursiveness

As we said, it was only to be hoped and expected that we could show that all μ-recursive functions are Turing-computable. What about the converse claim that the Turing-computable total functions are all μ-recursive?

This is in its way a much more substantial result. For Turing computability involves utterly 'free form' unstructured computation (at the level of 'machine code'): we place no restrictions at all on the way we stack up i-quadruples into a program, other than brute consistency. By contrast, μ-recursive functions are defined in terms of the algorithms that can be described by a higher-level computer language with just 'for' loops and 'do until' loops as the only programming structures. So we might well wonder whether every function which is computable using an arbitrarily organized Turing program can also be computed using only those two types of looping structure. Experience with the versatility of standard computer languages will probably lead us to expect a positive answer: but still, we do need a proof.

So let's now outline the needed argument for

Theorem 42.2 *All Turing-computable functions are μ-recursive.*

Proof strategy   Take the case where f is a monadic Turing-computable function. (The argument will generalize in the obvious way to many-place functions.)

Then, by hypothesis, some Turing program Π computes f. In other words, for each n, when Π is run from a standard initial configuration with n on the tape, it halts standardly with $f(n)$ on the tape.

So consider a run of program Π, where we execute the instructions in the next applicable i-quadruple after each tick of the clock: the clock keeps ticking, however, even if the program has halted.

At the j-th tick of the clock, and before the program has halted, the current *state-of-play* of the Turing computation is given by (i) a description of the contents of the tape; (ii) a specification of which cell is the currently scanned cell; (iii) the label for the i-quadruple we will need to execute next. Note that we start with only a finite number of occupied cells on the tape, and each step makes only one modification, so at every step there is still only a finite number of occupied cells; giving the description (i) is therefore always a finite task.

Suppose we use some kind of sensible Gödel-style numbering in order to encode the state-of-play at any step while the program is still running by a single code number s (where $s > 0$). Then we define the numerical *state-of-play function* $c(n, j)$ *for program* Π:

> If the computation which starts with input n is still running at time j, having reached the state-of-play coded s, then $c(n, j) = s$; otherwise, if the computation which starts with input n has halted at time j, $c(n, j) = 0$.

We can show that *the state-of-play function for a given Π is primitive recursive.* For just reflect that in running through the first j steps of a Turing computation, any searches for the next instruction are bounded by the length of the Turing program Π. So a computation evaluating $c(n, j)$ won't involve open-ended searches, i.e. it can be programmed up using 'for' loops, hence it evaluates a p.r. function. (We won't pause, however, to formally prove this result: the details are just horribly messy and there is neither pleasure nor enlightenment to be had in hacking through them.)

To continue: if the computation halts at step h, then for all $j \leq h$, $c(n, j) > 0$ and $c(n, Sh) = 0$. So $c(n, \mu z[c(n, Sz) = 0])$ gives the code describing the state-of-play – and in particular the contents of the tape – at the point where the computation halts (and by hypothesis, it always does). Therefore,

$$f(n) = decode(c(n, \mu z[c(n, Sz) = 0]))$$

where *decode* is a function that decodes a state-of-play description s and returns the number encoded in the output block of binary digits on the tape at the end of the computation. Assuming the Gödel numbering is sensible so that *decode* is also primitive recursive, it follows immediately that f is μ-recursive.   ⊠

If the overall proof strategy here also seems familiar, that's because we used the same basic idea in Section 38.4 when sketching a proof that the Ackermann–Péter function is μ-recursive.

42.4 Generalizing

We've defined a Turing machine as dealing with *binary* symbols (and blanks), using a *one-tape* work-space, moving its focus of operations *one cell* at a time, and also reading/writing *one cell* at a time. Innumerable variations are evidently possible! For example, we could use a larger repertoire of symbols, or we could consider a machine with more than one tape, or a machine than can move the scanned cell by any given number of cells. *But such changes don't make any difference to what our machines can compute* – i.e. they don't take us outside the class of μ-recursive functions. We've already argued for that informally in the case of changing the size of the symbol set (see Section 41.1); and we can similarly argue, case by case, that e.g. working with two tapes doesn't make a difference either, by sketching a way of transforming a program for a two-tape Turing machine into an equivalent Turing program of the original type.

However, such program-transformations can be *very* messy to effect. So note that the proof-strategy of the last section can be adapted to get us much nicer equivalence proofs. For suppose we describe a fancy new machine T. Then, for any sensible machine architecture, the corresponding coding function $c_T(n, j) = s$ will still be primitive recursive (where s is now a suitable code for describing the state-of-play at time j of the computation for input n on our modified machine T). Then by just the same reasoning it follows that the function computed is still μ-recursive, and so what T computes is just what can be computed by some regular Turing machine.

43   Halting and incompleteness

Our first main theorem in this chapter establishes the 'recursive unsolvability of the self-halting problem' for Turing machines. This is one of those pivotal results like the Diagonalization Lemma which at first sight can seem just an oddity but which entails a whole raft of important results. We use this theorem to establish (or re-establish) various claims about incompleteness and decidability.

We also prove a version of Kleene's Normal Form Theorem: this leads to yet another proof of incompleteness.

43.1   Two simple results about Turing programs

(a)   As a preliminary, let's note

> **Theorem 43.1** *We can effectively (and hence recursively) enumerate the Turing programs.*

Proof sketch   Use some system for Gödel-numbering sets of i-quadruples. For example, use powers of primes to code up single i-quadruples; then form the super g.n. of a sequence of codes-for-quadruples by using powers of primes again.

Now run through numbers $e = 0, 1, 2, \ldots$. For each e, take prime factors of e, then prime factors of their exponents. If this reveals that e is the super g.n. of a set of i-quadruples, then check that it is a consistent set and hence a Turing program (that is an effective procedure, and the search is bounded by the size of the set of i-quadruples). If e *is* the super g.n. of some Turing program Π, put $\Pi_e = \Pi$; otherwise put $\Pi_e = \Pi^*$ (where Π^* is some default favourite program). Then $\Pi_0, \Pi_1, \Pi_2, \ldots$ is an effectively generated list of all possible Turing programs (with many repetitions).

A labour-saving appeal to Church's Thesis then tells us that the set of Turing programs is r.e.   ⊠

(b)   In Section 42.3, we defined $c(n, j)$ as giving a Gödel-style code number for the state-of-play at the j-th step of a run of a given function-computing program Π with input n; $c(n, j)$ defaults to zero once the run has halted gracefully.

Now we generalize, and introduce the function $c'(e, n, j)$ which gives the code for the state-of-play at the j-th step of a run of the program Π_e in our standard enumeration, where the run starts with input n.

But note that previously we were always considering a program that computes a total one-place function, so it halts gracefully and delivers an output for every input n and never freezes. We now need to specify what happens to $c'(e, n, j)$ if the program Π_e runs out of instructions on input n. So let's say

328

$c'(e, n, j) = s$, where $s > 0$ is the code number describing the state-of-play as we enter the jth step in a run of the Turing program Π_e given the initial input n, unless either the computation has already halted gracefully in which case $c'(e, n, j) = 0$, or it has frozen in which case $c'(e, n, j) = c'(e, n, j - 1)$.

Note that, unlike c, the function c' is not regular: for many values of e and n, a run of the program Π_e given input n will not halt gracefully, so $c'(e, n, j)$ never hits zero as j increases.

We saw, by a rough-and-ready argument, that our original state-of-play function is p.r.: similarly,

Theorem 43.2 *The generalized state-of-play function c' is p.r.*

Proof sketch Decoding e to extract the program Π_e doesn't involve any unbounded searches; tracking the execution of Π_e on input n through j ticks of the clock doesn't involve any unbounded searches. So the computation involved in evaluating $c'(e, n, j)$ still involves no unbounded searches overall (the addition of the clause to cope with freezing plainly doesn't change that). So we are still evaluating a p.r. function.[1]   ⊠

43.2   The halting problem

(a)   Let $\Pi_0, \Pi_1, \Pi_2, \ldots$ be an effective enumeration of Turing programs. Then

The self-halting problem is to find an effective procedure that will decide, for any e, whether Π_e halts when set to work with its own index-number e as input.

We can immediately prove the following formal theorem:

Theorem 43.3 *The self-halting problem is not recursively solvable.*

Proof Let $h(e)$ be the characteristic function whose value is 0 if the machine number e halts with input e, and is otherwise 1.[2] Then, by definition, the self-halting problem is recursively solvable iff h is μ-recursive. Hence, by the equivalence established in the last chapter, the self-halting problem is recursively solvable iff there is a Turing machine H for computing the function h. We now show that there can be no such machine as H, which proves the theorem.

Suppose that there *is* a Turing machine H which computes h, and consider the result of 'chaining together' H with the trivial machine L which, when fed 0

---

[1]The difference between the original sort of state-of-play function c and our generalized function c' can be thought of like this. c is, in effect, a particular Turing program Π 'compiled' into a p.r. function: c' is a p.r. interpreter for arbitrary Turing programs.

[2]Recall once more, for us 0 serves as the truth-value *true*: see Section 2.2.

on its input tape, goes into an infinite loop and never halts but when fed 1 halts leaving the tape untouched. Call this composite machine D.

Like any Turing machine, D will appear in our enumeration of all the machines; it's the machine with program Π_d for some d. So next we ask the cunning question: does D halt when fed its own index number d as input? (And off we go on yet another 'diagonal' argument ...!)

Suppose D *does* halt on input d. Then, by definition of h, $h(d) = 0$. Since H, by hypothesis, computes h, this means that H must halt on input d and output 0. So chaining H together with the looper L gives us a composite machine which doesn't halt on input d. But that composite machine is D! So D doesn't halt. Contradiction.

Suppose then that D does *not* halt on input d. That means $h(d) = 1$. Since H computes h, this means that H must halt on input d and output 1. So chaining H together with the looper L gives us a composite machine which halts on input d. That composite machine is D again. So D halts after all. Contradiction.

Hence there can be no such composite machine as D. But L trivially exists. Which means that there can indeed be no such Turing machine as H. $\boxtimes$

And now we can appeal to Church's Thesis in interpretative mode. The recursive unsolvability of the self-halting problem shows that there is indeed no effective way of telling of an arbitrary Π_e whether it will halt on input e.

(b) Having proved that the *self*-halting problem isn't recursively solvable, we can go on to deduce that various other halting problems are also unsolvable. The general strategy is to suppose that the target problem P is recursively solvable; show this entails that the self-halting problem is recursively solvable too; then use our last theorem to deduce that P can't be recursively solvable after all; and then we appeal again to Church's Thesis to legitimate interpreting this result as showing that P isn't effectively solvable at all.

To take the simplest but also most important example, consider the problem which is usually simply called

> *The* halting problem: to find an effective procedure that will decide, for any e and n, whether Π_e halts when given input n.

There can't be such an effective procedure, however, because

Theorem 43.4 *The halting problem is not recursively solvable.*

Proof In Section 41.3 we defined a Turing copier which takes input e and produces output e, e. Now suppose that there is a machine H' which takes the inputs e, n and delivers a verdict on whether Π_e halts for input n. Chain together the copier and H', and we'd get a composite machine which takes input e and delivers a verdict on whether Π_e halts for input e. But we've just shown that there can be no such machine. So there is no such machine as H'.

Hence, the characteristic function $h(e, n)$ – whose value is 0 if the machine number e halts with input n, and is otherwise 1 – is not μ-recursive (else there would be a machine H' that computes it). Hence the halting problem isn't recursively solvable. ⊠

(c) Why are halting problems interesting in general? Well, for a start, because when we write Turing (or other) programs we'd like to be able to check that they work as advertised: we'd like to be able to check they don't get stuck in never-ending loops when they are supposed to halt and deliver an output. So it would be good if there were a general effective procedure for program-checking, which can at least determine whether a program Π_e halts when it is supposed to halt. And we've just shown that that's impossible.

43.3 The *Entscheidungsproblem* again

Next, here is a lovely result which shows another aspect of the significance of the unsolvability of the halting problem:

> **Theorem 43.5** *The recursive unsolvability of the halting problem entails that the* Entscheidungsproblem *is recursively unsolvable too.*

Proof Review the definition of the function c' in Section 43.1. Being a p.r. function, c' can be expressed and captured by Q by a four-place open wff $\mathsf{C}(\mathsf{x}, \mathsf{y}, \mathsf{z}, \mathsf{w})$.

So, for any e, $\mathsf{C}(\overline{\mathsf{e}}, \overline{\mathsf{e}}, \overline{\mathsf{j}}, 0)$ is true iff $c'(e, e, j) = 0$, i.e. just when Π_e with input e has halted by step j. Hence, since Q is p.r. adequate, $\mathsf{Q} \vdash \mathsf{C}(\overline{\mathsf{e}}, \overline{\mathsf{e}}, \overline{\mathsf{j}}, 0)$ whenever Π_e with input e has halted by step j. And if Π_e never halts with input e then, for each j, $\mathsf{Q} \vdash \neg\mathsf{C}(\overline{\mathsf{e}}, \overline{\mathsf{e}}, \overline{\mathsf{j}}, 0)$.

So now put $\mathsf{H}(\overline{\mathsf{e}}) =_{\mathrm{def}} \exists\mathsf{z}\mathsf{C}(\overline{\mathsf{e}}, \overline{\mathsf{e}}, \mathsf{z}, 0)$. It follows that, if Π_e eventually halts at some step j with input e, then $\mathsf{Q} \vdash \mathsf{H}(\overline{\mathsf{e}})$: and if it doesn't ever halt then – assuming Q is ω-consistent (which it is, being sound!) – $\mathsf{Q} \nvdash \mathsf{H}(\overline{\mathsf{e}})$.

Now let $\hat{\mathsf{Q}}$ again be the conjunction of the seven non-logical axioms of Q. Then $\hat{\mathsf{Q}} \to \mathsf{H}(\overline{\mathsf{e}})$ will be a logical theorem iff $\mathsf{Q} \vdash \mathsf{H}(\overline{\mathsf{e}})$ iff Π_e halts with input e.

Suppose we could recursively decide what's a first-order theorem. Then it would follow that we could recursively decide whether Π_e eventually halts with input e, for arbitrary e. Contraposing gives us our theorem.[3] ⊠

43.4 The halting problem and incompleteness

We already knew from Theorem 40.3 that the *Entscheidungsproblem* is not recursively solvable. But it is still worth highlighting our new Theorem 43.5, just

---

[3]A link between the halting problem and the *Entscheidungsproblem* was first made by Turing (1936, §11); see also Büchi (1962). However, the original version of the linking argument doesn't go via Q's p.r. adequacy and soundness, but depends upon more directly writing claims about programs and halting states as first-order wffs. You can find a textbook presentation of this line of proof in e.g. Boolos et al. (2002, ch. 11).

to emphasize how our limitative theorems hereabouts are all so very closely interconnected. This section and the next note more interconnections.

(a)   We've just seen that, using a Gödel-numbering scheme to code up facts about Turing machines, there will be a purely arithmetical L_A-sentence $H(\bar{e})$ which 'says' that the program Π_e halts when run on input e.

Now let's suppose, for reductio, that the truths of L_A are effectively enumerable. That means that there is an effectively computable function that given successive inputs 0, 1, 2 ... delivers as output a listing of (the Gödel numbers of) the truths of L_A.

So start effectively generating that listing. Since exactly one of $H(\bar{e})$ and $\neg H(\bar{e})$ is true, the Gödel number for one of them must eventually turn up – by hypothesis the effective enumeration lists (the Gödel numbers of) *all* the truths. So setting the enumeration going will be an effective way of deciding which of $H(\bar{e})$ and $\neg H(\bar{e})$ is true – wait and see which turns up! That gives us an effective way of solving the self-halting problem and determining whether the program Π_e halts when run on input e.

However, we now know that there is no effective way of solving the self-halting problem. So we can conclude that there is no effective way of enumerating the truths of L_A.

We can now either make a labour-saving appeal to Church's Thesis, or we can fairly easily rigorize the informal line of thought just sketched, to get:

> **Theorem 43.6** *The recursive unsolvability of the self-halting problem entails that $\mathcal{T}_A$ – True Basic Arithmetic, the set of truths of L_A – is not r.e.*

Given that the self-halting problem is indeed not recursively solvable, that yields the second proof we wanted for

> **Theorem 40.11** $\mathcal{T}_A$ *is not recursively enumerable.*

And this time – compare Section 40.8 – it is a proof that depends on considerations about step-by-step computations.[4]

(b)   And now a semantic incompleteness theorem very quickly follows, by a formal analogue of the argument in Chapter 6. For Theorem 40.6 tells us that the theorems of a recursively axiomatized formal theory *are* r.e. Hence, in particular, the theorems of a recursively axiomatized sound theory of arithmetic T framed in L_A are r.e.; hence T's theorems cannot be all of $\mathcal{T}_A$. But if φ is one of the truths in $\mathcal{T}_A$ which T can't prove, then T (being sound) can't prove $\neg\varphi$ either. Whence, by appeal to the recursive unsolvability of the self-halting problem, we have shown that

---

[4]There is also an alternative line of proof that runs more closely parallel to the informal argument in Section 6.2, but we won't spell this out. Enthusiasts can have the pleasure of consulting Cutland (1980, ch. 8).

Theorem 43.7 *A recursively axiomatized sound theory T whose language is L_A cannot be negation-complete.*

As a reality check, show that we can, by simple tweaks, also deduce that a recursively axiomatized sound theory T whose language *includes* L_A can't be negation-complete.

43.5 Another incompleteness argument

The alert reader might ask: in proving incompleteness by thinking about Turing machines halting, can we again drop the assumption that we are dealing with a *sound* theory of arithmetic and replace it e.g. with the weaker assumption that we are dealing with an ω-consistent theory?

We can! So, merely because it's rather fun if you like this sort of thing, let's re-prove a canonical version of the First Theorem. Here again is what we want to establish:

Theorem 22.3 *If T is a p.r. adequate, p.r. axiomatized theory whose language includes L_A, then there is a L_A-sentence φ of Goldbach type such that, if T is consistent then $T \nvdash \varphi$, and if T is ω-consistent then $T \nvdash \neg\varphi$.*

And this time we prove the result without constructing a canonical Gödel sentence or appealing to the Diagonalization Lemma.

Proof Recall the definition of $H(\bar{e})$ from Section 43.3, and now consider a Turing machine which, on input e, looks at values of m in turn, and tests whether m numbers a T-proof of $\neg H(\bar{e})$, i.e. for successive values of m it checks whether $Prf_T(m, \ulcorner\neg H(\bar{e})\urcorner)$, and it eventually halts when this p.r. relation holds or else it trundles on forever (evidently, we can program such a machine). This Turing machine, in short, tries to prove that Π_e doesn't halt with input e. For some s, it will be Π_s in our standard enumeration of machines. And – very predictably! – we first ask whether Π_s halts on input s.

If it does, then $T \vdash H(\bar{s})$ (because a p.r. adequate theory can prove each true $H(\bar{e})$ – see the proof of Theorem 43.5). But also, by the definition of Π_s, its halting implies that for some m, $Prf_T(m, \ulcorner\neg H(\bar{s})\urcorner)$, so $T \vdash \neg H(\bar{s})$. So, assuming T is consistent, Π_s *doesn't* halt on input s.

And now we can show (i) that if T is consistent, it doesn't prove $\neg H(\bar{s})$, and (ii) if T is ω-consistent, then it doesn't prove $H(\bar{s})$.

(i) First assume that $T \vdash \neg H(\bar{s})$. Then for some m, $Prf_T(m, \ulcorner\neg H(\bar{s})\urcorner)$, so Π_s would halt for input s. But we've just shown that Π_s doesn't halt for input s. So T doesn't prove $\neg H(\bar{s})$.

(ii) Now assume alternatively that $T \vdash H(\bar{s})$ i.e. $T \vdash \exists x C(\bar{s}, \bar{s}, x, 0)$. But since Π_s doesn't halt on input s, we have $c'(s, s, j) \neq 0$ for each j; and so $T \vdash \neg C(\bar{s}, \bar{s}, \bar{j}, 0)$

for each j. Which makes T ω-inconsistent. So if T is ω-consistent, it doesn't prove $\mathsf{H}(\bar{s})$.

Finally, you can readily confirm that the undecidable wff $\varphi =_{\text{def}} \neg\mathsf{H}(\bar{s})$ is equivalent to one of Goldbach type. $\boxtimes$

43.6   Kleene's Normal Form Theorem

That's the main business of this chapter done. But – now that we've reached this far – I can't resist adding a few more sections for enthusiasts. First, in this section, we'll prove a (version) of a very illuminating result due to Kleene (1936a). Then in the final section, we'll show that this result once more gives us another somewhat different route to incompleteness.

Still – very pretty though the arguments are! – you might well want to skip on a first reading.

(a)   Review the argument of Section 42.3. By exactly the same reasoning, if the Turing program Π_e in our enumeration in fact *does* compute a μ-recursive function $f_e(n)$ then we will have

$$f_e(n) = decode(c'(e, n, \mu z[c'(e, n, Sz) = 0])).$$

Now just for brevity, let's define two functions by composition as follows:

$$t(x, y, z) =_{\text{def}} c'(x, y, Sz),$$
$$u(x, y, z) =_{\text{def}} decode(c'(x, y, z)).$$

Both t and u are p.r. functions: hence – at least, given we've done the hard work of spelling out the definition of c' – we have established the following result:

> There is a pair of three-place p.r. functions t and u such that any one-place μ-recursive function can be given in the standard form
>
> $$f_e(n) = u(e, n, \mu z[t(e, n, z) = 0])$$
>
> for some value of e.

Which is *almost*, but not quite, Kleene's Normal Form theorem.

Note, by the way, that it is quite emphatically *not* being claimed that for *every* value of e our definition delivers a μ-recursive function. For t is not a regular function: for given values of e and n, $t(e, n, z)$ may never hit zero for any value of z. The claim is the other way around: for any f which is μ-recursive, there is a value of e such that $f = f_e$ (and for this e, $t(e, n, z)$ will always hit zero for some value of z).[5]

---

[5] Let's also note an interesting corollary of almost-Kleene. Imagine for a moment that we could replace the use of the *unbounded* minimization operator with *bounded* minimization. In other words, suppose that, associated with some function f_e we could find a corresponding p.r. function g_e, such that

(b) Here, mostly because we want to link up with standard presentations, is a more official version of Kleene's theorem for one-place functions:

> **Theorem 43.8** *There is a three-place p.r. function T and a one-place p.r. function U such that any one-place μ-recursive function can be given in the standard form.*
>
> $$f_e(n) =_{\text{def}} U(\mu z[T(e, n, z) = 0])$$
>
> *for some value of e.*

So the *only* difference in the official version is that, by a bit of not-quite-trivial juggling, we can get all the dependence on e and n packed into the function T, allowing U to be a one-place function.

This little difference between our almost-Kleene result and the real thing is a distinction which doesn't matter *at all* in most applications. However, for enthusiasts, we'd better derive the official theorem:

Proof Consider the four-place relation $R(e, n, r, j)$ which holds when both $t(e, n, j) = 0$ and $r = u(e, n, (\mu z \leq j)[t(e, n, z) = 0])$. When the Turing program Π_e successfully computes a monadic (total) function, then the R relation holds if the program, run with input n initially on the tape, halts gracefully with the result r on the tape, no later than time j. R is a p.r. relation since it is a conjunction of two p.r. relations (remember: *bounded* minimization keeps us within the domain of p.r. relations).

Now put $T(e, n, z) = 0$ when $(\exists r \leq z)(\exists j \leq z)(z = 2^r \cdot 3^j \wedge R(e, n, r, j))$ and put $T(e, n, z) = 1$ otherwise. Then T is p.r. (definition by cases). When the Turing program Π_e computes a monadic function, execution halts at the least value of z such that $T(e, n, z) = 0$. And we can extract the value of r, i.e. the result of the computation, by using the familiar factorizing function $exf(z, 0)$ which gives the exponent of π_0, i.e. 2, in the factorization of z. So put $U(z) =_{\text{def}} exf(z, 0)$. Then, as we wanted, $f_e(n) = U(\mu z [T(e, n, z) = 0])$, with U and T p.r. functions.[6] ⊠

$$f_e(n) =_{\text{def}} u(e, n, (\mu z \leq g_e(n))[t(e, n, z) = 0])$$

Then f_e would not just be recursive but be *primitive* recursive (see Section 14.7, Fact D). So, contraposing, if f_e is computable but not p.r., there can't be a p.r. function $g_e(n)$ which bounds the number of steps required in the Turing computation of $f_e(n)$. In Section 38.5, fn. 7, we noted that the *values* of a μ-recursive-but-not-p.r. function $f_e(n)$ don't have to grow explosively as n increases: but we now see that *lengths of computations* go wild.

[6] Again, note that T is *not* regular, so we are *not* saying that, for every e, we can read the equation $f_e(n) =_{\text{def}} U(\mu z[T(e, n, z) = 0])$ from right to left to define a μ-recursive function: the claim is just that, when f_e is μ-recursive there is an e which makes that equation hold.

We can of course readily generalize the argument to get a Normal Form theorem to cover many-place functions. The idea is that for any k there is a $k + 2$-place p.r. function T_k such that every k-place μ-recursive function $f(\vec{n})$ can be given in the form $U(\mu z [T_k(e, \vec{n}, z) = 0])$ for some e.

Note also that we *can* use a similar proof idea to establish Kleene's Normal Form theorem about recursive functions more directly, i.e. without going via the different idea of a Turing

43.7   A note on partial computable functions

To repeat, any total computable function is $f_e(n) =_{\text{def}} U(\mu z[T(e, n, z) = 0])$ for some e. Now define another 'diagonal' function:

$d(n) = f_n(n) + 1$ when $f_n(n)$ is defined (when $\exists z\, T(n, n, z) = 0$) and let $d(n)$ be undefined otherwise.

As with t, the function T is not regular – i.e. for given n, there may be no z such that $T(n, n, z) = 0$. So the function d is indeed not total. However, our definition gives us a way of effectively computing the value of $d(n)$ for those values for which it *is* defined. So d is a *partial computable* function.

Here is another definition:

A partial function f *has a recursive completion* if there is a μ-recursive (and so total) function g such that, for for all n where $f(n)$ is defined, $f(n) = g(n)$.

We now have a simple but telling result:

Theorem 43.9 *The function d has no recursive completion.*

Proof   Suppose g is μ-recursive and completes d. For some e, g is the function f_e. Since g is total, $g(e)$, i.e. $f_e(e)$, is defined. So $d(e)$, i.e. $f_e(e) + 1$, is defined. But g must agree with d when d is defined. So $f_e(e) = g(e) = d(e) = f_e(e) + 1$. Contradiction!   ⊠

In sum, you can't just take a partial computable function, 'fill the gaps' by now stipulating that the function should then take the value zero, and still get a computable function.[7]

43.8   Kleene's Theorem entails Gödel's First Theorem

We'll finish this chapter by noting a *wonderfully* pretty result which takes us back once again to our most central Gödelian concerns:

Theorem 43.10 *Kleene's Normal Form Theorem entails the First Incompleteness Theorem.*

Proof   Suppose that there is a p.r. axiomatized formal system of arithmetic S which is p.r. adequate, is ω-consistent (and hence consistent), and is negation-complete.

---

program. Just Gödel-number the computations which execute the function's recursive recipe: see e.g. Odifreddi (1999, pp. 90–96).

[7] Gap-filling-by-stipulation was, famously, Frege's preferred strategy for avoiding non-denoting terms in a logically perfect language. We now see that the strategy is far from innocuous: it can turn computable partial functions into non-computable ones.

Since S is p.r. adequate, there will be a four-place formal predicate T which captures the three-place p.r. function T that appears in Kleene's theorem. And now consider the following definition,

$$\overline{d}(n) = \begin{cases} U(\mu z[T(n,n,z) = 0]) + 1 & \text{if } \exists z(T(n,n,z) = 0) \\ 0 & \text{if } S \vdash \forall z \neg \mathsf{T}(\overline{n}, \overline{n}, z, 0) \end{cases}$$

We'll show that, given our assumptions about S, this well-defines a μ-recursive function.

Take this claim in stages. First, we need to show that our two conditions are exclusive and exhaustive:

i. The two conditions are mutually exclusive (so the double-barrelled definition is consistent). For assume that both (a) $T(n,n,k) = 0$ for some number k, and also (b) $S \vdash \forall z \neg \mathsf{T}(\overline{n}, \overline{n}, z, 0)$. Since the formal predicate T captures T, (a) implies $S \vdash \mathsf{T}(\overline{n}, \overline{n}, \overline{k}, 0)$. Which contradicts (b), given that S is consistent.

ii. The two conditions are exhaustive. Suppose the first of them doesn't hold. Then for every k, it isn't the case that $T(n,n,k) = 0$. Hence, for every k, $S \vdash \neg \mathsf{T}(\overline{n}, \overline{n}, \overline{k}, 0)$. By hypothesis S is ω-consistent, so we can't also have $S \vdash \exists z \mathsf{T}(\overline{n}, \overline{n}, z, 0)$. Hence by the assumption of negation-completeness we must have $S \vdash \neg \exists z \mathsf{T}(\overline{n}, \overline{n}, z, 0)$, which is equivalent to the second condition.

Which proves that, given our initial assumptions, our conditions well-define a total function $\overline{d}$.

Still given the same assumptions, $\overline{d}$ is effectively computable. Given n, just start marching through the numbers $k = 0, 1, 2, \ldots$ until we find the first k such that either $T(n,n,k) = 0$ (and then we put $\overline{d}(n) = U(\mu z[T(n,n,z) = 0]) + 1$), or else k is the super g.n. of a proof in S of $\forall z \neg \mathsf{T}(\overline{n}, \overline{n}, z, 0)$ (and then we put $\overline{d}(n) = 0$). Each of those conditions can be effectively checked to see whether it obtains – in the second case because S is p.r. axiomatized, so we can effectively check whether k codes for a sequence of expressions which is indeed an S-proof. And we've just shown that eventually one of the conditions must hold.

Now we make a labour-saving appeal to Church's Thesis, and conclude that $\overline{d}$ is μ-recursive. Which shows that our initial assumptions about S imply that the special function d defined at the beginning of the previous section has a recursive completion $\overline{d}$. But that is ruled out by Theorem 43.9!

So our assumptions about S can't all be true together. Hence it follows from Kleene's Normal Form Theorem that, if S is a p.r. axiomatized, p.r. adequate, ω-consistent theory, it can't also be negation-complete – which is the canonical form of Gödel's First Theorem again. ⊠

And, I'm rather tempted to add, if you don't find *that* argument a delight, then maybe you aren't quite cut out for this logic business after all!

44 The Church–Turing Thesis

We now bring together Church's Thesis and Turing's Thesis into the *Church-Turing Thesis*, explain what the Thesis claims when properly interpreted, and say something about its history and about its status.

44.1 Putting things together

Right back in Chapter 3 we stated *Turing's Thesis*: a numerical (total) function is effectively computable by some algorithmic routine if and only if it is computable by a Turing machine. Of course, we initially gave almost no explanation of the Thesis. It was only very much later, in Chapter 41, that we eventually developed the idea of a Turing machine and saw the roots of Turing's Thesis in his general analysis of the fundamental constituents of any computation.

Meanwhile, in Chapter 38, we introduced the idea of a μ-recursive function and noted the initial plausibility of *Church's Thesis*: a numerical (total) function is effectively computable by an algorithmic routine if and only if it is μ-recursive.

Then finally, in Chapter 42, we outlined the proof that a total function is Turing-computable if and only if it is μ-recursive. *Our two Theses are therefore equivalent.* And given that equivalence, we can now talk of

> *The Church–Turing Thesis* The effectively computable total numerical functions are the μ-recursive/Turing-computable functions.

Crucially, this Thesis links what would otherwise be merely technical results about μ-recursiveness/Turing-computability with intuitive claims about effective computability; and similarly it links claims about recursive decidability with intuitive claims about effective decidability. For example: it is a technical result that PA is not a recursively decidable theory. But what makes that theorem really significant is that – via the Thesis – we can conclude that there is no intuitively effective procedure for deciding what's a PA theorem. The Thesis is in the background again when, for example, we focus on a generalized version of the First Theorem, which says that any recursively axiomatized ω-consistent theory containing Q is incomplete. Why is *that* technical result significant? Because the Thesis directly links the idea of being recursively axiomatized to the idea of being an effectively axiomatized theory in the informal intuitive sense.

So, in sum, we repeatedly depend on the Thesis in giving interesting interpretative glosses to our technical results in this book. That's why it is so important to us. The Thesis, properly understood, is almost universally accepted. This chapter briefly explains why.

44.2 From Euclid to Hilbert

Any schoolchild, when first learning to calculate (e.g. in learning how to do 'long division'), masters a number of elementary procedures for computing the answers to various problems. And ever since Euclid, who e.g. gave an elegant and efficient routine for finding the greatest common divisor of two integers, mathematicians have developed more and more algorithms for tackling a wide variety of problems.[1]

These algorithms – to repeat our characterization of Section 3.1 – are finite sequential step-by-step procedures which can be set out in every detail in advance of being applied to any particular case. Every step is (or, by further breaking down, it can be made to be) 'small' so that it is readily executable by a calculator (human or otherwise) with limited cognitive resources; the steps don't require any ingenuity or mathematical acumen at all. The rules for moving from one step to the next are self-contained and fully determine the sequence of steps (they don't involve e.g. pausing to consult an outside oracle or to toss coins).[2] We also require that an algorithmic procedure is to deliver its output after a finite number of computational steps.

Such algorithms are evidently great to have when we can get them. It is no surprise, then, that algorithms to deal with an ever-widening domain of problems have been sought over two millennia. And mathematicians had – we may reasonably suppose – a pretty clear idea of the kind of thing they were looking for before they had ever heard of μ-recursiveness or Turing machines. The idea of an algorithm was simply taken for granted, and wasn't subject to any close analysis. Even in the foundational ferment of the first quarter of the last century, there seems to have been very little explicit discussion. And surely not because the idea was thought too foggy to take very seriously, but for exactly the opposite reason. Compared with the profound worries about the infinite generated by e.g. the set-theoretic paradoxes and the intuitionistic critique of classical mathematics, the idea of a finite, algorithmic, step-by-step procedure must have seemed relatively clear and unproblematic.

Consider again Hilbert and Ackermann's classic *Grundzüge der theoretischen Logik* (1928). In §11, as we've noted before, the authors famously pose the

---

[1]For a generous sample of cases, ancient and more modern, see Chabert (1999).

[2]Determinism is built into the classical conception. Nowadays, we also talk of 'randomized' and 'probabilistic' and 'quantum' algorithms. These are perhaps natural enough extensions of the original idea; however, they assuredly *are* extensions. More generally, perhaps it is true to say:

> In fact the notion of algorithm is richer these days than it was in Turing's days. And there are algorithms ... not covered directly by Turing's analysis, for example, algorithms that interact with their environments, algorithms whose inputs are abstract structures, and geometric or, more generally, non-discrete algorithms. (Blass and Gurevich, 2006, p. 31)

But our concern here is with the articulation of the classical idea of deterministic step-by-step routines of the kind that the early founding fathers had in mind in arriving at the concept of an effective procedure.

Entscheidungsproblem, i.e. the problem of deciding for an arbitrary sentence of first-order logic whether it is valid or not. They note that the corresponding decision problem for propositional logic is easily solved – we just use a truthtable: in this case, at any rate, we have a 'well-developed algebra of logic' where, 'so to speak, an arithmetic treatment' is possible. So now,

> The decision problem for first-order logic presents itself. ... The decision problem is solved when one knows a process which, given a logical expression, permits the determination of its validity/satisfiability in a finite number of operations. ... We want to make it clear that for the solution of the decision problem a process needs to be given by which [validity] can, in principle, be determined (even if the laboriousness of the process would make using it impractical). (Hilbert and Ackermann, 1928, §11)

The quest, then, is for an effective procedure that works – at least in principle, given world enough and time – to deliver a verdict in a finite number of steps by means of a quasi-arithmetical computation. In sum, the *Entscheidungsproblem* – 'the chief problem of mathematical logic', as they call it – is the problem of finding an algorithm for deciding validity. And *Hilbert and Ackermann plainly took themselves to have set a well-posed problem* that needed only the bare minimum of elucidation which we have just quoted.

Of course, it is one thing to pose a decision problem, and quite another thing to solve it. And there is a crucial asymmetry here between positive and negative cases. Hilbert and Ackermann note, for example, that the decision problem for monadic first-order logic (i.e. logic with only one-place predicates) *does* have a positive solution: see Löwenheim (1915). Now, to show that Löwenheim's procedure solves the restricted decision problem, we just need (i) to see that it is algorithmic, and (ii) to prove that it always delivers the right verdict. And step (i) requires only that we recognize an algorithm when we see one; we don't need e.g. to have command of any general story about necessary conditions for being an algorithm. It is similar for other decision problems with positive solutions: we just need to be able to say 'Look, here's an algorithm that does the trick'.

Suppose, on the other hand, that we begin to suspect that a certain decision problem is unsolvable and want to confirm that conjecture. For example, let's re-run history and suppose that we have tried but failed to discover a general positive solution for the *Entscheidungsproblem*. We find some solutions for limited cases, extending Löwenheim's work, but the suspicion begins to dawn that there isn't a general method for deciding validity for arbitrarily complex first-order sentences. We therefore set out to prove that the decision problem is unsolvable – i.e. prove that there is *no* algorithm which decides first-order validity across the board. Obviously, we can't establish this by an exhaustive search through possible algorithms, one at a time, since there are unlimitedly many of those. Hence, to get our negative proof, we'll need some way of proving facts about the class of *all* possible algorithms. So *now* we will need some clean finite characterization of the infinite class of algorithms.

In headline terms, then, the situation is this: our implicit pre-theoretical grasp of the idea of a step-by-step computation is enough for a good understanding of what the *Entscheidungsproblem* is. But to show that it is unsolvable we need more; we need an explicit story about general features of all algorithmic procedures in order to show that no procedure with those features can solve the problem. Or to quote Kleene:

> [The] intuitive notion of a computation procedure, which is real enough to separate many cases where we know we do have a computation procedure before us from many others where we know we don't have one before us, is vague when we try to extract from it a picture of all possible computable functions. And we must have such a picture, in exact terms, before we can hope to prove that there is no computation procedure at all for a certain function, ... Something more is needed for this. (Kleene, 1967, p. 231)

We will need to say more in the next chapter, however, about whether 'vague' is *le mot juste* here.

44.3   1936 and all that

Faced with the *Entscheidungsproblem* and the growing suspicion that it was unprovable, Church, Kleene, Gödel and Turing developed a variety of accounts of what makes for an effectively computable function (and hence what makes for an effective decision procedure).[3] The drama was played out in Princeton (USA) and Cambridge (England).

(a)   To start with, in the early 1930s in Princeton, Alonzo Church and his then students Stephen Kleene and Barkley Rosser developed the so-called 'λ-calculus'. After some false starts, this proved to be a powerful foundational system: Church first conjectured in 1934 that every effectively calculable function is 'λ-definable'.

However, we won't pause to explain what this means. For – at least until it was taken up much later by theoretical computer scientists – the λ-calculus didn't win over a great number of enthusiasts, and a couple of other approaches to computability very quickly came to occupy centre stage.[4]

First, in lectures he gave during his visit to Princeton in 1934, Gödel outlined the idea of a so-called *general recursive* function, drawing on work by Jacques

---

[3]To be absolutely explicit about the connection: the Gödel-numbering trick turns the decision problem about validity into a decision problem about the corresponding numerical property of numbering-a-valid-wff; and the characteristic function trick turns this numerical decision problem into a question about the algorithmic computability of a corresponding function. We'll have to radically truncate the historical story in what follows, but I hope in not too misleading a way. For many more details, see Davis (1982), Gandy (1988) and Sieg (1997).

[4]But see, for example, Trakhtenbrot (1988) and Barendregt (1997).

Herbrand. Consider how we compute values of the Ackermann–Péter function p (see Section 38.3). We are given some fundamental equations governing p; and then we work out the value of e.g. $p(2, 1)$ by repeated substitutions in these equations. Roughly speaking, the idea of a general recursive function is the idea of a function whose values can similarly be uniquely generated by repeated substitutions into initial equations (where these equations aren't now restricted to primitive recursive definitions but might allow e.g. double recursions).

But again we won't go into further details here.[5] For by 1936, Kleene was highlighting the neat concept of a μ-*recursive* function which we met in Chapter 38 – i.e. the idea of a function definable by primitive recursion and minimization. And it almost immediately became clear that in fact the λ-definable total functions are the same functions as Gödel's general recursive total functions which are the same as the μ-recursive functions.[6]

This convergence of approaches was enough to convince Church of the soundness of his original conjecture – a version of Church's Thesis, as we now think of it – identifying calculability with λ-definability. So in his classic 1936 paper, he writes:

> We now define the notion, already discussed, of an effectively calculable function of positive integers by identifying it with the notion of a recursive function of positive integers (or of a λ-definable function of positive integers). This definition is thought to be justified by the considerations which follow [most notably, the equivalence results], so far as a positive justification can ever be obtained for the selection of a formal definition to correspond to an intuitive notion. (Church, 1936b, p. 356)

Kleene had already been persuaded: he reports that – when Church originally conjectured that the Thesis is true – he 'sat down to disprove it by diagonalizing out of the class of the λ-definable functions'; and finding that this couldn't be done he 'became overnight a supporter'.[7] Gödel, however, was more cautious: his recollection was that he was initially 'not at all convinced' that all effective computations were covered by his characterization of general recursiveness.[8]

(b) Meanwhile, in 1935 in Cambridge, Turing was working quite independently on his account of computation.

We have already sketched this in Section 41.1. To repeat, but now more in his own words, Turing invites us to 'imagine the operations performed by the [human] computer to be split up into "simple operations" which are so elementary that it is not easy to imagine them further divided.'[9] Take our computing agent

---

[5]See Gödel (1934, §9). For a modern introduction to 'Herbrand–Gödel' computability, as it is also called, see for example Mendelson (1997, §5.5) or Odifreddi (1999, pp. 36–38).

[6]The key equivalence results were published in Kleene (1936a,b).

[7]Kleene (1981, p. 59). For the idea of 'diagonalizing out' see Section 38.6.

[8]See the letter from Gödel to Martin Davis as quoted by Kleene (Gödel, 1986, p. 341).

[9]This, and the following quotations in this paragraph, come from Turing (1936, §9).

to be writing on squared paper (though the two-dimensional character of the usual paper we use for hand computation is inessential: we can 'assume that the computation is carried out on one-dimensional paper, i.e. on a tape divided into squares'). Then, Turing continues, we can suppose that a *single* symbol is being observed at any one time: if the computing agent 'wants to observe more, he must use successive observations.'[10] Likewise, '[w]e may suppose that in a simple operation not more than one symbol is altered. Any other changes can be split into simple changes of this kind.' Further, we may, without loss of generality, 'assume that the squares whose symbols are changed are always "observed" squares.' Having read and written in some square, we then need to move to work on a new square. The human computer needs to be able to recognize the squares he has to jump to next. And Turing claims that 'it is reasonable to suppose that these can only be squares whose distance from ... the immediately previously observed squares does not exceed a certain fixed amount' (without loss of generality we can suppose the computer gets to new squares stepwise, jumping one square at a time, as in our presentation in Chapter 41). As each simple operation is completed, the human computer moves to some new 'state of mind' which encapsulates determinate instructions about what to do next by way of changing the contents of a square and jumping to a new square (or cell, to revert to our earlier terminology). In sum:

> It is my contention that these operations [i.e. reading/writing cells on a tape, moving the scanned cell, and jumping to a new instruction] include all those that are used in the computation of a number. (Turing, 1936, §1)

And *if* all that is right, then it is immediate that any function which is effectively computable via an algorithm will be computable by a Turing machine. The converse, however, is surely secure: any Turing-computable function is computable by us (had we but world enough, and time), by following the program steps. Whence Turing's Thesis.

(c) Turing learnt of the Princeton research on λ-definability and recursiveness just as he was about to send off his own classic paper for publication. In response, he added a key appendix outlining a proof of the equivalence of Turing-computability with λ-definability. And thus the two strands of work that we've just outlined came together. Still only 24, Turing then left Cambridge to continue his research at Princeton.

(d) It would be wrong to say that in 1936 Church was still defending his Thesis *merely* because a number of different but apparently 'empirically adequate' characterizations of computability had turned out to be equivalent, while in contrast Turing based *his* Thesis on a bottom-up analysis of the very idea of

---

[10]In our presentation, we took the fundamental symbols to be just '0' and '1'. Turing was more generous. But for the reasons we gave, working with (say) 27 basic symbols rather than two in the end makes no odds.

computation. For a start, Church takes over Gödel's 1934 idea of general recursiveness, and that can be thought of as an attempt to locate the essence of a computation in the repeated manipulation of equations. And in his paper, Church also gives what is often referred to as his 'step-by-step argument' for the Thesis, which is related to the argument which we gave in Section 38.5.

Still, Turing's analysis does seem to dig deeper. It convinced Church: if he was somewhat guarded in 1936 (saying that his own account of computability is justified 'so far as a positive justification can ever be obtained'), he is quite emphatic a year later:

> It is ... immediately clear that computability, so defined [by Turing], can be identified with (especially, is no less general than) the notion of effectiveness as it appears in certain mathematical problems (various forms of the *Entscheidungsproblem*, various problems to find complete sets of invariants in topology, group theory, etc., and in general any problem which concerns the discovery of an algorithm). (Church, 1937, p. 42)

Gödel was similarly convinced. In a 'Postscriptum' added in 1964 to a reprint of his 1934 lectures, he writes:

> Turing's work gives an analysis of the concept of 'mechanical procedure' (alias 'algorithm' or 'computation procedure' or 'finite combinatorial procedure'). This concept is shown to be equivalent with that of a 'Turing machine'. (Gödel, 1934, pp. 369–370)

And Gödel remarks in a footnote that 'previous equivalent definitions of computability' (and he refers to Church's account of λ-definability and his own treatment of general recursiveness) 'are much less suitable for our purpose' – the purpose, that is, of giving a 'precise and unquestionably adequate definition' of concepts like undecidability. The thought, then, is that Turing's analysis *shows* that algorithmic computability is 'unquestionably' Turing computability (while λ-definability and recursiveness earn their keep as alternative characterizations of computability because of the theorems that show them to be equivalent to Turing computability).

44.4   What the Church–Turing Thesis is and is not

(a)   It is important to note that there are three levels of concept that are in play here when we talk about computation.

1. At the *pre-theoretic* level – and guided by some paradigms of common-or-garden real-world computation – there is a loose cluster of inchoate ideas about what we can compute with paper and pencil, about what a computer (in the modern sense) can compute, and about what a mechanism more generally might compute.

2. Then at what we might call the *proto-theoretic* level we have, inter alia, one now familiar way of picking out strands from the pre-theoretic cluster while idealizing away from practical limitations of time or amount of paper, giving us the notion of an effectively computable function. So here some theoretic tidying has already taken place, though the concept still remains somewhat vaguely characterized (what makes a step in a step-by-small-step algorithmic procedure 'small' enough to be admissible?).

3. Then at the *fully theoretic* level we have tightly characterized concepts like the concept of a μ-recursive function and the concept of a Turing-computable total function.

It would be quite implausible to suppose that the inchoate pre-theoretic cluster of ideas at the first level pins down anything very definite. No, the Church–Turing Thesis sensibly understood, in keeping with the intentions of the early founding fathers, is a view about the relations between concepts at the second and third level. The Thesis kicks in *after* some proto-theoretic work has been done. The claim is that the functions that fall under the proto-theoretic idea of an effectively computable function are just those that fall under the concept of a μ-recursive function and under the concept of a Turing-computable total function. NB: the Thesis is a claim about the *extension* of the concept of an effectively computable function.

(b)   There are other strands in the pre-theoretic hodgepodge of ideas about computation than those picked up in the idea of effective computability: in particular, there's the idea of what a machine can compute and we can do some proto-theoretic tidying of that strand too. But the Church–Turing Thesis is *not* about this idea. It must not be muddled with the entirely different claim that a physical machine can only compute recursive functions – i.e. the claim that any possible computing mechanism (broadly construed) can compute no more than a Turing machine.[11] For perhaps there could be a physical set-up which somehow or other is *not* restricted to delivering a result after a finite number of discrete, deterministic steps, and so is enabled to do more than any Turing machine. Or at least, if such a 'hypercomputer' is impossible, that certainly can't be established merely by arguing for the Church–Turing Thesis.

Let's pause over this important point, and explore it just a little further. We have seen that the *Entscheidungsproblem* can't be solved by a Turing machine. In other words, there is no Turing machine which can be fed (the code for) an arbitrary first-order wff, and which will then decide, in a *finite* number of steps, whether it is a valid wff or not. Here, however, is a simple specification for a non-Turing hypercomputer that could be used to decide validity.

Imagine a machine that takes as input the (Gödel number for the) wff φ which is to be tested for validity. It then starts effectively enumerating (numbers for) the theorems of a suitable axiomatized formal theory of first-order logic. We'll

---

[11]See Copeland (2008) for much more on this point.

suppose our computer flashes a light if and when it enumerates a theorem that matches φ. Now, our imagined computer *speeds up* as it works. It performs one operation in the first second, a second operation in the next half second, a third in the next quarter second, a fourth in the next eighth of a second, and so on. Hence after two seconds it has done an infinite number of tasks, thereby enumerating and checking *every* theorem to see if it matches φ! So if the computer's light flashes within two seconds, φ is valid; if not, not. In sum, we can use our wildly accelerating machine to decide validity, because it can go through an *infinite* number of steps in a finite time.

Now, you might very reasonably think that such accelerating machines are a mere philosophers' fantasy, physically impossible and not to be taken seriously. But actually it isn't quite as simple as that. For example, we can describe space-time structures consistent with General Relativity which apparently have the following feature. We could send an 'ordinary' computer on a trajectory towards a spacetime singularity. According to its own time, it's a non-accelerating computer, plodding evenly along, computing forever and never actually reaching the singularity. But according to us – such are the joys of relativity! – it takes a finite time before it vanishes into the singularity, accelerating as it goes. Suppose we set up our computer to flash us a signal if, as it enumerates the first-order logical theorems, it ever reaches φ. We'll then get the signal within a bounded time just in case φ is a theorem. So our computer falling towards the singularity can be used to decide validity.

Now, there are quite fascinating complications about whether this fanciful story actually works within General Relativity.[12] But no matter. The important point is that the issue of whether there could be this sort of Turing-beating physical set-up – where (from our point of view) an infinite number of steps are executed – has nothing to do with the Church–Turing Thesis properly understood. For that is a claim about effective computability, about what can be done in a *finite* number of steps following an algorithm.

44.5   The status of the Thesis

So the question whether the Church–Turing Thesis is true is *not* an issue about the limits of all possible machines (whatever exactly that means). The question is: are the functions computable-in-principle by step-by-small-step, finite, deterministic processes exactly the μ-recursive/Turing-computable functions? The Church–Turing Thesis gives a positive answer, and no serious challenge has ever been successfully mounted.[13]

So the Thesis – when construed as we are construing it – is more or less universally believed. But we have already seen intimations of two different assessments

---

[12]For discussion of the ingenious suggestion and its pitfalls, see Earman (1995, ch. 4).

[13]Which isn't to say that there haven't been attempts, and some of these failures can be instructive. See e.g. Kalmár (1959) and the riposte by Kleene (1987).

of its status, modest and bold. The more modest view (which perhaps echoes that of the earlier Church) can be expressed as follows:

> Various attempts to characterize the class of effectively computable functions have converged on the same class of recursive functions.[14] No one has ever succeeded in describing a computable function that isn't recursive. All this weight of evidence therefore warrants our adopting recursiveness/Turing-computability at least as an 'explication' of our intuitive concept of effective computability – i.e. as a fruitful, simple, clean *replacement* for our proto-theoretic concept.[15] Hence we can accept the Thesis, though not as a statement of the bald truth but rather as recording a *decision* about how best to locate a sharply defined class of functions in the area initially gestured to by our still somewhat vague proto-theoretic concept.

The bolder stance (inspired by Turing, and seemingly adopted by the later Church and by Gödel) maintains that

> Further reflection on the very notion of an effective calculation together with mathematical argument shows that the effectively calculable functions can be none other than the μ-recursive/Turing-computable ones, and so the Thesis is a demonstrably true claim about the coextensiveness of our proto-theoretic and formal concepts.

Now, despite Turing's work and Gödel's consequent endorsement of the bolder stance, the modest view seems to have been dominant both in the passing comments of mathematicians and in philosophical discussions of the Church–Turing Thesis. As it happens, I think we *can* be bolder, and explain why in the next chapter. However, don't be distracted by those contentious arguments: *accepting the Thesis in a modest spirit is quite enough for our purposes in this book*. For what we really care about is linking up the technical results about e.g. recursive decidability with claims about what is effectively decidable in the intrinsically interesting intuitive sense. And so long as we accept the Thesis as a working assumption, that's enough to make the link, whatever its philosophical status is.

---

[14] For accessible reviews of a number of formal definitions of computability in addition to μ-recursiveness and Turing-computability, see Cutland (1980, chs. 1 and 3) and Odifreddi (1999, ch. 1). Note particularly the idea of a register machine, which idealizes the architecture of a real-world computer.

[15] For more on the idea of an explication, see Carnap (1950, ch. 1).

45 Proving the Thesis?

An algorithm, we said, is a sequential step-by-step procedure which can be fully specified in advance of being applied to any particular input. Every minimal step is to be 'small' in the sense that it is readily executable by a calculator with limited cognitive resources. The rules for moving from one step to the next must be entirely determinate and self-contained. And an algorithmic procedure is to deliver its output, if at all, after a finite number of computational steps. The Church–Turing Thesis, as we are interpreting it, is then the claim that a numerical function is effectively computable by such an algorithm iff it is μ-recursive/ Turing-computable (note, we continue to focus throughout on *total* functions).

The Thesis, to repeat, is *not* a claim about what computing 'machines' can or can't do. Perhaps there can, at least in principle, be 'machines' that out-compute Turing machines – but if so, such hypercomputing set-ups will not be finitely executing algorithms (see Section 44.4).

And as we also stressed, it is enough for our wider purposes that we accept the Thesis's link between effective computability by an algorithm and μ-recursiveness/Turing computability; we don't have to take a particular stance on the *status* of the Thesis. But all the same, it is very instructive to see how we might go about following Turing (and perhaps Gödel) in defending a bolder stance by trying to give an informal proof that the intuitive and formal concepts are indeed coextensive. So in this chapter I attempt such a demonstration.

It should be clearly signalled that the core argument is contentious. Still, it is perhaps good to be reminded as we get towards the end of this book that, when it comes to questions about the interpretative gloss that we put on technical results in logic, things aren't always as black-and-white as textbook presentations can make them seem. Take our discussions here at least as a provocation to further thought and exploration.

45.1 Vagueness and the idea of computability

(a)   Recall the project. In Section 44.4, we distinguished three levels of concepts of computability: (1) the inchoate pre-theoretical level guided by paradigms of common-or-garden real-world computation; (2) the proto-theoretic notion of effective computability; (3) the formal concepts of μ-recursiveness, Turing computability, and so on.

Now, the move from (1) to (2) involves a certain exercise in conceptual sharpening. And there is no doubt an interesting story to be told about the conceptual dynamics involved in reducing the amount of 'open-texture',[1] getting rid of some of the imprecision in our initial inchoate concept – for this exercise isn't just an *arbitrary* one. However, it plainly would be over-ambitious to claim that in refining our inchoate concept and homing in on the idea of effective computability we are simply explaining what we were clearly talking about from the beginning, at the first level. And that isn't any part of the claim in this chapter.

Rather, the Church-Turing Thesis (read boldly) is that, once we have arrived at the *second*, considerably more refined though still somewhat vague, concept of an *effective* computation, *then* we in fact have a concept which pins down the same unique class of functions as the third-level concepts.

Now, in the literature we repeatedly find claims like this:

> It is important to realize that the [Church–Turing] thesis is not susceptible of proof: it is an unsubstantiable claim .... (Bridges, 1994, p. 32)

But why so? The most usual reason given is some version of the following:

> Since our original notion of effective calculability of a function (or of effective decidability of a predicate) is a somewhat vague intuitive one, the thesis cannot be proved. (Kleene, 1952, p. 317)

But what kind of vagueness is in question here, and why is it supposed to block all possibility of a proof?[2]

Let's consider two initial ways of elaborating the sort of claim Kleene makes, deploying the ideas of what I'll call *borderline-vagueness-in-extension* and *poly-vagueness* respectively. I will argue that neither interpretation leads to a good argument against the unprovability of the Church-Turing Thesis.

(b) The vague concept of a tall man allows the possibility of *borderline cases*, so its extension needn't be sharply bounded. We can line up a series of men, shading from tall to non-tall by imperceptibly tiny stages: there is no sharp boundary to the class of tall men.[3]

Now suppose that it is claimed that the vague concept of an effectively computable function similarly allows the possibility of borderline cases so its extension too is not sharply bounded. The picture is that we can likewise line up a series of functions shading gradually from the effectively computable to the non-computable, and again there is no sharp boundary to be found.

If that picture is right, then the extension of 'effectively computable function' is blurry. But the extension of 'recursive function' is of course entirely sharp.

---

[1]The phrase 'open texture' is due to Friedrich Waismann (1945). Waismann's work has been rather neglected of late; but see Shapiro (2006a) and the Appendix to Shapiro (2006b).

[2]We could readily give twenty quotations making much the same claim.

[3]I won't here consider the so-called epistemic theory of vagueness according to which there really *is* a sharp boundary to the class of tall men, but we don't and can't know where it is: see the Introduction to Keefe and Smith (1999). Enthusiasts for the epistemic theory will have to rewrite this subsection to accord with their preferred way of thinking of vagueness.

So the two concepts don't strictly speaking have the same extension, and the Church–Turing biconditional can't be strictly true.[4]

We have, however, no compelling reason to suppose that this picture *is* right. To be sure, our initial characterization of a step-by-step algorithm used vague talk (e.g. we said that the steps should be 'small'). *But note that it just doesn't follow from the fact that the idea of an* algorithm *is somewhat vague that there can be borderline cases of* algorithmically computable *functions.*

Compare: I wave an arm rather airily and say, 'The men over there are great logicians'. The sense of 'over there' is vague; yet I may determinately refer to none other than Gödel, Church and Kleene, if they are the only men in the vicinity I sweepingly gesture towards (so on any reasonable sharpening of 'over there' in the context, I pick out the same men). Likewise, even if informal talk of effectively computable functions is in some sense an imprecise verbal gesture, it could be the case that this gesture picks out a quite determinate class of functions (i.e. the only natural class in the vicinity, which is located by any reasonable sharpening of the idea of a function calculable by an algorithm proceeding by small steps).

In a slogan, then: vagueness in the *sense* of 'effectively computable' does not necessarily make for vagueness in the *extension* – and the Church–Turing Thesis is a claim about extensions.[5]

So there is no simple argument from *conceptual* vagueness to some supposed borderline vagueness in the *extension* of 'effectively computable function'. And now we can just be blunt. If that extension did suffer from borderline vagueness then, as we said, there could be a series of functions which – as we march along from one function to the next – takes us from the plainly algorithmically-computable-in-principle to the plainly not-computable-even-in-principle via borderline cases. But as far as I know no one has ever purported to describe such a series. I suggest that's because there *isn't* one.

(c) Consider next the discussion of the concept of a polyhedron in Imre Lakatos' wonderful *Proofs and Refutations* (1976). Lakatos imagines a class examining the Euler conjecture that, for any polyhedron, $V - E + F = 2$.[6] And, after some discussion of Cauchy's argument in support of the conjecture, the student Alpha suggests a counterexample. Take a solid cube with a cubic 'hole' buried in the middle of it. This has 12 faces (six outside faces, six inside), 16 vertices, and 24 edges, so in this case $V - E + F = 4$.

Has Alpha described a genuine counterexample? 'Polyhedron' is already a proto-theoretical term, defined along the lines of 'a solid bounded by polygonal faces'. But does Alpha's example count as a solid in the intended sense? Well,

---

[4] Compare: 'The *extension* of "effectively computable" is vague and … "recursive" sharpens it, which is the main reason [the Thesis] is important and was introduced by Church in the first place.' (Nelson, 1987, p. 583, my emphasis.)

[5] For us, at any rate, it would be a mistake to write, e.g., 'Church's thesis is the proposal to identify an intuitive notion with a precise, formal, definition' (Folina, 1998, p. 311): it isn't the *notions* but their *extensions* which are being identified.

[6] V, of course, is the number of vertices, E the number of edges, F the number of faces.

our initial practice arguably just doesn't settle whether Alpha's hollowed-out cube falls under this concept of a polyhedron or not; so the concept could be sharpened up in different ways, quite consistently with the implicit informal rules we'd mastered for applying and withholding the concept to 'ordinary' cases. To put it another way: Alpha's example reveals that our initial talk about polyhedra fails to pick out a single mathematical 'natural kind' – we can legitimately disambiguate the term in more than one way, and we *need* to disambiguate it before we can prove or disprove Euler's conjecture.

Let's say that a term is 'poly-vague' if it ambiguously locates more than one mathematical kind (though each kind might be precisely bounded).[7] And whether or not 'polyhedron' is a good example, the general phenomenon of proto-theoretical mathematical concepts that can be rigorized in more than one way is surely incontestable. So is 'effectively computable' another one? If it is, then our proto-theoretic talk here fails to pick out a single mathematical 'natural kind', and – before disambiguation – it will therefore be indeterminate what the Church–Turing Thesis says, and hence the Thesis as it stands will be unprovable.

But *is* 'computable function' poly-vague in this way? Well, we have no reason whatsoever to suppose that there is more than one mathematically natural class of total functions in the vicinity picked out by the intuitive notion of an effectively computable function (once we have done our rough-and-ready proto-theoretic clarification of the idea of computing-by-following-an-algorithm). As we've said before, all our attempts to define effective computability famously point to exactly the same class of μ-recursive/Turing-computable functions. So we certainly *can't* take the claim that 'effectively computable function' is poly-vague as a starting point for arguing about the Church–Turing Thesis.

45.2 Formal proofs and informal demonstrations

In sum, we can't take either vagueness-in-extension or poly-vagueness for granted here. So what *are* we to make of Kleene's blunt claim that the idea of an effectively computable function is a vague intuitive one?

We seem to be left with nothing much more than the truism that our intuitive, proto-theoretical notion is intuitive and not fully theoretical (i.e. to grasp it doesn't involve grasping an explicit and sharply formulated general theory of what makes for a computation).

So consider again Kleene's expressed pessimism about whether we can 'extract' from our proto-theoretical notion a clearly defined account suitable for theoretical purposes (see the quotation at the end of Section 44.2). Kleene seems to be taking it as obvious that there aren't enough constraints governing our

---

[7]The unfriendly might think it is wrong to think of this as involving any kind of vagueness, properly so called, and that 'confused' would be a better label than 'poly-vague' – see Camp (2002). In a more friendly spirit, you might think that what we are talking about again is a kind of 'open texture' – see Shapiro (2006a). But let's not fuss about how exactly to describe the situation, for the point I need to make isn't sensitively dependent on such details.

proto-theoretical notion – constraints which anyone who has cottoned on to the notion can be brought to acknowledge – which together with mathematical argument will suffice to establish that the computable functions are the recursive ones. But that claim is *not* obvious. It needs to be defended as the conclusion of some arguments: it can't just be asserted as an unargued presumption.

Here is the same presumption at work again, this time with a couple of added twists:

> How can we in any wise demonstrate [that the effectively calculable functions are those computable by each of the precise definitions that have been offered]? Ultimately only in some formal system where the vague intuitive concept 'calculable function' would have to be made precise before it could be handled at all, so that in any case the vague intuitive idea would be eliminated before we began, so we would certainly fail to demonstrate anything about it at all. (Steen, 1972, p. 290)

The added assumptions here are that (i) formal systems can't handle vague concepts, and (ii) any genuine demonstration must 'ultimately' be in some formal system.

But neither assumption is compelling. (i) There are in fact a number of competing, non-classical, ways of formally representing the semantics of vague concepts, if we want to be explicit in modelling their vagueness. But waive that point.[8] Much more importantly, (ii) formalization doesn't somehow magically conjure proofs where there were none before. Formalization enforces honesty about what assumptions and inferential rules are being relied on, enabling us to expose suppressed premisses and inferential fallacies, to avoid trading on ambiguities, and so forth. We thereby push to their limits the virtues of explicitness and good reasoning that we hope to find in common-or-garden mathematical arguments. But those common-or-garden arguments can perfectly well involve good reasoning and sound demonstrations *before* we go formal.

Here are two examples. First, take the diagonal argument for what we cheerfully labelled Theorem 14.2, i.e. the result that not all effectively computable functions are primitive recursive.[9] Surely, by any reasonable standards, that argument counts as a perfectly good proof, even though one of the concepts it involves is 'a vague intuitive idea'.

Second, take the claim endorsed by almost all those who say that the Church-Turing Thesis is not provable, namely that it would be disprovable if false.[10] The thought is that if we could find a clear case of an intuitively effectively

---

[8] For some details, see e.g. the editors' Introduction to Keefe and Smith (1999).

[9] I was hardly stepping out of line in calling this result a 'theorem'! See, for just one example, 'Theorem 3.11' of Cohen (1987, §3.6): 'There is an *intuitively computable* function which is not primitive recursive' (my emphasis).

[10] For dissent, see Folina (1998) – but this again depends on the unargued presumption that any genuine proof must be translatable into a formal proof involving no vague concepts.

computable function which is provably not recursive, then that would decisively settle the matter. But that again would be a proof involving the application of an intuitive unformalized notion.

The point here is worth highlighting. For note that the argument for Theorem 14.2, for example, didn't invoke mere plausibility considerations. It was, let's say, *a demonstration in informal mathematics*. We might distinguish, then, three levels of mathematical argument – mere plausibility considerations, informal demonstrations, and ideally formalized proofs (or truncated versions thereof).[11] Some philosophers write as if the important divide has ideally formalized proofs on one side, and everything else on the other, pieces of informal mathematics and mere plausibility considerations alike. But that's simply not an attractive view: it misrepresents mathematical practice, and also pretends that the formal/informal distinction is much sharper than it really is.[12] Moreover, it certainly doesn't draw the line in a way that is relevant to our discussion of the status of the Church–Turing Thesis. The question we set ourselves is whether we can do better than give quasi-empirical plausibility considerations. If we *can* support the Thesis using arguments that have something like the demonstrative force of our diagonal argument to show that there are computable functions which aren't p.r., then this is enough to support a *bold* stance on the Thesis (in the sense of Section 44.5).

45.3 Squeezing arguments – the very idea

We have seen that it is not easy to find a sense of 'vague' in which it is clear at the very outset both that (i) our informal notion of an effectively computable function is vague and that (ii) this kind of vagueness must prevent our demonstrating that the Church–Turing Thesis is true. Which, of course, doesn't show that such a demonstration is possible, but at least it means that we shouldn't give up too soon on the project of looking for one.

So let's consider how we can show that a proto-theoretical, intuitive, informally characterized concept *I is* co-extensive with some explicitly defined

---

[11]One of the nicest examples of a plausibility consideration I know concerns Goldbach's conjecture that every even number greater than 2 is the sum of two primes. Just being told that no one has yet found a counterexample among the even numbers so far examined is not at all persuasive (after all, there are well-known examples of arithmetical claims, e.g. about the proportion of numbers up to n that are primes, which hold up to some utterly enormous number, and then fail). However, if you do a graphical analysis of the distribution of the pairs of primes that add to the successive even numbers you get a remarkable, fractal-like, pattern called the 'Goldbach Comet' which reveals a lot of entirely unexpected structure (I won't reproduce it here: an internet search will reveal some nice examples). And this suddenly makes Goldbach's conjecture look a *lot* more plausible. It no longer looks like a stand-alone oddity, but seems as if it should have interesting interconnections with a rich body of related propositions about the distribution of primes. Though that does indeed make it seem all the more puzzling that the conjecture has resisted proof.

[12]For emphatic resistance to exaggerating the formal/informal distinction, see Mendelson (1990).

concept. Here, outlined in very schematic form, is one type of argument that would deliver such a co-extensiveness result.

Suppose firstly that we can find some precisely defined concept S such that falling under concept S is certainly and uncontroversially a sufficient condition for falling under the concept I. So, when e is some entity of the appropriate kind for the predications to make sense, we have

K1. If e is S, then e is I.

Now suppose secondly that we can find another precisely defined concept N such that falling under concept N is similarly an uncontroversial necessary condition for falling under the concept I. Then we also have

K2. If e is I, then e is N.

In terms of extensions, therefore, we have

i. $|S| \subseteq |I| \subseteq |N|$,

where $|X|$ is the extension of X. In this way, the extension of I – vaguely gestured at and indeterminately bounded though that might be – is at least sandwiched between the determinately bounded extensions of S and N.

So far, though, so unexciting. It is no news at all that even the possibly fuzzy extensions of paradigmatically vague concepts can be sandwiched between those of more sharply bounded concepts. The extension of 'tall' (as applied to men) is uncontroversially sandwiched between those of 'over five foot' and 'over seven foot'.

But now suppose, just suppose, that in a particular case our informal concept I gets sandwiched in this way between such sharply defined concepts S and N, but we can *also* show that

K3. If e is N, then e is S.

In the sort of cases we are going to be interested in, S and N will be precisely defined concepts from some rigorous theory. So in principle, the possibility is on the cards that the result (K3) could actually be a theorem of the relevant mathematical theory.

But in that case, we'd have

ii. $|S| \subseteq |I| \subseteq |N| \subseteq |S|$,

so the inclusions can't be proper. What has happened, then, is that the theorem (K3) squeezes together the extensions $|S|$ and $|N|$ which are sandwiching the extension $|I|$, and so we have to conclude

iii. $|S| = |I| = |N|$.

In sum, the extension of the informally characterized concept I is now revealed to be none other than the identical extension of both the sharply circumscribed concepts S and N.

All this, however, is merely schematic. Are there any plausible cases of informal concepts I where this sort of *squeezing argument* can be mounted, and where we can show in this way that the extension of I is indeed the same as that of some sharply defined concept(s)?

45.4 Kreisel's squeezing argument

(a)   There is certainly one attractive example, due to Kreisel (1967), to whom the general idea of such a squeezing argument is due. His argument is worth knowing about anyway, so let's pause to explain it.

Take the entities being talked about to be arguments couched in a given regimented first-order syntax with a classical semantics. Here we mean of course arguments whose language has the usual truth-functional connectives, and whose quantifiers are understood, in effect, as potentially infinitary conjunctions and disjunctions. And now consider the concept I_L, the informal notion of being valid-in-virtue-of-form for such arguments. We will give a squeezing argument to show that this is coextensive with model-theoretic validity.

(b)   But first we should note that again, as with concepts of computability, there are three levels of concept of validity to think about.

1. Pre-theoretically, there's the idea of a logically watertight argument (the idea met at the very outset of a first logic course, shaped by some simple paradigms and arm-waving explanations).

2. Then, at the proto-theoretical level, there is the notion of an argument being valid-in-virtue-of-form.

3. Then, fully theoretically, we have a nice clean Tarskian definition of validity, at least for suitably regimented languages.

It would again be quite implausible to think that the pre-theoretical 'intuitive' notion of valid consequence has enough shape to it to pin down a unique extension. If you suppose otherwise, start asking yourself questions like this. Is the intuitive notion of consequence constrained by considerations of relevance? – are *ex contradictione quodlibet* inferences (which argue from a contradiction to an arbitrary conclusion) fallacious? When can you suppress necessarily true premisses and still have an inference which is intuitively valid? What about the inference 'The cup contains some water; so it contains some H_2O molecules'? That necessarily preserves truth, if we agree with Kripke (1980, Lecture 3): but is it valid in the intuitive sense? – if not, just why not?

In thinking about validity at the proto-theoretical level, we need to fix on responses to such initial questions. One route forward takes us to the idea of validity-in-virtue-of-form. We informally elucidate this concept by saying that an argument is valid in this sense if, however we spin the interpretations of the non-logical vocabulary, and however we pretend the world is, it's never the case that the argument's premises would come out true and its conclusion false. The 'form' of the argument, i.e. what is left fixed when we spin the interpretations, is enough to guarantee truth-preservation.

Of course, that explication takes us some distance from the inchoately pre-theoretical.[13] But it is still vague and informal: it's the sort of loose explanation we give a few steps into an introductory logic course. In particular, we've said nothing explicitly about where we can look for the 'things' to build the interpretations which the account of validity generalizes over. For example, just how big a set-theoretic universe can we call on? – which of your local mathematician's tall stories about wildly proliferating hierarchies of 'objects' do you actually take seriously enough to treat as potential sources of structures that we need to care about? If you do cheerfully buy into set-theory, what about allowing domains of objects that are even bigger than set-sized? Our informal explication just doesn't begin to speak to such questions.

But no matter; informal though the explication is, it does in fact suffice to pin down a unique extension for I_L, the concept of validity-in-virtue-of-form for first-order sentences. Here's how.

(b)   Take S_L to be the property being provable in your favourite proof system for classical first-order logic. Then, for any argument α,

L1. If α is S_L, then α is I_L.

That is to say, the proof system is classically sound: if you can formally deduce φ from some bunch of premises Σ, then the inference from Σ to φ is valid according to the elucidated conception of validity-in-virtue-of-form. That follows by an induction on the length of the proofs, given that the basic rules of inference are sound according to our conception of validity, and chaining inference steps preserves validity. Their evident validity in that sense is, after all, the principal reason why classical logicians accept the proof system's rules in the first place!

Second, let's take N_L to be the property of having no countermodel in the natural numbers. A countermodel for an argument is, of course, an interpretation that makes the premises true and conclusion false; and a countermodel in the natural numbers is one whose domain of quantication is the natural numbers, where any constants refer to numbers, predicates have sets of numbers as their extensions, and so forth. Now, even if we are more than a bit foggy about the limits to what counts as legitimate re-interpretations of names and predicates as mentioned in our informal explication of the idea of validity, we must surely recognize at least this much: if an argument does have a countermodel in the

---

[13]It now warrants *ex contradictione quodlibet* as a limiting case of a valid argument. And given that 'water' and 'H$_2$O' are bits of non-logical vocabulary, that means that the inference 'The cup contains water; so it contains H$_2$O' is of course not valid in virtue of form.

natural numbers then the argument certainly can't be valid-in-virtue-of-its-form in the informal sense. Contraposing,

L2. If α is I_L, then α is N_L.

So the intuitive notion of validity-in-virtue-of-form (for inferences in our first-order language) is sandwiched between the notion of being provable in your favourite system and the notion of having no arithmetical counter-model. Hence we have

i′. $|S_L| \subseteq |I_L| \subseteq |N_L|$.

But now, of course, it is a standard theorem that

L3. If α is N_L, then α is S_L.

That is to say, if α has no countermodel in the natural numbers, then α can be deductively warranted in your favourite classical natural deduction system.[14] So (L3) squeezes the sandwich together. We can conclude, therefore, that

iii′. $|S_L| = |I_L| = |N_L|$.

In sum, take the relatively informal notion I_L of a first-order inference which is valid-in-virtue-of-its-form (explicated as sketched): then our proto-theoretic assumptions about that notion constrain it to be coextensive with each of two sharply defined, mutually coextensive, formal concepts.

Which, I claim, is a very nice result. But whether you buy the details of that argument or not, it illustrates one *sort* of reasoning we might use to argue about computability. So the next question is whether we can turn a similar squeezing trick when the relevant entities are total numerical functions f, the informal concept I is that of effective computability, and the sandwiching concepts pin down recursiveness/Turing-computability.

45.5 The first premiss for a squeezing argument

The first premiss of a squeezing argument for computability is surely secure.

C1. If a function f is S_C (μ-recursive), then it is I_C (effectively computable in the intuitive sense).

For if f is μ-recursive it is Turing-computable, so there is a Turing-program which computes it; and a human calculating agent can in principle follow the entirely determinate algorithmic instructions in that program and therefore –

---

[14]Recall: the downward Löwenheim–Skolem theorem tells us that if there's any counter-model to α, then there is a countermodel whose domain is some or all the natural numbers. So if α has no countermodel in the natural numbers it can have no countermodel at all, so by completeness it is deductively valid.

given world enough and time – compute the value of f for any given input. So the function is effectively computable in the intuitive sense. Which establishes the first premiss.

45.6 The other premisses, thanks to Kolmogorov and Uspenskii

Now for the really contentious section! To complete the squeezing argument, we need to find a formally framed condition N_C which is weak enough to be a necessary condition for algorithmic computability, yet strong enough to entail μ-recursiveness (remember: we are concentrating on total functions). Can we find such a condition?

(a)   With a bit of regimentation, we can think of Turing's epoch-making 1936 paper as gesturing towards a suitable N_C and giving us the beginnings of a defence of the second and third premisses for a squeezing argument.

As we've seen, Turing notes various features that would *prevent* a procedure from counting as algorithmic in the intuitive sense. For example, an algorithmic procedure can't require an infinity of distinct fundamental symbols: 'if we were to allow an infinity of symbols, then there would be symbols differing to an arbitrarily small extent' (given they have to be inscribed in finite cells), and then the difference between symbols wouldn't be recognizable by a limited computing agent. For similar reasons, the computing agent can't 'observe' more than a limited amount of the workspace at one go. And computations shouldn't involve arbitrarily large jumps around the workspace which can't be reduced to a series of smaller jumps – a bound is set by cognitive resources of the computing agent, who needs to be able to recognize where to jump to next.[15]

Now put together the requirements of a finite alphabet, restricted local action on and movement in the workspace, etc. with whatever other similar constraints we can muster. We will get a – formally specifiable – composite necessary condition N_C for being calculable by an acceptably algorithmic procedure. In short, we will have something of the form

C2. If f is I_C, then f is N_C.

Then, to complete the squeeze, we need to be able to prove that any total function which is N_C is computable by a standard Turing machine/is μ-recursive. That is, we need to prove

C3. If f is N_C, then f is S_C.

Turing's remarks about how we can break down more general kinds of computation into small steps, and hence don't lose generality in concentrating on what

---

[15] For more on the same lines, see Turing (1936, §9).

we now think of as standard Turing machines (which e.g. change just one cell at a time), can be read as intimating the possibility of this sort of result.

There's now bad news and good news. The bad news is that Turing himself isn't clear about exactly how to fill in the necessary condition N_C in order to complete the argument. And in so far as he is clear, his way of spelling out N_C is surely too strong to look uncontentious. For example, when we do real-life computations by hand, we often insert temporary pages as and when we need them, and equally often throw away temporary working done earlier. In other words, our workspace doesn't have a fixed form: so when we are trying to characterize intuitive constraints on computation we shouldn't assume straight out – as Turing does – that we are dealing with computations where the 'shape' of the workspace stays fixed once and for all.

The good news is that this and other worries can be quieted by appeal to the very general condition for algorithmic computation given in 1958 by A. N. Kolmogorov and V. A. Uspenskii in their paper 'On the definition of an algorithm' (English translation 1963). Note, however, that we do *not* need to accept the claim implicit in their title, namely that the Kolmogorov–Uspenskii (KU) account gives necessary *and* sufficient conditions for being an algorithm: for our purposes, necessity is enough. I'll return to this point.

My claim, then, will be that *when the necessary condition N_C is identified with being-computable-by-a-KU-algorithm, the second and third premisses of the Turing-style squeezing argument are demonstrably true*. So in the rest of this section, we will spell out the linked ideas of a KU-algorithm and KU-computability (very slightly modified from Kolmogorov and Uspenskii's original version). We defend premiss (2) with N_C understood in terms of KU-computability. We then indicate briefly how the corresponding technical result (3) is proved, and so complete the squeeze.

(b) For Turing's reasons,

    i. We still take the alphabet of symbols which any particular algorithm works on to be finite.

But we now start relaxing Turing's assumptions about the shape of the workspace, to get the most general story possible.

First, though, a terminological point. We want a way of referring to the *whole* collection of 'cells' that contain data – whether they are in some general area where computations can take place, or in some reserved areas for passive memory storage ('memory registers'). In the case of Turing machines which lack such reserved areas, it was natural enough to use the term 'workspace' for the whole collection of cells. But now that we are moving to a more general setting – and given that 'workspace' naturally contrasts with 'reserved memory' – we had better use a different term to make it clear that we intend comprehensive coverage. I suggest *dataspace* – which has the added advantage of having a more abstract ring to it, and so doesn't too readily suggest a simple spatial arrangement.

Since we are going to allow the extent of the dataspace to change as the computation goes along, we can take it to consist of a finite network of 'cells' at every stage, adding more cells as we need them. So,

ii. The dataspace at any stage of the computation consists of a finite collection of 'cells' into which individual symbols are written (we can assume that there is a special 'blank' symbol, so every cell has some content). But now generalizing radically from the 'tape' picture, we'll allow cells to be arranged in any network you like, with the only restriction being that there is some fixed upper limit (which can be different when implementing different algorithms) on the number of immediate 'neighbours' we can get to from any given cell. Being 'neighbours' might be a matter of physical contiguity, but it doesn't have to be: a cell just needs to carry *some* kind of 'pointer' to zero or more other cell(s). Since the algorithm will need to instruct us to operate on cells and certain neighbours and/or move from one cell (or patch of cells) to some particular neighbour(s), we'll need some system for differentially labelling the 'pointers' from a cell to its various neighbours.

For vividness, you can depict cells as vertices in a directed graph, with vertices being linked to their neighbours by 'colour-coded' arrows (i.e. directed edges): the 'colours' are taken from a given finite palette, and the arrows linking one cell to its immediate neighbours are all different colours.[16] And why put an upper bound on the size of the colour palette (and so put an upper bound on the number of different arrows leading out from a given vertex)? Well, consider a finite computing agent with fixed 'on board' cognitive resources who is trying to follow program instructions of the kind that require recognizing and then operating on or moving around a group of cells linked up by particular arrows: there will be a fixed bound on the number of discriminations the agent can make.

Subject to those constraints, the dataspace can now be structured in any way you like (it needn't be equivalent to a tape or even to an n-dimension 'sheet' of paper): *we only require that the network of cells is locally navigable by a computing agent with fixed cognitive resources.*

Next, we've said all along that an algorithm should proceed by 'small' steps. So we can certainly lay down the following weak requirements:

iii. At every stage in the computation of a particular algorithm, a patch of the dataspace of at most some fixed bounded size is 'active'.

---

[16] For enthusiasts: here we have slightly modified Kolmogorov and Uspenskii's original treatment. They treat the dataspace as an *undirected* graph; putting it in our terms, if there is an arrow of colour c from vertex v_1 to v_2, then there is an arrow of colour c from vertex v_2 to v_1. Hence, for them, the restriction to a bounded number of 'arrows out' from a vertex implies a similar restriction to the number of 'arrows in'. But this is unnecessary, and is probably unwelcome if we want maximal generality, as is in effect shown by the independent work in Schönhage (1970, 1980).

iv. The next step of the computation operates only on the active area, and leaves the rest of the dataspace untouched.

Without loss of generality, we can think of the 'active' area of dataspace at any point in the computation to be the set of cells that are no more than n arrows away from some current focal vertex, where for a given algorithmic procedure n again stays fixed throughout the computation. Why keep n fixed? Because the maximum attention span of a limited cognitive agent stays fixed as he runs the algorithm (he doesn't get smarter!). However, we will otherwise be ultra-liberal and allow the bound n to be as large as you like.[17]

Now for perhaps the crucial radical relaxation of the Turing paradigm:

v. A single computing step allows us to replace a patch of cells in the active area of the dataspace with particular contents and a particular pattern of internal arrows by a new collection of cells (of bounded size) with new contents and new internal arrows.

So, at least internally to the active area of the dataspace, we can not only fiddle with the contents of the cells at vertices, but also change the local arrangement of coloured arrows (while preserving incoming arrows from outside the active area). As announced, then, the shape of the dataspace itself is changeable – and the dataspace can grow as needed as we replace a small patch by a larger patch with extra cells and new interlinkings.

Having worked on one patch of the dataspace, our algorithm needs to tell us which patch to work on next. And, for Turing's reasons,

vi. There is a fixed bound on how far along the current network of cells the focal vertex of the active patch shifts from one step of the algorithm to the next.

But again, we will be ultra-relaxed about the size of that bound, and allow it to be arbitrarily large for any given algorithm.

Thus far, then, we have described the dataspace and general mode of operation of a KU-algorithm. We now need to define the character of the algorithm itself:

vii. First there can be an initial set of instructions which sets up and structures a patch of dataspace, which 'writes in' the numerical input to the algorithm in some appropriate way. We might, for example, want to set up something with the structure of a register machine, with some initial numerical values in different registers. Or – as with our computation of the Ackermann–Péter function – we might start by writing into the dataspace an equation we are going to manipulate. Rather differently, we might want to start by writing a higher-level program into the dataspace, and then we will use the rest of the algorithm as an interpreter.

---

[17] Get the order of the quantifiers right here! – we are only making the weak claim that for any particular algorithm there is some bound on the size of its active dataspace as it runs; we aren't saying that there has to be some *one* bound which obtains across all algorithms.

viii. The body of a KU-algorithm then consists in a finite consistent set of instructions for changing clumps of cells (both their contents and interlinkings) within the active patch and jumping to the next active patch. Without loss of generality, we can follow Turing in taking these instructions as in effect labelled lines of code, giving bunches of conditional commands of the form 'if the active patch is of type P, then change it into a patch of type P'/move to make a different patch P'' the new active patch; then go on to execute line q_j/halt'. The instructions are to be implemented sequentially, one line at a time.

ix. Finally, we will need to specify how we read off numerical output from the configuration of cells if and when we receive a 'halt' instruction after a finite number of steps.

We can then offer the following natural definition:

x. A (monadic, total) function $f(n)$ is *KU-computable* if there is some KU-algorithm which, when it operates on n as input, delivers $f(n)$ as output.

The generalization to many-place $f(\vec{n})$ is obvious.

Now, we've given conditions (i) to (x) in a rough and informal style. But it should be clear enough how to go about developing the kind of fully formal abstract characterization articulated in detail by Kolmogorov and Uspenskii. Let's not go into that, however: the important thing is to grasp the basic conception.

(c)   The great generality of the KU story means that it surely covers *far* too many procedures to count as giving an analysis of the intuitive notion of an algorithm. What we ordinarily think of as algorithms proceed, we said, by 'small' steps; but KU-algorithms can proceed by very large operations on huge chunks of dataspace. But we needn't worry about that at all. The only question we need to focus on is: could the KU story possibly cover *too little*? Well, how could a proposed algorithmic procedure for calculating some function *fail* to be covered by the KU specification?

The KU specification involves a conjunction of requirements (finite alphabet, logically navigable workspace, etc.). So for a proposed algorithmic procedure to fail to be covered, it must falsify one of the conjuncts. But how? By having (and using) an infinite number of primitive symbols? Then it isn't usable by a limited computing agent like us (and we are trying to characterize the idea of an algorithmic procedure of the general type that agents like us could at least in principle deploy). By making use of a different sort of dataspace? But the KU specification only requires that the space has *some* structure which enables the data to be locally navigable by a limited agent. By not keeping the size of active patch of dataspace bounded? But algorithms are supposed to proceed by the repetition of 'small' operations which are readily surveyable by limited agents. By not keeping the jumps from one active patch of dataspace to the next active patch limited? But again, a limited agent couldn't then always jump to the next patch 'in one go' and still know where he was going. By the program that governs the updating of the dataspace having a different form? But KU-algorithms are entirely freeform; there is no more generality to be had.

The claim is that the very modest restrictions on finiteness of alphabet and bounded locality of operation in the dataspace are compulsory for any algorithm; otherwise, the KU specification imposes no significant restrictions.[18] So, as Kolmogorov and Uspenskii (1963, p. 231) themselves asserted, 'any algorithm is essentially subsumed under the proposed definition'. Hence, as we want:

C2. If f is I_C (effectively computable), then f is N_C (KU-computable).

(d)   We need not pause too long over the last premiss of the squeezing argument, i.e. over the technical result

C3. If f is N_C (KU-computable), then f is S_C (μ-recursive).

We get this by arithmetization once again – i.e. we use just the same kind of coding argument that we've used in outline twice before, first in Section 38.4 to show that the Ackermann–Péter function is μ-recursive, and then Section 42.3 to show that Turing-computable total functions are all μ-recursive. And indeed, we already implied by a rather handwaving argument in Section 38.5 that this kind of coding argument looked as if it could be used to give a general defence of Church's Thesis. So, in brief:

Sketch of a proof sketch   Suppose the KU-algorithm A computes a total monadic function $f(n)$ (generalizing to many-place functions is routine). Define a function $c(n, j)$ whose value suitably codes for the state of play at the j-th step of the run of A with input data n – i.e. the code describes the configuration and contents of the dataspace and locates the active patch. And let the state-of-play code default to zero once the computation has halted, so the computation halts at step number $\mu z[c(n, Sz) = 0]$. Therefore the final state of play of the computation has the code $c(n, \mu z[c(n, Sz) = 0])$. And hence, using a decoding function d to extract the value computed at that state of play, we get

$$f(n) = d(c(n, \mu z[c(n, Sz) = 0])).$$

But d will (fairly trivially) be primitive recursive. So, as usual with this type of proof, all the real work goes into showing that $c(n, j)$ is p.r. too. That result is basically ensured by the fact that the move from one state of play to the next is always boundedly local (so the transition from $c(n, j)$ to $c(n, Sj)$ can be computed using only 'for' loops). Hence $f(n)$ is μ-recursive.   ⊠

The squeeze is therefore complete! *We have shown that the (total) computable functions are just the μ-recursive ones.*

---

[18]That's right, at any rate, given our small modification of their original story, as remarked in fn. 16 above.

(e)   A reality check. At the beginning of subsection (c) we said 'Such is the great generality of the KU story, it probably covers far too many procedures to count as giving an analysis of the intuitive notion of an algorithm'. But in subsection (d) we have just argued that the KU-computable functions are exactly the μ-recursive functions and hence, by the squeezing argument, are exactly the effectively computable functions. Those claims might look to be in some tension; it is important to see that they are not.

To repeat, the idea of a *KU-algorithm* may well be far too generous to capture the intuitive notion of an algorithm – i.e. more *procedures* count as KU-algorithms than count as mechanical, step-by-small-step, procedures in the ordinary sense. But quite consistently with this, the *KU-computable functions* can be exactly those functions which are algorithmically computable by intuitive standards. That's because – by the third premiss of the squeezing argument – any function that might be computed by a KU-algorithm which operates on 'over-large' chunks of dataspace (i.e. a KU-algorithm which is too wild to count as an intuitive algorithm) is also tamely computable by a standard Turing machine.

45.7   The squeezing argument defended

I will consider three responses, the first two very briefly, the third – due to Robert Black (2000) – at greater length.

(a)   One response is to complain that the argument illegitimately pre-empts the possible future development of machines that might exploit relativistic or other yet-to-be-discovered physical mysteries in order to trump Turing machines.

But it should be clear by now that such complaints are beside the point. We need, as before, to sharply distinguish the Church–Turing Thesis proper – which is about what can be computed by finite step-by-step algorithmic procedures – from a claim about what might perhaps be computed by exploiting physical processes structured in some other way.

(b)   A second response is to complain that the argument illegitimately pre-empts the future development of mathematics: 'The various familiar definitions of computable functions (e.g. in terms of λ-definability and Turing machines) are radically different one from another. We can't second-guess how future research might go, and predict the new sort of procedures that might be described. So how do we know in advance that another definition won't sometime be offered that can't be regimented into KU form?'

Well, true, we can't predict how mathematics develops. New paradigms for abstract 'computing' processes may be discovered (mathematicians are always generalizing and abstracting in radical ways). We certainly aren't in the business of second-guessing such developments. We are only making the conceptual point that, whatever ideas emerge, they won't count as ideas of algorithmic calculation in the classic sense if they don't cleave to the basic conception of step-by-small-step local manipulations in a dataspace that can be navigated by limited agents. But that conception is all we built into the idea of a KU algorithm. So, to repeat, the claim is: if a procedure is properly describable as algorithmic in the traditional sense, it will be redescribable as a KU algorithm.

(c) I am in considerable agreement with Black's excellent discussion; but we part company at the very last step in our responses to Kolmogorov and Uspenskii. Towards the end of his paper, he writes:

> Given the extreme generality of [their] definition of locality, I think it is fair to say that we here have a rigorous proof of Church's thesis *if we can assume that bounded attention span is built into the intuitive notion of effective computation.* However, this is a rather big 'if'. ... [I]t is unclear why the idealization which allows a potentially infinite passive memory (the unlimited supply of Turing-machine tape) should not be accompanied by a corresponding idealization allowing unlimited expansion of the amount of information which can be actively used in a single step. (Black, 2000, p. 256, his emphasis)

However, the reason for not making the second idealization in fact seems pretty clear.

Let's start, though, with a reminder about the first idealization. The sense in which we 'allow a potentially infinite passive memory' is that we are simply *silent* about the extent of the dataspace (other than, in the KU specification, assuming that the occupied space is finite at any stage). In the same way, we are silent about how many steps a successful algorithmic calculation may take. And the reason for the silence is the same in both cases. An algorithmic procedure is to be built up from small steps which can be followed by a cognitive agent of limited accomplishment, on the basis of a limited amount of local information (the overall state of the dataspace may well, in the general case, be beyond the agent's ken). Hence, so far as the agent is concerned, the state and extent of the non-active portion of the dataspace at any moment has to be irrelevant, as is the extent of the past (and future!) of the computation. We count an algorithm as in good standing so long as it keeps issuing instructions about what to do locally at the next step (irrespective of the history of the computation or the state of play beyond the active part of the dataspace).

Of course, the real-world implementation of a particular algorithmic procedure may well run up against limitations of external physical resources. But if we can't keep working away at the algorithm because we run out of paper or run out of time, then we – so to speak – blame the poverty of the world's resources rather than say that there is something intrinsically at fault with our step-by-step procedure as an in-principle-computable algorithm.

Now, in the KU story, there is another kind of silence – a silence about just how smart a computing agent is allowed to be. Which means that, for *any* n, we allow KU-algorithms that require the computing agent to have an attention span of 'radius' n (i.e. that require the agent to deal at one go with a patch of cells linked by up to n arrows from some current focal cell). So in that sense, we *do* allow an 'unlimited amount of information' to be 'actively used in a single step'. And to be sure, that looks like a decidedly generous – indeed quite wildly over-generous – interpretation of the idea that an algorithm should proceed by 'small' steps: so we might very well wonder whether every KU-algorithm will count as an algorithm in the intuitive sense. But, as we said before, no matter. The claim that concerns us is the converse one that a procedure that *isn't* a KU-algorithm won't count as an algorithm by intuitive standards – a claim which of course gets the more secure the wider we cast the KU net.

What the KU story *doesn't* allow, though, are procedures which require the capacities of the computing agent to expand in the course of executing a given algorithm (so that on runs of the algorithm for different inputs, the agent has to 'take in at a glance' ever bigger patches of dataspace, without any limit). Any KU algorithm is a finite list of instructions, fixed in advance, so there will be a maximum amount of data space the agent is called upon to process in a single instruction of that algorithm. But isn't this just fine? The intuitive idea of an idiot-proof algorithm – which always proceeds by *small* steps, accessible to limited agents – surely rules out that the steps should get ever bigger, without a particular limit, requiring ever more cognitive resources from the computing agent. So Black's worry is unfounded: in fact, the *absence* of a limit on the size of the dataspace and the *presence* of an algorithm-dependent limit on 'attention span' in the KU story are both natural concomitants of the same fundamental assumption that an algorithm has to be processed by a computing agent of limited internal cognitive capacity, doing limited things locally.

45.8 To summarize

So I have argued that the Church–Turing Thesis, which links some of our earlier technical results to intuitive claims about axiomatized theories, decidability, etc., itself seems susceptible to an intuitively compelling demonstration.

But be that as it may. Whatever its exact status, the Thesis is quite secure enough for us to lean on in interpreting technical results: and *that* is sufficient for all our theorems earlier in the book to have their advertised deep interest.

46 Looking back

Let's finish by taking stock one last time. At the end of the last Interlude, we gave a road-map for the final part of the book. So we won't repeat the gist of that detailed local guide to recent chapters; instead, we'll stand further back and give a global overview. And let's concentrate on the relationship between our various proofs of incompleteness. Think of the book, then, as falling into four main parts:

(a)   The first part (Chapters 1 to 8), after explaining various key concepts, proves two surprisingly easy incompleteness theorems. Theorem 6.3 tells us that if T is a *sound* effectively axiomatized theory whose language is *sufficiently expressive*, then T can't be negation-complete. And Theorem 7.2 tells us that we can weaken the soundness condition and require only consistency if we strengthen the other condition (from one about what T can express to one about what it can prove): if T is a *consistent* effectively axiomatized theory which is *sufficiently strong*, then T again can't be negation-complete.

Here the ideas of being sufficiently expressive/sufficiently strong are defined in terms of expressing/capturing enough effectively decidable numerical properties or relations. So the *arguments* for our two initial incompleteness theorems depend on a number of natural assumptions about the intuitive idea of effective decidability. And the *interest* of those theorems depends on the assumption that being sufficiently expressive/sufficiently strong is a plausible desideratum on formalized arithmetics. If you buy those assumptions – and they are intuitively attractive ones – then we have proved Gödelian incompleteness without tears (and incidentally, without having to construct any 'self-referential' sentences). But it isn't very satisfactory to leave things like that, given the undischarged assumptions. And much of the ensuing hard work over the nearly three hundred pages that follow is concerned with avoiding those assumptions, one way or another.

(b)   The core of the book (Chapters 10 to 30) proves incompleteness again, without relying on informal assumptions about a theory's being sufficiently expressive/sufficiently strong, and without relying on the idea of effective decidability at all. Two ideas now drive the proofs – ideas that are simple to state but inevitably rather messy to prove. (i) The first idea involves the *arithmetization of syntax* by Gödel-numbering: we show that key numerical relations like *m codes for a* PA *proof of n* are primitive recursive, and similarly for any p.r. axiomatized theory. (ii) The other idea is that PA and even Q are *p.r. adequate* theories: that is to say, they can express/capture all p.r. functions and relations. With these two ideas in place, the rest of the argument is then relatively straightforward. We can use Gödel's method of explicitly constructing a particular sentence G which, via coding, 'says' of itself that it isn't provable. We can then show that PA is incomplete if sound, since it can't prove its Gödel sentences. And again, we can weaken the semantic assumption of soundness to a syntactic assumption, this time the assumption of ω-consistency. If PA is ω-consistent, then it is incomplete.

Further, the dual arguments for PA's incompleteness then generalize in easy ways. By the semantic argument, *any* sound theory which contains Q and which is sensibly axiomatized (more carefully: is p.r. axiomatized) is incomplete: moreover there are undecidable sentences which are Π_1 sentences of arithmetic. In fact, by the syntactic argument, *any* ω-consistent p.r. axiomatized theory which contains Q is incomplete, whether it is sound or not. Or, to generalize to the result closest to Gödel's own First Incompleteness Theorem, Theorem 22.3: if T includes the language of basic arithmetic, can capture all p.r. functions, is p.r. axiomatized, and is ω-consistent, then there are Π_1 arithmetical sentences undecidable in T. And to improve these last results, Rosser's construction then tells us how to replace the assumption of ω-consistency with plain consistency. (The Diagonalization Lemma which we prove en route also delivers some other results, like Tarski's Theorem on the arithmetical undefinability of arithmetic truth.)

(c)   The next part of the book (Chapters 31 to 37) looks at The Second Incompleteness Theorem. PA can't prove the sentence Con which arithmetizes in a natural way the claim that PA is consistent. A fortiori, PA cannot prove that stronger theories are consistent either. The proof of this result then generalizes from PA to any theory which contains Q plus a little induction.

The Second Theorem reveals again that there are undecidable sentences like Con which aren't 'self-referential', thus reinforcing a point that emerged in the first part of the book: incompleteness results aren't somehow irredeemably tainted with self-referential paradox. And where the First Theorem sabotages traditional logicist ambitions, the Second Theorem sabotages traditional versions of Hilbert's Programme.

(d)   The final part of the book (Chapters 38 to 45) returns to the approach to incompleteness we initially explored in the first part. But we now trade in the informal notion of effective decidability for the idea of *recursive* decidability (or for the provably equivalent idea of being-decidable-by-a-Turing-machine). We can then use results about Turing machines to re-prove incompleteness, still using Cantor-like diagonalization tricks but now without going via the Diagonalization Lemma, to get formal analogues of our informal theorems Theorem 6.3 and Theorem 7.2. And finally, for fun, we also proved the First Theorem again by invoking Kleene's Normal Form Theorem.

But in trading informal talk of effectiveness for precise talk of recursiveness do we change the subject from our original concerns with effectively axiomatized theories, etc.? The Church–Turing Thesis assures us that we don't, and the book concludes by defending the Thesis.

In sum, that all gives us a number of different routes to the pivotal Gödelian incompleteness results. Our discussions are certainly not the end of the story: there are other routes too, some of a different character again, and not involving the usual kind of diagonalization tricks. They must remain a story for another day. But at least we've made a start . . .

Further reading

(a)   Let's begin with five recommendations for parallel reading (fuller publication details are in the bibliography).

1. For a beautifully clear introduction, presenting and rounding out some of the logical background we assume in this book, and also giving a very nice proof of incompleteness, Christopher Leary's *A Friendly Introduction to Mathematical Logic* is hard to beat.

2. George Boolos and Richard Jeffrey's *Computability and Logic* (3rd edition) covers most of the same ground as this book, and more besides, but does things in a different order – dealing with the general theory of computability before exploring Gödelian matters. There are also significantly expanded later editions, with John Burgess as a third author: but many readers will prefer the shorter earlier version.

3. Richard L. Epstein and Walter Carnielli's *Computability* also discusses computation in general before covering the incompleteness theorems; this is attractively written with a lot of interesting historical asides.

4. Raymond Smullyan's terse *Gödel's Incompleteness Theorems* is deservedly a modern classic; it should particularly appeal to those who appreciate elegant conciseness.

5. Even briefer is the *Handbook* essay on 'The Incompleteness Theorems' by Craig Smoryński, though it still manages to touch on some issues beyond the scope of this book.

(b)   So where next? There are many pointers in the footnotes scattered through this book. So I'll confine myself here to mentioning readily available books which strike me as being, in their different ways, particularly good. First, a group very directly concerned with Gödelian issues:

6. Raymond Smullyan's *Diagonalization and Self-Reference* examines in detail exactly what the title suggests.

7. In his wonderful *The Logic of Provability*, George Boolos explores in depth the logic of the provability predicate (the modal logic of our '$\Box$').

8. For more on what happens when we add sequences of consistency statements to expand an incomplete theory, and much else besides, see Torkel Franzén, *Inexhaustibility*.

Next, a couple of fine books that explore Peano Arithmetic and its variants in more depth:

9. Richard Kaye's *Models of Peano Arithmetic* will tell you – among many other things – more about 'natural' mathematical statements which are independent of PA.

10. Petr Hájek and Pavel Pudlák's *Metamathematics of First-Order Arithmetic* is encyclopaedic but still surprisingly accessible.

As we saw in the last part of the book, incompleteness results are intimately related to more general issues about computability and decidability. For more on those issues, here are a few suggestions:

11. For a brief and very accessible overview, see A. Shen and N. K. Vereshchagin, *Computable Functions*.

12. Nigel Cutland's *Computability* is deservedly a modern classic, with much more detail yet also remaining particularly accessible.

13. A more recent text, also quite excellent, is S. Barry Cooper, *Computability Theory*.

14. Another splendid book – with more historical and conceptual asides than the others – is Piergiorgio Odifreddi, *Classical Recursion Theory*.

15. For much more on the Church-Turing Thesis, though of rather variable quality, see Olszewski et al. (eds.), *Church's Thesis after 70 Years*.

Finally, we mentioned second-order arithmetics in Chapter 29. For an exploration-in-depth of the framework of second-order, see the indispensable

16. Stewart Shapiro, *Foundations without Foundationalism*.

Bibliography

Gödel's papers are identified by their dates of first publication; but translated titles are used and references are to the versions in the *Collected Works*, where details of the original publications are to be found. Similarly, articles or books by Frege, Hilbert etc. are identified by their original dates, but page references are whenever possible to standard English translations.

Ackermann, W., 1928. Zum Hilbertschen Aufbau der reelen Zahlen. *Mathematische Annalen*, 99: 118–133. Translated as 'On Hilbert's construction of the real numbers' in van Heijenoort 1967, pp. 495–507.

Aigner, M. and Ziegler, G. M., 2004. *Proofs from The Book*. Berlin: Springer, 3rd edn.

Avigad, J. and Feferman, S., 1998. Gödel's functional ('Dialectica') interpretation. In S. R. Buss (ed.), *Handbook of Proof Theory*, pp. 337–406. Amsterdam: Elsevier Science B.V.

Balaguer, M., 1998. *Platonism and Anti-Platonism in Mathematics*. New York: Oxford University Press.

Barendregt, H., 1997. The impact of the Lambda Calculus in logic and computer science. *Bulletin of Symbolic Logic*, 3: 181–215.

Benacerraf, P., 1967. God, the Devil and Gödel. *The Monist*, 51: 9–32.

Bernays, P., 1930. Die Philosophie der Mathematik und die Hilbertsche Beweistheorie. *Blätter für deutsche Philosophie*, 4: 326–367. Translated as 'The philosophy of mathematics and Hilbertian proof-theory' in Mancosu 1998, pp. 234–265.

Bezboruah, A. and Shepherdson, J. C., 1976. Gödel's second incompleteness theorem for Q. *Journal of Symbolic Logic*, 41: 503–512.

Black, R., 2000. Proving Church's Thesis. *Philosophia Mathematica*, 8: 244–258.

Blass, A. and Gurevich, Y., 2006. Algorithms: a quest for absolute definitions. In Olszewski et al. (2006), pp. 24–57.

Boolos, G., 1993. *The Logic of Provability*. Cambridge: Cambridge University Press.

Boolos, G., 1998. *Logic, Logic, and Logic*. Cambridge, MA: Harvard University Press.

Boolos, G., Burgess, J., and Jeffrey, R., 2002. *Computability and Logic*. Cambridge: Cambridge University Press, 4th edn.

Boolos, G. and Jeffrey, R., 1989. *Computability and Logic*. Cambridge: Cambridge University Press, 3rd edn.

Bridge, J., 1977. *Beginning Model Theory*. Oxford Logic Guides 1. Oxford: Clarendon Press.

Bridges, D. S., 1994. *Computability: A Mathematical Sketchbook*. New York: Springer-Verlag.

Büchi, J. R., 1962. Turing machines and the *Entscheidungsproblem*. *Mathematische Annalen*, 148: 201–213.

Buss, S. R., 1994. On Gödel's theorems on lengths of proofs I: number of lines and speedups for arithmetic. *Journal of Symbolic Logic*, 39: 737–756.

Buss, S. R., 1998. Proof theory of arithmetic. In S. R. Buss (ed.), *Handbook of Proof Theory*, pp. 79–147. Amsterdam: Elsevier Science B.V.

Camp, J. L., 2002. *Confusion: A Study in the Theory of Knowledge*. Cambridge, MA: Harvard University Press.

Cantor, G., 1874. Über eine Eigenschaft des Inbegriffs aller reelen algebraischen Zahlen. *Journal für die reine und angewandte Mathematik*, 77: 258–262. Translated as 'On a property of the set of real algebraic numbers' in Ewald 1996, Vol. 2, pp. 839–843.

Cantor, G., 1891. Über eine elementare Frage der Mannigfaltigkeitslehre. *Jahresbericht der Deutschen Mathematiker-Vereinigung*, 1: 75–78. Translated as 'On an elementary question in the theory of manifolds' in Ewald 1996, Vol. 2, pp. 920–922.

Carnap, R., 1934. *Logische Syntax der Sprache*. Vienna: Springer. Translated into English as Carnap 1937.

Carnap, R., 1937. *The Logical Syntax of Language*. London: Kegan Paul Trench, Trubner & Co.

Carnap, R., 1950. *Logical Foundations of Probability*. Chicago: Chicago University Press.

Chabert, J.-L. (ed.), 1999. *A History of Algorithms*. Berlin: Springer.

Chihara, C. S., 1973. *Ontology and the Vicious Circle Principle*. Ithaca: Cornell University Press.

Church, A., 1936a. A note on the *Entscheidungsproblem*. *Journal of Symbolic Logic*, 1: 40–41.

Church, A., 1936b. An unsolvable problem of elementary number theory. *American Journal of Mathematics*, 58: 345–363.

Church, A., 1937. Review of Turing 1936. *Journal of Symbolic Logic*, 2: 42–43.

Cohen, D. E., 1987. *Computability and Logic*. Chichester: Ellis Horwood.

Cooper, S. B., 2004. *Computability Theory*. Boca Raton, Florida: Chapman and Hall/CRC.

Copeland, B. J. (ed.), 2004. *The Essential Turing*. Oxford: Clarendon Press.

Copeland, B. J., 2008. The Church-Turing Thesis. In E. N. Zalta (ed.), *The Stanford Encyclopedia of Philosophy*. Fall 2008 edn. URL http://plato.stanford.edu/ archives/fall2008/entries/church-turing/.

Craig, W., 1953. On axiomatizability within a system. *Journal of Symbolic Logic*, 18: 30–32.

Curry, H. B., 1942. The inconsistency of certain formal logics. *Journal of Symbolic Logic*, 7: 115–117.

Cutland, N. J., 1980. *Computability*. Cambridge: Cambridge University Press.

Davis, M., 1982. Why Gödel didn't have Church's Thesis. *Information and Control*, 54: 3–24.

Dedekind, R., 1888. *Was sind und was sollen die Zahlen?* Brunswick: Vieweg. Translated in Ewald 1996, Vol. 2, pp. 790–833.

Detlefsen, M., 1986. *Hilbert's Program*. Dordrecht: D. Reidel.

Earman, J., 1995. *Bangs, Crunches, Whimpers, and Shrieks: Singularities and Acausalities in Relativistic Spacetimes.* New York: Oxford University Press.

Enderton, H. B., 2002. *A Mathematical Introduction to Logic.* San Diego: Academic Press, 2nd edn.

Epstein, R. L. and Carnielli, W. A., 2000. *Computability: Computable Functions, Logic, and the Foundations of Mathematics.* Wadsworth, 2nd edn. 3rd edition 2008 from Advanced Reasoning Forum.

Ewald, W. (ed.), 1996. *From Kant to Hilbert.* Oxford: Clarendon Press.

Fairtlough, M. and Wainer, S. S., 1998. Hierarchies of provably recursive functions. In S. R. Buss (ed.), *Handbook of Proof Theory*, pp. 149–207. Amsterdam: Elsevier Science B.V.

Feferman, S., 1960. Arithmetization of metamathematics in a general setting. *Fundamenta Mathematicae*, 49: 35–92.

Feferman, S., 1984. Kurt Gödel: conviction and caution. *Philosophia Naturalis*, 21: 546–562. In Feferman 1998, pp. 150–164.

Feferman, S., 1998. *In the Light of Logic.* New York: Oxford University Press.

Feferman, S., 2000. Why the programs for new axioms need to be questioned. *Bulletin of Symbolic Logic*, 6: 401–413.

Feferman, S., 2006. Are there absolutely unsolvable problems? Gödel's dichotomy. *Philosophia Mathematica*, 14: 134–152.

Field, H., 1989. *Realism, Mathematics and Modality.* Oxford: Basil Blackwell.

Fisher, A., 1982. *Formal Number Theory and Computability.* Oxford: Clarendon Press.

Fitch, F. B., 1952. *Symbolic Logic.* New York: Roland Press.

Flath, D. and Wagon, S., 1991. How to pick out the integers in the rationals: An application of number theory to logic. *The American Mathematical Monthly*, 98: 812–823.

Folina, J., 1998. Church's Thesis: prelude to a proof. *Philosophia Mathematica*, 6: 302–323.

Franzén, T., 2004. *Inexhaustibility: a Non-Exhaustive Treatment.* Association for Symbolic Logic: Lecture Notes in Logic 16. Wellesley, MA: A. K. Peters.

Franzén, T., 2005. *Gödel's Theorem: An Incomplete Guide to its Use and Abuse.* Wellesley, MA: A. K. Peters.

Frege, G., 1882. Über die wissenschaftliche Berechtigung einer Begriffsschrift. *Zeitschrift für Philosophie und philosophische Kritik*, 81: 48–56. Translated as 'On the scientific justification of a conceptual notation' in Frege 1972, pp. 83–89.

Frege, G., 1884. *Die Grundlagen der Arithmetik.* Breslau: Verlag von Wilhelm Koebner. Translated as Frege 1950.

Frege, G., 1891. *Über Funktion und Begrif.* Jena: Hermann Pohle. Translated as 'Function and concept' in Frege 1984, pp. 137–156.

Frege, G., 1893. *Grundgesetze der Arithmetik*, vol. I. Jena: Hermann Pohle.

Frege, G., 1903. *Grundgesetze der Arithmetik*, vol. II. Jena: Hermann Pohle.

Frege, G., 1950. *The Foundations of Arithmetic.* Oxford: Basil Blackwell.

Frege, G., 1964. *The Basic Laws of Arithmetic.* Berkeley and Los Angeles: University of California Press.

Frege, G., 1972. *Conceptual Notation and related articles.* Oxford: Clarendon Press. Edited by Terrell Ward Bynum.

Frege, G., 1984. *Collected Papers.* Oxford: Basil Blackwell.

Friedman, H., 2000. Normal mathematics will need new axioms. *Bulletin of Symbolic Logic*, 6: 434–446.

Gallier, J. H., 1991. What's so special about Kruskal's theorem and the ordinal Γ_0? A survey of some results in proof theory. *Annals of Pure and Applied Logic*, 53: 199–260.

Gandy, R., 1988. The confluence of ideas in 1936. In R. Herken (ed.), *The Universal Turing Machine*, pp. 55–111. Oxford: Oxford University Press.

Gentzen, G., 1935. Untersuchungen über das logische Schliessen. *Mathematische Zeitschrift*, 39: 176–210, 405–431. Translated as 'Investigations into logical deduction' in Szabo 1969, pp. 68–131.

Gentzen, G., 1936. Die Widerspruchsfreiheit der reinen Zahlentheorie. *Mathematische Annalen*, 112: 493–565. Translated as 'The consistency of elementary number theory' in Szabo 1969, pp. 132–213.

Gentzen, G., 1937. Der Unendlichkeitsbegriff in der Mathematik. Unpublished lecture, translated as 'The concept of infinity in mathematics' in Szabo 1969, pp. 223–233.

Gentzen, G., 1938. Neue Fassung des Widerspruchsfreiheitsbeweises für die reine Zahlentheorie. *Forschungen zur Logik*, 4: 19–44. Translated as 'New version of the consistency proof for elementary number theory' in Szabo 1969.

Giaquinto, M., 2002. *The Search for Certainty.* Oxford: Clarendon Press.

Gödel, K., 1929. *Über die Vollständigkeit des Logikkalküls.* Ph.D. thesis, University of Vienna. Translated as 'On the completeness of the calculus of logic' in Gödel 1986, pp. 60–101.

Gödel, K., 1931. Über formal unentscheidbare Sätze der *Principia Mathematica* und verwandter Systeme I. *Monatshefte für Mathematik und Physik*, 38: 173–198. Translated as 'On formally undecidable propositions of *Principia Mathematica* and related systems I' in Gödel 1986, pp. 144–195.

Gödel, K., 1932. Über Vollständigkeit und Widerspruchsfreiheit. *Ergebnisse eines mathematischen Kolloquiums*, 3: 12–13. Translated as 'On consistency and completeness' in Gödel 1986, pp. 234–237.

Gödel, K., 1933. The present situation in the foundations of mathematics. Unpublished lecture. In Gödel 1995, pp. 45–53.

Gödel, K., 1934. On undecidable propositions of formal mathematical systems. Mimeographed lecture notes, taken by S. C. Kleene and J. B. Rosser. In Gödel 1986, pp. 346–371.

Gödel, K., 1936. Über die Lange von Beweisen. *Ergebnisse eines mathematischen Kolloquiums*, 7: 23–24. Translated as 'On the length of proofs' in Gödel 1986, pp. 396–399.

Gödel, K., 1951. Some basic theorems on the foundations of mathematics and their implications. The 'Gibbs Lecture'. In Gödel 1995, pp. 304–323.

Gödel, K., 1958. Über eine bisher noch nicht benützte Erweiterung des finiten Standpunktes. *Dialectica*, 12. Translated as 'On a hitherto unutilized extension of the finitary standpoint' in Gödel 1990, pp. 241–251.

Gödel, K., 1972. Some remarks on the undecidability results. In Gödel 1990, pp. 305–306.

Gödel, K., 1986. *Collected Works, Vol. 1: Publications 1929–1936*. New York and Oxford: Oxford University Press.

Gödel, K., 1990. *Collected Works, Vol. 2: Publications 1938–1974*. New York and Oxford: Oxford University Press.

Gödel, K., 1995. *Collected Works, Vol. 3: Unpublished Essays and Lectures*. New York and Oxford: Oxford University Press.

Gödel, K., 2003a. *Collected Works, Vol. 4: Correspondence A–G*. Oxford: Clarendon Press.

Gödel, K., 2003b. *Collected Works, Vol. 5: Correspondence H–Z*. Oxford: Clarendon Press.

Goldrei, D., 1996. *Classic Set Theory*. Boca Raton, Florida: Chapman and Hall/CRC.

Goodstein, R. L., 1944. On the restricted ordinal theorem. *Journal of Symbolic Logic*, 9: 33–41.

Grandy, R. E., 1977. *Advanced Logic for Applications*. Dordrecht: D. Reidel.

Hájek, P. and Pudlák, P., 1993. *Metamathematics of First-Order Arithmetic*. Berlin: Springer.

Hale, B. and Wright, C., 2001. *The Reason's Proper Study*. Oxford: Clarendon Press.

Harrington, L. A., Morley, M. D., Ščedrov, A., and Simpson, S. G. (eds.), 1985. *Harvey Friedman's Research on the Foundations of Mathematics*. Amsterdam: North-Holland.

Hart, W. D. (ed.), 1996. *The Philosophy of Mathematics*. Oxford Readings in Philosophy. Oxford: Oxford University Press.

Heck, R., 2007. The logic of Frege's Theorem. Forthcoming in a *Festschrift* for Crispin Wright.

Hedman, S., 2004. *A First Course in Logic*. Oxford: Oxford University Press.

Henkin, L., 1952. A problem concerning provability. *Journal of Symbolic Logic*, 17: 160.

Hilbert, D., 1918. Axiomatisches Denken. *Mathematische Annalen*, 78: 405–15. Translated as 'Axiomatic thought' in Ewald 1996, Vol. 2, pp. 1107–1115.

Hilbert, D., 1926. Über das Unendliche. *Mathematische Annalen*, 95: 161–90. Translated as 'On the infinite' in van Heijenoort 1967, pp. 369–392.

Hilbert, D. and Ackermann, W., 1928. *Grundzüge der theoretischen Logik*. Berlin: Springer. 2nd edn. 1938, translated as *Principles of Mathematical Logic*, New York: Chesea Publishing Co., 1950.

Hilbert, D. and Bernays, P., 1934. *Grundlagen der Mathematik, Vol I*. Berlin: Springer.

Hilbert, D. and Bernays, P., 1939. *Grundlagen der Mathematik, Vol II*. Berlin: Springer.

Hintikka, J. (ed.), 1969. *The Philosophy of Mathematics*. Oxford: Oxford University Press.

Hodges, W., 1997. *A Shorter Model Theory*. Cambridge: Cambridge University Press.

Hunter, G., 1971. *Metalogic*. London: Macmillan.

Isaacson, D., 1987. Arithmetical truth and hidden higher-order concepts. In The Paris Logic Group (ed.), *Logic Colloquium '85*. Amsterdam: North-Holland. Page references are to the reprint in Hart 1996.

Isles, D., 1992. What evidence is there that 2^{65536} is a natural number? *Notre Dame Journal of Formal Logic*, 33: 465–480.

Jeroslow, R. G., 1973. Redundancies in the Hilbert-Bernays derivability conditions for Gödel's second incompleteness theorem. *Journal of Symbolic Logic*, 38: 359–367.

Kalmár, L., 1959. An argument against the plausibility of Church's Thesis. In A. Heyting (ed.), *Proceedings of the Colloquium held at Amsterdam, 1957*, pp. 72–80. Amsterdam: North-Holland.

Kaye, R., 1991. *Models of Peano Arithmetic*. Oxford Logic Guides 15. Oxford: Clarendon Press.

Keefe, R. and Smith, P. (eds.), 1999. *Vagueness: A Reader*. Cambridge, MA: MIT Press.

Ketland, J., 1999. Deflationism and Tarski's paradise. *Mind*, 108: 69–94.

Ketland, J., 2005. Deflationism and the Gödel phenomena: reply to Tennant. *Mind*, 114: 75–88.

Kirby, L. and Paris, J., 1982. Accessible independence results for Peano arithmetic. *Bulletin of the London Mathematical Society*, 14: 285–293.

Kleene, S. C., 1936a. General recursive functions of natural numbers. *Mathematische Annalen*, 112: 727–742.

Kleene, S. C., 1936b. λ-definability and recursiveness. *Duke Mathematical Journal*, 2: 340–353.

Kleene, S. C., 1936c. A note on recursive functions. *Bulletin of the American Mathematical Society*, 42: 544–546.

Kleene, S. C., 1952. *Introduction to Metamathematics*. Amsterdam: North-Holland Publishing Co.

Kleene, S. C., 1967. *Mathematical Logic*. New York: John Wiley.

Kleene, S. C., 1981. Origins of recursive function theory. *Annals of the History of Computing*, 3: 52–67.

Kleene, S. C., 1987. Reflections on Church's Thesis. *Notre Dame Journal of Formal Logic*, 28: 490–498.

Kolata, G., 1982. Does Gödel's theorem matter to mathematics? *Science*, 218: 779–780. Also in Harrington et al., 1985.

Kolmogorov, A. N. and Uspenskii, V. A., 1963. On the definition of an algorithm. *American Mathematical Society Translations*, 29: 217–245.

Kreisel, G., 1957. Abstract: A refinement of ω-consistency. *Journal of Symbolic Logic*, 22: 108–109.

Kreisel, G., 1965. Mathematical logic. In T. L. Saaty (ed.), *Lectures on Modern Mathematics*, vol. III, pp. 95–195. New York: John Wiley.

Kreisel, G., 1967. Informal rigour and completeness proofs. In I. Lakatos (ed.), *Problems in the Philosophy of Mathematics*, pp. 138–171. Amsterdam: North-Holland. Reprinted in Hintikka 1969.

Kreisel, G., 1980. Kurt Gödel. *Biographical Memoirs of Fellows of the Royal Society*, 26: 149–224.

Kreisel, G., 1990. Review of Gödel's 'Collected Works, Volume II'. *Notre Dame Journal of Formal Logic*, 31: 602–641.

Kretzmann, N. and Stump, E., 1988. *The Cambridge Translations of Medieval Philosophical Texts: Vol. 1, Logic and the Philosophy of Language*. Cambridge: Cambridge University Press.

Kripke, S. A., 1980. *Naming and Necessity*. Oxford: Basil Blackwell.

Lakatos, I., 1976. *Proofs and Refutations*. Cambridge: Cambridge University Press.

Leary, C. C., 2000. *A Friendly Introduction to Mathematical Logic*. New Jersey: Prentice Hall.

Lindström, P., 2003. *Aspects of Incompleteness*. Natick, MA: A. K. Peters, 2nd edn.

Löb, M. H., 1955. Solution of a problem of Leon Henkin. *Journal of Symbolic Logic*, 20: 115–118.

Löwenheim, L., 1915. On possibilities in the calculus of relatives. In van Heijenoort 1967, pp. 232–251.

Lucas, J. R., 1961. Minds, machines and Gödel. *Philosophy*, 36: 112–127.

Lucas, J. R., 1996. Minds, machines and Gödel: A retrospect. In P. J. R. Millican and A. Clark (eds.), *Machines and Thought: The Legacy of Alan Turing*, pp. 103–124. Oxford: Oxford University Press.

Mac Lane, S., 1986. *Mathematics: Form and Function*. New York: Springer-Verlag.

Mancosu, P. (ed.), 1998. *From Brouwer to Hilbert*. Oxford: Oxford University Press.

Mancosu, P., Zach, R., and Badesa, C., 2008. The development of mathematical logic from Russell to Tarski, 1900-1935. In L. Haaparanta (ed.), *The Development of Modern Logic*. Oxford University Press.

Martin, R. M., 1943. A homogeneous system for formal logic. *Journal of Symbolic Logic*, 8: 1–23.

Martin, R. M., 1949. A note on nominalism and recursive functions. *Journal of Symbolic Logic*, 14: 27–31.

Mellor, D. H. and Oliver, A. (eds.), 1997. *Properties*. Oxford University Press.

Mendelson, E., 1964. *Introduction to Mathematical Logic*. Princeton, NJ: van Nostrand.

Mendelson, E., 1990. Second thoughts about Church's thesis and mathematical proofs. *Journal of Philosophy*, 87: 225–233.

Mendelson, E., 1997. *Introduction to Mathematical Logic*. Boca Raton, Florida: Chapman and Hall, 4th edn.

Meyer, A. R. and Ritchie, D., 1967. Computational complexity and program structure. Tech. Rep. RC-1817, IBM.

Milne, P., 2007. On Gödel sentences and what they say. *Philosophia Mathematica*, 15: 193–226.

Moschovakis, Y., 2006. *Notes on Set Theory*. New York: Springer, 2nd edn.

Mostowski, A., 1952a. On models of axiomatic systems. *Fundamenta Mathematicae*, 39: 133–158.

Mostowski, A., 1952b. *Sentences Undecidable in Formalized Arithmetic: An Exposition of the Theory of Kurt Gödel.* Amsterdam: North-Holland Publishing Co.

Myhill, J., 1952. A derivation on number theory from ancestral theory. *Journal of Symbolic Logic,* 17: 192–197.

Nelson, E., 1986. *Predicative Arithmetic.* Mathematical Notes, 32. Princeton, NJ: Princeton University Press.

Nelson, R. J., 1987. Church's Thesis and cognitive science. *Notre Dame Journal of Formal Logic,* 28: 581–614.

Odifreddi, P., 1999. *Classical Recursion Theory: The Theory of Functions and Sets of Natural Numbers.* Studies in logic and the foundations of mathematics, vol. 125. Amsterdam: North-Holland, 2nd edn.

Olszewski, A., Woleński, J., and Janusz, R. (eds.), 2006. *Church's Thesis after 70 Years.* Frankfurt a.M.: Ontos Verlag.

Parikh, R., 1971. Existence and feasibility in arithmetic. *Journal of Symbolic Logic,* 36: 494–508.

Paris, J. and Harrington, L. A., 1977. A mathematical incompleteness in Peano Arithmetic. In J. Barwise (ed.), *Handbook of Mathematical Logic,* pp. 1133–1142. Amsterdam: North-Holland.

Parsons, C., 1980. Mathematical intuition. *Proceedings of the Aristotelian Society,* 80: 145–168.

Parsons, C., 1998. Finitism and intuitive knowledge. In M. Schirn (ed.), *The Philosophy of Mathematics Today,* pp. 249–270. Oxford: Oxford University Press.

Peano, G., 1889. *The Principles of Arithmetic.* In van Heijenoort 1967, pp. 85–97.

Penrose, R., 1989. *The Emperor's New Mind.* Oxford: Oxford University Press.

Penrose, R., 1994. *Shadows of the Mind.* Oxford: Oxford University Press.

Péter, R., 1934. Über den zusammenhang der verschiedenen Begriffe der rekursiven Funktionen. *Mathematische Annalen,* 110: 612–632.

Péter, R., 1935. Konstruktion nichtrekursiver Funktionen. *Mathematische Annalen,* 111: 42–60.

Péter, R., 1951. *Rekursive Funktionen.* Budapest: Akadémiai Kiadó. Translated as Péter 1967.

Péter, R., 1967. *Recursive Functions.* New York: Academic Press.

Pohlers, W., 1989. *Proof Theory.* Lecture Notes in Mathematics 1407. Berlin: Springer-Verlag.

Potter, M., 2000. *Reason's Nearest Kin.* Oxford: Oxford University Press.

Potter, M., 2004. *Set Theory and its Philosophy.* Oxford: Oxford University Press.

Presburger, M., 1930. Über die Vollständigkeit eines gewisse Systems der Arithmetik ganzer Zahlen, in welchem die Addition als einzige Operation hervortritt. In F. Leja (ed.), *Comptes-rendus du I Congrès des Mathématiciens des Pays Slaves, Varsovie 1929,* pp. 92–101. Translated as Presburger 1991.

Presburger, M., 1991. On the completeness of a certain system of arithmetic of whole numbers in which addition occurs as the only operation. *History and Philosophy of Logic,* 12: 225–233.

Pudlák, P., 1998. The length of proofs. In S. R. Buss (ed.), *Handbook of Proof Theory*, pp. 547–638. Amsterdam: Elsevier Science B.V.

Putnam, H., 1960. Minds and machines. In S. Hook (ed.), *Dimensions of Mind*, pp. 148–179. New York: New York University Press. In Putnam 1975, pp. 362–385.

Putnam, H., 1975. *Mind, Language and Reality: Philosophical Papers, Vol. 2*. Cambridge: Cambridge University Press.

Quine, W. V., 1940. *Mathematical Logic*. Cambridge, MA: Harvard University Press.

Raatikainen, P., 2003. Hilbert's program revisited. *Synthese*, 137: 157–177.

Ramsey, F. P., 1925. The foundations of mathematics. *Proceedings of the London Mathematical Society*, 25: 338–384. Page references are to the reprint in Ramsey 1990.

Ramsey, F. P., 1990. *Philosophical Papers*. Cambridge: Cambridge University Press. Edited by D. H. Mellor.

Rautenberg, W., 2006. *A Concise Introduction to Mathematical Logic*. New York: Springer, 2nd edn.

Robinson, J., 1949. Definability and decision problems in arithmetic. *Journal of Symbolic Logic*, 14: 98–114.

Robinson, R., 1952. An essentially undecidable axiom system. In *Proceedings of the International Congress of Mathematicians, Cambridge, Mass., 1950*, vol. 1, pp. 729–730. Providence, R.I.

Russell, B., 1902. Letter to Frege. In van Heijenoort 1967, pp. 124–125.

Russell, B., 1903. *The Principles of Mathematics*. London: George Allen and Unwin.

Russell, B., 1908. Mathematical logic as based on the theory of types. *American Journal of Mathematics*, 30: 222–262. Reprinted in Russell 1956.

Russell, B., 1956. *Logic and Knowledge: Essays 1901–1950*. London: George Allen and Unwin.

Russell, B. and Whitehead, A. N., 1910–13. *Principia Mathematica*. Cambridge: Cambridge University Press.

Schönhage, A., 1970. Universelle Turing Speicherung. In J. Dörr and G. Hotz (eds.), *Automatentheorie und Formal Sprachen*, pp. 369–383. Mannheim: Bibliogr. Institut.

Schönhage, A., 1980. Storage modification machines. *SIAM Journal on Computing*, 9: 490–508.

Shapiro, S., 1991. *Foundations without Foundationalism: A Case for Second-Order Logic*. Oxford Logic Guides 17. Oxford: Clarendon Press.

Shapiro, S., 1998. Incompleteness, mechanism and optimism. *Bulletin of Symbolic Logic*, 4: 273–302.

Shapiro, S., 2006a. Computability, proof, and open-texture. In Olszewski et al. (2006), pp. 420–455.

Shapiro, S., 2006b. *Vagueness in Context*. Oxford: Clarendon Press.

Shen, A. and Vereshchagin, N. K., 2003. *Computable Functions*. Rhode Island, USA: American Mathematical Society.

Shepherdson, J. C. and Sturgis, H. C., 1963. Computability of recursive functions. *Journal of the Association for Computing Machinery*, 10: 217–255.

Shoenfield, J. R., 1967. *Mathematical Logic*. Reading, MA: Addison-Wesley.

Sieg, W., 1997. Step by recursive step: Church's analysis of effective calculability. *Bulletin of Symbolic Logic*, 3: 154–180.

Simmons, K., 1993. *Universality and the Liar*. Cambridge: Cambridge University Press.

Simpson, S. G., 1991. *Subsystems of Second Order Arithmetic*. Berlin: Springer.

Skolem, T., 1923. Begründung der elementaren arithmetik .... *Videnskapsselskapets skrifter, I. Mathematisk-naturvidenskabelig klasse, no. 6*. Translated as 'The foundations of elementary arithmetic established by means of the recursive mode of thought, without the use of apparent variables ranging over infinite domains' in van Heijenoort 1967, pp. 303–333.

Skolem, T., 1930. Über einige Satzfunktionen in der Arithmetik. *Skrifter utgitt av Det Norske Videnskaps-Akademi i Oslo, I. Mathematisk-naturvidenskapelig klasse*, 7: 1–28. Reprinted in Skolem 1970, pp. 281–306.

Skolem, T., 1970. *Selected Works in Logic*. Oslo: Universitetsforlaget.

Smoryński, C., 1977. The incompleteness theorems. In J. Barwise (ed.), *Handbook of Mathematical Logic*, pp. 821–865. Amsterdam: North-Holland.

Smoryński, C., 1982. The varieties of arboreal experience. *Mathematical Intelligencer*, 4: 182–189. Also in Harrington et al., 1985.

Smullyan, R. M., 1992. *Gödel's Incompleteness Theorems*. Oxford: Oxford University Press.

Smullyan, R. M., 1994. *Diagonalization and Self-Reference*. Oxford: Clarendon Press.

Steen, S. W. P., 1972. *Mathematical Logic*. Cambridge: Cambridge University Press.

Stoljar, D. and Damnjanovic, N., 2012. The deflationary theory of truth. In E. N. Zalta (ed.), *The Stanford Encyclopedia of Philosophy*. Summer 2012 edn. URL http://plato.stanford.edu/archives/sum2012/entries/truth-deflationary/.

Szabo, M. E. (ed.), 1969. *The Collected Papers of Gerhard Gentzen*. Amsterdam: North-Holland.

Tait, W. W., 1981. Finitism. *Journal of Philosophy*, 78: 524–546. Reprinted in Tait 2005.

Tait, W. W., 2002. Remarks on finitism. In W. Sieg, R. Sommer, and C. Talcott (eds.), *Reflections on the Foundations of Mathematics: Essays in Honor of Solomon Feferman*, pp. 410–419. Association for Symbolic Logic and A K Peters. Reprinted in Tait 2005.

Tait, W. W., 2005. *The Provenance of Pure Reason. Essays in the Philosophy of Mathematics and Its History*. New York: Oxford University Press.

Takeuti, G., 1987. *Proof Theory*. Amsterdam: North-Holland.

Tarski, A., 1933. *Pojecie Prawdy w Jezykach Nauk Dedukcyjnych*. Warsaw. Translated into English in Tarksi 1956, pp. 152–278.

Tarski, A., 1956. *Logic, Semantics, Metamathematics*. Oxford: Clarendon Press.

Tarski, A., Mostowski, A., and Robinson, R., 1953. *Undecidable Theories*. Amsterdam: North-Holland Publishing Co.

Tennant, N., 2002. Deflationism and the Gödel phenomena. *Mind*, 111: 551–582.

Tennant, N., 2005. Deflationism and the Gödel phenomena: reply to Ketland. *Mind*, 114: 89–96.

Tourlakis, G., 2002. A programming formalism for PR. URL http://www.cs.yorku.ca/~gt/papers/loop-programs.ps.

Tourlakis, G., 2003. *Lectures in Logic and Set Theory*. Cambridge: Cambridge University Press.

Trakhtenbrot, B. A., 1988. Comparing the Church and Turing approaches: two prophetical messages. In R. Herken (ed.), *The Universal Turing Machine*, pp. 603–630. Oxford University Press.

Turing, A., 1936. On computable numbers, with an application to the *Entscheidungsproblem*. *Proceedings of the London Mathematical Society*, 42: 230–265. In Copeland 2004, pp. 58–90.

van Dalen, D., 1994. *Logic and Structure*. Berlin: Springer-Verlag, 3rd edn.

van Heijenoort, J. (ed.), 1967. *From Frege to Gödel*. Cambridge, MA: Harvard University Press.

Velleman, D. J., 1994. *How to Prove It*. Cambridge: Cambridge University Press.

Visser, A., 1989. Peano's smart children: a provability logical study of systems with built-in consistency. *Notre Dame Journal of Formal Logic*, 30: 161–196.

von Neumann, J., 1927. Zur Hilbertschen Beweistheorie. *Mathematische Zeitschrift*, 26: 1–46.

Waismann, F., 1945. Verifiability. *Proceedings of the Aristotelian Society, Supplementary Volume*, 19: 119–150.

Wang, H., 1974. *From Mathematics to Philosophy*. London: Routledge and Kegan Paul.

Wilkie, A. J. and Paris, J., 1987. On the scheme of induction for bounded arithmetic formulas. *Annals of Pure and Applied Logic*, 35: 261–302.

Willard, D. E., 2001. Self-verifying axiom systems, the incompleteness theorem, and related reflection principles. *Journal of Symbolic Logic*, 66: 536–596.

Wright, C., 1983. *Frege's Conception of Number as Objects*. Aberdeen: Aberdeen University Press.

Yaqub, A., 1993. *The Liar Speaks the Truth*. New York and Oxford: Oxford University Press.

Zach, R., 2003. Hilbert's program. In E. N. Zalta (ed.), *The Stanford Encyclopedia of Philosophy*. URL http://plato.stanford.edu/archives/fall2003/entries/hilbert-program/.

Zach, R., 2005. Review of Potter 2000. *Notre Dame Journal of Formal Logic*, 46: 503–513.

Zach, R., 2006. Hilbert's program then and now. In D. Jacquette (ed.), *Philosophy of Logic*, pp. 411–447. Amsterdam: North-Holland.

Index

Entries for *technical terms* and for *theorems* give the principal location(s) where you will find them explained. Entries for *names* link to not-merely-bibiliographical occurrences.